The Jewel on the Mountaintop

Fifty Years of the European Southern Observatory

Claus Madsen

Acknowledgement:
As a title, *The Jewel on the Mountaintop* is derived from a speech by Swiss Ambassador Jean-Pierre Keusch at the inauguration of the 3.5-metre New Technology Telescope (NTT) on 6 February 1990. The NTT was, in many ways, a precursor to the Very Large Telescope, which — staying with the metaphor — constitutes the true jewel on the mountaintop.

Cover Photo:
Dawn at Paranal — the first rays of the Sun are reflected by the telescope enclosures at Paranal. (Photo: Gerd Hüdepohl)

ADRIAAN BLAAUW

In Memoriam

Table of Contents

Author's Preface

In early 1988, my supervisor at ESO, Richard West, called me to his office, where I met with Adriaan Blaauw, who had served as ESO's second Director General, succeeding Otto Heckmann. Richard explained that Adriaan had begun to work on a series of articles about the early years of the organisation at the suggestion of ESO's then Director General Harry van der Laan. These articles had originally been intended for *The ESO Messenger,* but had soon turned into a book project, and Richard asked me to assist Adriaan in particular with pictures. Richard had served under Adriaan and it was clear that great respect and mutual sympathy existed between the two. Richard wanted to extend all the help he could to Adriaan in his new undertaking. During the months to come I did my best to help, but the task soon changed from one simple job among many to a deeply fascinating assignment that enabled me to obtain a deeper insight into ESO's history as told not just by an insider, but by someone of the impeccable academic and human stature that has always characterised Adriaan. It was an intellectual feast that I have never forgotten.

I had joined ESO only eight years earlier as a member of what was then called the ESO Sky Atlas Laboratory. The main task of the laboratory was to produce the ESO Survey of the southern skies, based on large photographic glass plates obtained with the ESO 1-metre Schmidt telescope at the La Silla Observatory and later also, together with the UK's Science Research Council (SRC), the ESO/SRC Deep Surveys, a task that also involved the UK Schmidt Telescope at Siding Spring in New South Wales, Australia. All in all, the task of the laboratory took more than 20 years to complete. My own task was peripheral to the main job, but it included developing techniques to enhance the exploitation of the tremendous scientific treasures that were contained in these glass plates. Furthermore, I had become interested in, and made a bit of a name for myself in, wide-field imaging techniques and their application to astronomy. But when Adriaan began his book project my professional path had taken a twist. In 1986, ESO established an information service and Richard, who had been in charge of the laboratory, was appointed head of this new service. He asked me whether I would like to follow him into this new endeavour, and this enabled me to pursue another interest of mine, the public awareness of science and, more generally, the interaction between science and society, topics that were to determine my continued career for the decades to come.

Adriaan's book was published in 1991. Although it was a history tracing the earliest years of ESO, it appeared to me as almost eerily timely, for at the time ESO was passing through a rough patch as we struggled with the challenges of the recently approved Very Large Telescope (VLT) project. His precise description of the difficulties in the 1950s and 1960s provided both food for thought and a good understanding of the challenges ahead — but also of the ingenuity, tenacity and immense dedication displayed by everyone who had been associated with the great adventure that we now label rather summarily ESO. This is perhaps the most important lesson learnt from what we now call an adventure and the "five-star success" (Flensted Jensen, 2002) that has enabled the strong comeback of European astronomy: success does not come by itself. Rather it is the result of a mechanism that can foster, absorb and bring to fruition great ideas, wherever they may be. Many battles were fought, many mountains climbed (also literally), many hurdles overcome — some which are widely known, others which will remain known to only a few. Some which deserve scrutiny because lessons can be learnt from them, others that deserve little more than to disappear in the mist of forgotten times.

Meanwhile, many years have passed. In December 2009, ESO's Director General, Tim de Zeeuw asked me if I would be interested in writing a new book about ESO's fascinating history — focussing on the more recent period that Blaauw for obvious reasons could not cover. I gladly accepted the offer and the challenge. I have been privileged to follow ESO's evolution for more than 30 years, and in connection with the 40th anniversary actually produced a video about ESO's history. With this background, and in preparation for this book, I have carried out extensive interviews with key actors, both from inside and outside the organisation, trying to obtain a more comprehensive understanding of ESO's role and evolution. These interviews and conversations have been invaluable and I am more than pleased to acknowledge the many people who have given me their time, and their voices, telling me their version of the ESO story. A complete list of interviewees is given in Appendix 3.

The interviews have complemented the treasure trove of information found in *The ESO Messenger*, the quarterly magazine that has been published since 1974 (and now amounting to a staggering 10 000 pages), the Annual Reports, minutes of the ESO Council and Committee meetings and a myriad of other publications from ESO. Of course, with such rich material, the challenge has been that of selection. It is the hard job, but also the prerogative of the author, to make the choices. I have done so knowing that others might have chosen differently.

This book sees ESO's development as falling into four distinct phases: Catching Up (1962–1980), Years of Experimentation (1980–1990), The Breakthrough (up until the VLT's inauguration and science operations, and also first fringes at the Very Large Telescope Interferometer [VLTI]), and finally, Towards New Horizons (up until 2007[1]), when ESO began to set the research agenda through new and unique projects. This is a rough division, of course. The reality is that while one can observe a chronological development — and a logical evolution both of the organisation and the scientific and societal context in which it has operated — the process was anything but linear or straightforward. Rather, it was an iterative one, and with various elements compounding and interacting with each other. In this sense a simple timeline is an inadequate guide for a full comprehension of ESO in all its dimensions — for science, for the scientists, for technology, for the Member States and for their citizens. It is therefore hoped that the excursions that I have made from the exact timeline help the reader, rather than cause confusion. Also, while it is hoped that members of the astronomical community will read this book with interest, it is aimed at a wider audience, for the ESO story is not just interesting in its own right. It is both a symbol of, and an example for, a period during which Europe struggled to regain its former position as a continent of science through a process of cooperation and integration — ultimately within a framework known as the European Research Area (ERA). Many astronomical ideas and concepts are therefore explained in a way that makes them accessible to readers with a variety of backgrounds.

In his book, *Europe's Quest for the Universe*, Lodewijk Woltjer summarises the main elements behind the advances in modern ground-based astronomy, to which ESO has so richly contributed. He writes: *"The trio that ... has dominated progress was increased sensitivity, increased angular resolution and increased wavelength coverage.... To the above we have to add a fourth: cost reduction."* But there is a fifth element, which is perhaps the single most important one: the human one. This is therefore also a story about people — how many individuals from many different cultures and walks of life came together to contribute to the ESO project; mostly working hand-in-hand, though sometimes also struggling to find common ground — each thinking and acting within his or her own reference system — yet at the end of the day found constructive and forward-looking solutions. This is perhaps the essence of ESO's success: its ability to rally world-class people and top-notch material resources in pursuit

[1] The year 2007 coincides with the departure of ESO's sixth Director General, Catherine Cesarsky. Although some important developments that took place after 2007, but which originated earlier, will be mentioned, the author does not attempt to describe ESO's history beyond 2007.

a common and fundamental goal for humankind — to obtain a deeper and better understanding of the Universe in which we live and of which we are a part.

This is now the time for acknowledgements and thanks. I have already mentioned my "interview victims", several of whom provided most valuable comments as the manuscript progressed. But others have helped. They include my long-time collaborators and friends: Hans Hermann Heyer, who helped me with photographs, Herbert Zodet provided sound recordings of past interviews, Uta Grothkopf supplied ESO's publication statistics and helped me to find many papers and logbooks, Jean-Michel Bonneau was a competent source as regards ESO budget and finance questions, Ferdinando Patat and his team provided statistical information about observing proposals. I must also thank Elisabeth Völk for her invaluable help in (re-)establishing contact with former colleagues, and for many fruitful conversations about ESO's long history, and Mafalda Martins, who did the graphical layout of the book. I also want to thank everyone who provided photos for the book (as indicated in the captions). Note that images without a specific photo credit are courtesy of ESO. Last, but certainly not least, I wish to acknowledge the work of and thank the four people, who agreed to review the final version of the manuscript: Bob Fosbury, Henrik Grage, George Miley and Richard West.

This book is dedicated to the memory of Adriaan Blaauw. My last visit to him took place in July 2010, when I conducted my final interview with him about his long association with ESO. He also provided most valuable comments on the early draft manuscript of this book. Even at the age of 96, he was not only able to draw on his rich memories in a clear and succinct manner, but having visited both ESO Headquarters and Paranal Observatory recently, it was also evident that he retained a keen interest and pride in the further development of the organisation. "*My dream now,*" he confided to me, "*is to live to see ESO celebrate its 50th anniversary in 2012.*" Sadly, that wish would not be fulfilled. On 1 December 2010, he passed away.

Despite his great age, Blaauw travelled to Chile to visit Paranal in February 2010. Here he is seen enjoying the Paranal sunset together with Tim de Zeeuw, ESO's current Director General. (Photo: Ueli Weilenman)

Prologue

The Hinge: The VLT

"OK for everybody on the mountain. Thank you very much: We have the First Light!"

Massimo Tarenghi, Paranal Observatory,
Monday, 25 May 1998, 22:00 hours local time.

I had the good fortune to be on Paranal on 25 May 1998, the night of first light for the first VLT Unit Telescope, Antu. Together with my colleague Herbert Zodet and a small team from Swiss television, I was supposed to produce a piece for the international television news, to be distributed by satellite. I vividly remember the tense atmosphere in the control room as night fell — soon to be transformed into moments of complete elation once the first exposures had been provisionally examined. People react very differently in such situations: Jason Spyromilio spontaneously broke into a beautiful dance of joy. Massimo Tarenghi kept his carefully controlled composure, but he was clearly happy too. For a split second, he gave away his true feelings as only an astronomer can: *"Oh-point-five!"* he cried triumphantly to Jörg Eschwey, the site construction manager, referring to the image resolution of 0.5 arcseconds. We did a short interview with him in the control room immediately afterwards, and although he clad his statement in modest words, his deep sense of satisfaction was plain.

The observing team left the control room for half an hour or so for a midnight champagne party, the only time ever that observations were interrupted for reasons other than adverse weather conditions or technical problems. I stayed behind because I had promised to give a telephone interview to the early morning programme on German radio. As I was put through to the studio at WDR, I heard the usual pop song and a traffic alert concerning jams on the highways in the Cologne area. Then the studio host announced that he was now connected live to the Paranal Observatory, where the world's biggest telescope had just been put to use for the first time. And then he fired off his question: *"So, what did you see...?"* The public mass media have their own dynamics — the VLT had been on its way for more than 20 years, it was more than ten years since the project had been formally approved, and thousands of individuals had been involved in realising this dream — and then there were just a few seconds to explain the outcome. I honestly don't recall what I said, but probably not much more than a minute later, it was over; the next song was played, followed by the weather forecast for North-Rhine-Westphalia. Showers, I believe.

Massimo Tarenghi (right) talking to the author at the VLT console shortly after first light, with Herbert Zodet behind the camera. (Photo: Peter Gray)

We did the news piece for television on our return to Santiago, uploaded it to the satellite for worldwide distribution the following day and then attended the first light press conference at ESO's Santiago Office. I left for Europe in the late evening and, as usual, the in-flight entertainment included a news programme, and suddenly I found myself, with some 300 other passengers, watching the VLT first light: happy faces, Jason's joyous dance, a view of the planetary nebula NGC 6302 (the Bug Nebula, one of the first light press pictures), and an excited Rolf Kudritzki, of the Universitäts-Sternwarte München, praising the first pictures at the Garching press conference.

The public mass media indeed have their own dynamics — catch a few photons from outer space, and everybody watches.... For me, it was the end of a string of intensely hectic days. I closed my eyes.

≈

Before that, for years, we had toured the ESO Member States with exhibitions and presentations about the VLT. At a technology fair in 1997, a very important representative of his government's ministry for research literally stormed into our stand. He must have had an awful day. It was late and he had clearly only come because one of his subordinates had implored him to stop by. He was in an extreme hurry and exuded an unholy mixture of exhaustion, frustration and stress. He nodded briskly and went up to a model of the VLT. *"So this is the VLT,"* he barked, *"this [expletive] thing that is costing us so much money...."* In such circumstances, prudence would normally dictate a cautious reply, but he caught me off guard, and it was the end of a long day for me too. So, instead of seeking refuge in diplomacy, I shot back: *"If you think this is expensive, what are you going to say, when this telescope delivers the first picture of a planet outside the Solar System?"* He looked at me, completely stunned, and then became very friendly. Nonetheless I felt uncomfortable that night, feeling that I had committed the unforgivable sin of overselling. Yet only seven years later, the VLT did exactly that: it delivered the first direct image of an exoplanet.

There are many tales to be told about the VLT and some of them will be mentioned in this book. The VLT represented the dream of a generation of astronomers and it marked an impressive comeback for European astronomy after decades of taking second place to scientists across the Atlantic. How did ESO achieve this great success and where would ESO go after this magnificent achievement? What did it mean for science? What did all of this mean for Europe? And what did it mean for the proverbial man in the street, whose tax money had enabled the realisation of this dream?

Searching for these answers, we will embark on a historic journey that began 45 years before the magical moment of first light, that 25th day of May 1998....

Part I: Catching Up

Ont signé:

Prof. O. Heckmann
Directeur de l'Observatoire de Hambourg

Prof. A. Unsöld
Directeur de l'Observatoire de Kiel

Dr. P. Bourgeois
Directeur de l'Observatoire royal de Belgique

Dr A. Couder
Astronome de l'Observatoire de Paris

Prof. A. Danjon
Directeur de l'Observatoire de Paris

Prof. R. O. Redman
Directeur de l'Observatoire de Cambridge

Prof. J. H. Oort
Directeur de l'Observatoire de Leyde

Prof. P. Th. Oosterhoff
Astronome de l'Observatoire de Leyde

Prof. P. J. van Rhijn
Directeur du Laboratoire Astronomique "Kapteyn"
Groningue

Prof. B. Lindblad
Directeur de l'Observatoire de Stockholm

Prof. K. Lundmark
Directeur de l'Observatoire de Lund

Prof. K. G. Malmquist
Directeur de l'Observatoire d'Uppsala

Signatories to the 1954 declaration.

Chapter I-1

The Oldest Science

"The situation in the thirties was largely that Europe — European astronomers, in a way, left big telescope astronomy to the Americans."

Bengt Strömgren[1]

Astronomy has often been described as the oldest of the sciences. It is evident that most cultures have studied the heavens and that these studies have deeply influenced people's lives both in practical terms and in their world view.

Without prejudice to other cultures, however, it seems fair to state that for several centuries, astronomy, as a science, was closely associated with Europe, especially following the Renaissance and the Enlightenment. Copernicus, Tycho and Kepler stand out as early examples, though still tied to the pre-telescope era. The great breakthrough enabled by the invention of the telescope and its use for astronomical observations by Galileo, initiated an increasingly dynamic evolution of European astronomy. The names of Adams, Le Verrier, Halley, Messier, Rømer, Lacaille, the Cassinis, the Herschels, Laplace, Lomonosov, Bode, Piazzi, Struve, Bessel, Fraunhofer, and many others grace the history of astronomy. Newton stands out by establishing the mathematical framework that has held for centuries. At the beginning of the 20th century, scientists such as Hertzsprung, Eddington, Oort and others were expanding our knowledge about the world, but names like Hoyle, Bethe and von Weizsäcker also deserve mention, as they signal the gradual transition of astronomy in the direction of astrophysics, from a taxonomical and phenomenological description of the world to a physical understanding of it. The advent of nuclear physics, Einstein's theories of relativity and the strange world of quantum physics also set their mark on the science of astronomy as much as they were inspired by it.

[1] Interview with Bengt Strömgren by Lillian Hoddeson and Gordon Baym, 13 May 1976. (Niels Bohr Library and Archives, courtesy: The American Institute of Physics).

For astronomers in Europe in the first half of the 20th century, the main focus was the study of the stellar system known as the Milky Way galaxy, or simply the Galaxy, and the individual stars therein. As Adriaan Blaauw expressed it, at the time "*the Universe was the Universe filled with the stars….*" Their studies ranged from objects in the Solar System (including asteroids and comets), individual stars (including double and variable stars) to the shape, structure and kinematics of the Milky Way. Whilst the attention of observational astronomy in Europe was on our own galaxy, interest in modern-day cosmology — the study of the large-scale properties of the Universe as we know it today — received a boost in the US with Edwin P. Hubble's 1925 paper, based on his observations with the 100-inch Hooker reflector in California. This paper provided empirical evidence that some of the fuzzy patches in the sky — nebulae — were stellar systems far away from, and independent of, our galaxy. Hubble's 1929 paper went further and suggested that these systems were independent galaxies and moving away from us. This discovery set US astronomy on a quite different path from that pursued by European astronomers, even though scientists in Europe, such as Alexander Friedman, Georges Lemaître[2], Willem de Sitter, Arthur Eddington, and Otto Heckmann had undertaken seminal theoretical work in cosmology, which today forms one of the main areas in astronomical research worldwide.

≈

Where was astronomy in the mid-1950s? The main focus of many astronomers was on stellar evolution and on understanding the internal structure of the stars and their atmospheres, supported by progress in nuclear physics. Stellar spectral classification was important in determining the ages, masses and temperatures of stars, greatly helped by theoretical modelling, which in turn was enabled by technological advances in the field of computing. Astronomers had also developed a fairly good understanding of the Milky Way system as a whole, with its different stellar populations and star formation taking place in the spiral arms. Radio astronomy had enabled the mapping of neutral hydrogen (HI) in the Milky Way, and there was an emerging understanding also of how spiral structures form from discs of gas and dust. So, for stellar astronomy, a coherent if rudimentary and incomplete picture was

[2] Two years before Hubble's famous paper, Lemaître had published his ideas about the expanding Universe — not dissimilar to those of Hubble's — in the *Annales de la Société scientifique de Bruxelles* (Block, 2011). The philosophical roots of the idea of galaxies as islands in the Universe, however, can be traced back to Immanuel Kant's work *The General History of Nature and Theory of the Heavens*, published in 1755.

beginning to emerge, although many questions were still open. Many details of star formation, internal stellar structure, chemical abundances and their relationship to stellar evolution were all poorly understood, as were late stellar evolutionary stages. The situation was even less clear when it came to extragalactic astronomy. The dispute about the reality of the Big Bang took centre stage; with proponents, such as George Gamow, arguing that the Universe was expanding — and thus ever changing — while opponents, such as Fred Hoyle[3], countered that the Universe could exist within a general steady state, expanding, perhaps, but with new matter being constantly created to replenish it. With time, the Big Bang theory gained widespread acceptance. Establishing the age of the Universe was a related question, and here too the ideas were quite rudimentary, with estimates ranging between 10 and 20 billion years. But, as so often, searching for answers led to new questions. Radio astronomy would soon lead to the discovery of mysterious radio sources such as quasars, which served as probes in the study of the intervening space. The advent of space flight would soon open up potent windows for observations in previously inaccessible wavelength domains, such as gamma and X-rays. The combination of ever more sophisticated observational techniques and a better general theoretical understanding provided for a dynamic development of the discipline in a continuous interplay between observations, theoretical work and the development of new technologies. New and powerful telescopes — both in the optical and radio domains — would play an important role, but in the mid-1950s, all the large telescopes were located in the northern hemisphere.

While differences existed between astronomy as conducted in Europe and in the United States of America, by the middle of the 20th century this reflected the hard reality that the twin scientific communities were confronted by very different circumstances. Thanks to philanthropy, American astronomers had been able to construct large new telescopes in sites that were favourable to astronomical observations — under clear skies and on mountaintops far away from cities — beginning in 1918 with the 100-inch Hooker reflector on Mount Wilson and culminating in 1948 with the 200-inch Hale telescope on Mount Palomar in southern California. European astronomers were not so fortunate. It is fruitless to speculate to what extent this was a result of a different scientific emphasis, but it is clear that the maelstrom of the world wars and totalitarian ideologies that engulfed Europe had dire consequences for all European science, including astronomy.

[3] Fred Hoyle was arguably one of the most pronounced opponents of the Big Bang theory. Ironically, to ridicule the idea, he was also the man who coined the term.

But Europe's relative weakness in the field was a problem for astronomy as a whole. As Otto Heckmann, ESO's first Director, wrote in his book *Sterne, Kosmos, Weltmodelle*: *"American astronomy, based on large instruments, seemed destined to remain a monologue, even though fruitful science demands dialogue, yes even controversy."* (Heckmann, 1976).

So, in the early 1950s, a group of astronomers undertook an important initiative to reverse the decline of European astronomy. The vehicle for change and turnaround was to build new and competitive observational facilities.

In the southern hemisphere.

Chapter I-2

Returning from the Abyss

"Il faut faire l'Europe."

Charles Fehrenbach[1]

The initiative by the astronomers, however unique it may have appeared, reflected the prevailing *zeitgeist*: a European desire to rise from the ashes of the Second World War and an understanding that this had to happen within a broader European framework. This understanding was underpinned by the Cold War chill (the Warsaw Pact being signed in 1954), causing western European countries to move closer together, Germany's arduous efforts to re-join the community of civilised, democratic nations and, of course, by the gigantic economic and social challenges that faced each of Europe's nation states at the time. Food rationing in the United Kingdom, for example, was only lifted in the summer of 1954. In France the Fourth Republic was entangled in protracted colonial wars and saw no less than eight governments between 1953 and 1958. At the same time, two social trends gained strength: the drive towards European cooperation and the notion that science and technology constituted the key to a better future. Following one of the most horrific and bitter conflicts in the history of humankind, post-war European cooperation was not built on love. It was built on a desire to avoid the mistakes of the past that had caused the conflict. This meant reaching out to former enemies, and it is therefore not surprising that European integration became a project driven primarily by an intellectual and visionary elite — in politics, but also in science[2]. In the political arena, European cooperation was first suggested by Winston Churchill in his famous speech on 19 September 1946 in Zurich, but was given a more concrete form on 9 May 1950 by the French Foreign Minister Robert Schuman, initiating the process that would later lead to the Treaties of Rome and, ultimately, to the creation of the European Union. It should be mentioned that post-war cooperation between the countries of

[1] Quoted in Blaauw, 1991.

[2] By contrast, when the International Council of Scientific Unions was created after the First World War, it was not possible to include scientists from what was then known as the Central Powers, Germany and Austria.

western Europe was nurtured by the United States of America and science and technology were fields that presented themselves as fertile ground for cross-border collaboration. As the historian of science John Krige has pointed out, the US saw this as an important element of its *"foreign-policy objectives in the European theatre"* at the time (Krige, 2005). So it was a member of the US delegation to the fifth UNESCO General Conference in June 1950, Nobel Laureate Isidor I. Rabi, who encouraged European collaboration in nuclear physics. On 1 July 1953, the European Council for Nuclear Research (Conseil Européen pour la Recherche Nucléaire, CERN) opened a convention establishing a European Organisation for Nuclear Research for signature, and thus began a research facility that would evolve to become the world leader in particle physics. This convention also inspired and paved the way for a series of other European collaborative projects in science. ESO would become the second such initiative, with its convention based on that of CERN. Also in ESO's case, gentle US encouragement played an important role, both during the initial discussions and later on, although there is no direct evidence that this was led by wider strategic considerations, as was the case for CERN. Thus, in the spring of 1953, Walter Baade — a German astronomer who had emigrated before the war and was now working at the Palomar and Mount Wilson Observatories — visited Leiden[3]. In discussions with Jan Oort, then Director of the Leiden Observatory (Sterrewacht Leiden) and arguably one of the most respected European astronomers, the idea of a European observatory was born — an idea that rapidly caught on. Soon thereafter, in June a group of astronomers began to consider it, using the opportunity of a conference on Galactic research, held in Groningen[4], and by the beginning of 1954 it was given concrete form, with a formal statement signed in Leiden on 26 January of that year by 12 leading astronomers from Belgium, France, Germany, the Netherlands, Sweden and the United Kingdom. The main elements of the statement, including the scientific rationale, would later find their way into the convention: an observatory was to be established in the southern hemisphere, comprising a large telescope (3-metre class) and a Schmidt survey telescope (thus imitating the highly successful combination at Mount Palomar), jointly supported by several European countries. However, the statement also declared that participation should be restricted to neighbouring countries *("le nombre des participants a quelque pays voisins formant un groupe restreint")*.

[3] Baade had worked at the Hamburger Sternwarte in the 1920s before joining the Palomar and Mount Wilson Observatories in 1931. In the mid-1930s he had contemplated a return to Europe to assume the directorship of the Hamburger Sternwarte. Before his migration, Baade had been awarded a one-year Rockefeller fellowship, sponsored by Harlow Shapley. He was the first German astronomer to be accepted for a fellowship after the Great War.

[4] International Astronomical Union (IAU) Symposium No. 1.

Participants at the IAU Conference No. 1 in Vosbergen (Groningen), from left to right: Guillermo Haro, Bertil Lindblad, Wilhelm Becker, Richard Stoy, Walter Baade, Otto Heckmann, Vladimir Kourganoff, Jöran Ramberg, Jan Hendrik Oort, Lukas Plaut, Carl Schalén, Adriaan Blaauw, William Morgan, Harold Spencer-Jones, Laura Nassau, Pieter Oosterhoff, Jason Nassau, Priscilla Bok, Pieter van Rhijn, Petr Grigorevich Kulikovsky, Boris Vasilevich Kukarkin, Viktor Ambartsumian, Pavel Petrovich Paranago and Oleg Melnikov. (Courtesy: J. Merkelijn-Katgert Leiden/IAU)

≈

The choice of the southern hemisphere had historical justification and recognised that the southern skies, being relatively unexplored and at the same time featuring objects of particular interest, offered rich hunting grounds for Europe's astronomers. With easy access to the centre of the Milky Way and a unique view of the twin Magellanic Clouds, the southern skies could provide the stage for the comeback of European observational astronomy that its astronomers so coveted.

European astronomers had shown interest in southern hemisphere astronomy quite early. In 1750, Nicolas Louis de Lacaille observed the southern skies from the Cape of Good Hope, determining the positions of nearly 10 000 stars. Later, in 1834, John Herschel set up his observatory at Wynberg in the Western Cape Province, with the purpose of extending the catalogues compiled by his father, William Herschel, to cover objects in the southern skies. John Herschel recorded 2307 nebulae and 2102 double stars, which were added to the General Catalogue of Nebulae and Clusters. By the mid-20th century, astronomers from Belgium, Germany, Ireland, the Netherlands, Sweden, and the United Kingdom had access to observational facilities in South Africa, as had the Americans (Harvard). Aside from the many interesting objects in the southern skies, in South Africa the astronomers found observational conditions that were vastly superior to those in Europe, with clear skies and truly dark sites.

≈

The Leiden Statement was a clear manifestation of the strong desire by Europe's astronomers to regain their position at the forefront of astromical research. Within months, they had transformed a vague idea into a specific project proposal. Possibly intoxicated by the speed of events, they also stated that there was *"no task more urgent" ("il n'y a pas de tâche plus urgente")* than to establish this new observatory[5]. Alas, it was to take eight years, before the formal agreement, in the shape of an intergovernmental agreement, was signed, and no less than 22 years before the large telescope, called for in the statement, would point to the stars for the first time. One of the important lessons from the creation of ESO is that the key to success lay in the ability of Europe's astronomers to undertake concerted actions in pursuit of clearly defined goals. Following in the footsteps of particle physicists, astronomers pioneered formally organised cooperation in science in Europe. It was hardly a coincidence. Astronomers have a long tradition of cross-border collaboration. It can in fact be argued that by its very nature few sciences are as amenable to international collaboration as astronomy. But pooling resources also meant addressing the rather more tricky issue of finding the right balance between national projects and joint activities. This was perhaps more of an issue for the larger countries, and notably for France and the UK, both of whom at some level might have had the capacity to go it alone. Tackling this issue was bound to be difficult and to take time, and for a while it led to rather different decisions in those two countries.

At the same time, the astronomers' impatience was understandable, for European astronomy was steadily ceding ground to the US. Discussing the state of European

[5] As we shall see later, to avoid losing time, the idea was by and large to copy telescopes in the US rather than to start independent development projects (Heckmann, *ibid.*).

astronomy in 1950s, Gustav Tammann wrote: *"There were top achievements [in Europe], but they came from single individuals. The average situation compared with the United States was desolate. Every young PhD student, particularly when he was interested in observations, had only one aim: to find at least a temporary position in the US."* (Tammann, 1995).

≈

The long wait ahead of Europe's astronomer was, obviously, not known to them as they set about to create the basis for their new observatory already towards the end of the following year — in 1955. Hardly surprisingly, given both the historic and contemporary links, the Leiden Statement had identified South Africa as the place to locate the observatory. During the coming years, the efforts focussed on finding the right location there, while also creating the necessary financial basis and legal framework for an observatory. In his book about ESO's early history, Adriaan Blaauw (Blaauw, 1991) has provided an in-depth account of these efforts and the details shall not be repeated here. It suffices to mention that having looked at a number of locations in various parts of the country, interest gradually focussed on two locations in the Great Karoo semi-desert, which covers a vast area roughly between Port Elizabeth and Cape Town: Zeekoegat, at 1000 metres elevation, located 80 kilometres south of the town of Beaufort West, and Klawervlei, 35 kilometres northwest of it, offering three particular sites with elevations between 1490 and 1970 metres. Site tests were initially carried out with the help of 25-centimetre reflecting telescopes, with visual assessment of the diffraction rings and, using small refractors, with the help of photoelectric measurements, to determine atmospheric extinction. Basic meteorological data were recorded as well. Later, measurements of temperature variations at different elevations above ground were used to better understand local conditions regarding turbulence and its effect on what astronomers call "seeing", i.e. the blurring and twinkling of stellar images caused by atmospheric effects. To assist the scientists, young people with practical skills were sought. An announcement in *De Verkenner*, the Dutch Boy Scouts' magazine, explained that *"during the next two years, ... simple astronomical observations are to be made at a number of places in South Africa. Since the observers will have to work and live under fairly primitive conditions, scouts [are sought] to take these measurements. Leaders or former scouts, who would like at least four months of adventurous life in South Africa can [obtain] further information... ."* Among the volunteers were two young Dutchmen, Jan Doornenbal and

Albert Bosker[6]. After the South African campaign had ended, both rejoined ESO in 1965. Doornenbal later became Head of the Mechanical Group at La Silla, but left in 1969, whereas Bosker would serve ESO for 38 years, ending his career as Deputy Administrator at Paranal. In South Africa, observers worked on a sunset-to-sunrise work plan for 25 consecutive nights, followed by six days off. As André Muller, a Dutch astronomer who was to play a key role at ESO during the early years, reported, *"for the participants of the ESO Seeing Expedition 1961 in the Union of South Africa: With the exception of these five nights, there will be no opportunity for outings, whatsoever!"* Site testing in remote areas is a tough job, which requires almost superhuman stamina, dedication and discipline. It involves relatively sophisticated technology, yet it takes place under the most primitive general conditions and furthermore, in this case, cost was becoming an issue[7].

Photo left: André Muller and Paul McSharry at the Flathill observing site. At the end of an observing period, all the observers met to compare results. Photo right: Albert Bosker and McSharry with a 25-centimetre Danjon seeing telescope. (Photos: Jan Doornenbal)

To boost the ESO undertaking and instil renewed enthusiasm by doing real science, Charles Fehrenbach, then director of the Observatoire de Marseille, proposed installing a refractor, the so-called Grand Prisme Objectif (GPO) at Zeekoegat. The GPO was a double astrograph with twin four-metre-long telescope tubes. One was equipped with a prism and, by rotating the prism between exposures, it was possible to obtain radial velocities — the velocity with which an object moves along the

[6] Dedicated boy scouts as they were, once in South Africa they started a local Boy Scout Troop, which stood out by being interracial — unheard of in South Africa at the time, but tolerated locally.

[7] At the time costs had to be covered out of national research agency budgets.

line of sight — with an accuracy of about 5 km/s[8]. The GPO, which became operational in 1961, was used thus to measure the radial velocities of stars, a project fully within mainstream European research of the time. In a sense, the GPO enjoyed the honour of being the first ESO telescope and, as it turned out, it came close to being so twice. First in South Africa and later, at ESO's site in South America where it followed hot on the heels of the 1-metre photometric telescope, which, in a provisional dome, was the first telescope to become operational there. Fehrenbach's suggestion was not only motivated by scientific reasons. He also saw this as a move to secure continued French support for the ESO project, at a time of great uncertainty in that country[9]. In an article in *The ESO Messenger* he later summarised the situation: *"During these bad years (1953–1960), the United Kingdom definitely withdrew from this project and the French authorities did not feel themselves to be very much involved. The French delegation was no longer authorised to participate, but the astronomers did not want to give up the project and they thought of building a French station for studying the Magellanic Clouds, the operation of which would be in the framework of ESO."* (Fehrenbach, 1981). German astronomers also installed a 40-centimetre telescope at the Klawerlei site to carry out photometric measurements. It was used both for site testing and for photometric observations of the Milky Way.

The site tests ceased in early 1963 and they ended with a surprise. As Blaauw put it in his book, perhaps slightly tongue in cheek: *"At the end, bewilderment and consent."*

≈

Meanwhile, in Europe, much effort went into organisational matters. It should be remembered that ESO did not yet exist as a legal entity and thus all activities depended on the goodwill of national institutes and agencies. The original cost estimate foresaw a capital outlay of 2.5 million US dollars and an operational budget of 100 000 US dollars per annum. This estimate was soon revised upwards to 3.5 million and 135 000 US dollars respectively. By 1959, the expected capital need had grown to 5 million US dollars. And as site tests were underway, their cost, too, went

[8] It is interesting to compare this number with the performance of the instrumentation in use nowadays. The HARPS spectrograph at La Silla, for example, provides an accuracy better than 60 cm/s. For spectrographs for the next generation of telescopes much higher performance is planned.

[9] Preoccupied with its domestic problems, the position of France *vis-à-vis* ESO was unclear for a long time. In his memoires, Otto Heckmann alludes to discussions among the remaining four countries — Germany, Belgium, the Netherlands and Sweden as to whether to proceed without French participation. Luckily, it did not become necessary.

up. The cost of the final site-testing period of one and a half years had increased tenfold over the cost of the preceding time period of similar duration. The growing cost meant that the original thought that ESO could be established as a joint facility between national organisations was no longer viable. Stronger commitment was needed and that could only come from the national governments. Blaauw later described the situation: *"Many of the delegates from government side had been people who had a leading position in science policy in their own countries, and we have just been fortunate ... that the people in these important positions of science policy have been so broad-minded that they said a European effort, also from the government point of view, is the one that is the thing of the future."* Without prejudice to others who helped behind the scenes, the names of two civil servants stand out: Henk Bannier (of the Netherlands), who first argued for an intergovernmental agreement and Gösta Funke (of Sweden). Both later became presidents of the ESO Council. The consequence was the decision, in 1960, to pursue a solution involving an international convention and the setting up of an intergovernmental organisation. It was fortuitous. This particular legal framework has proved to be invaluable in ensuring the stability and long-term strength of the organisation — essential as it embarked on the long road towards facilitating a recovery for European astronomy. But the group that had fought so hard to get ESO off the ground was not yet at the goal. In 1959, the United Kingdom withdrew from the project, giving preference to collaboration within the British Commonwealth — arguably a most serious setback for the ESO project as it struggled to stay afloat. Fortunately, at the same time, the Ford Foundation (of the US) declared its willingness to support a European observatory under certain conditions — notably that at least four countries were ready to commit to the project[10, 11]. The grant of one million US dollars, 20% of the estimated capital cost, was undoubtedly instrumental in bringing the discussions regarding ESO to a successful conclusion. On 5 October 1962, representatives of the five founding countries gathered in Paris to sign the ESO Convention. The Convention would come into force upon parliamentary ratification of the fourth signatory. On 17 January 1964,

[10] The original statutes of the Ford Foundation allowed support for initiatives to strengthen world peace, freedom and democracy as well as education. In the late 1950s, the Foundation expanded its portfolio temporarily to include science and engineering. Heckmann attributes the Ford grant to ESO to the diplomacy of Bertil Lindblad and Jan Oort, arguing that support for this project would stimulate the early efforts of European integration in general (Heckmann, *ibid.*).

[11] The Ford grant not only ensured the necessary capital base for ESO, but may also have played a role in finally securing French participation. The grant was discussed with Jean Monnet and the French Minister of Finance, Antoine Pinay and it is believed that the topic was also raised with the President of the Republic, Charles de Gaulle. This was not the last time that ESO affairs became a topic for discussion in the highest political circles in France.

the final hurdle was passed, as France deposited its instrument of ratification. But by that time a series of decisive operational decisions had already been taken, determining the future of the organisation.

ESO was a child of the European integration process, but it also played its own role in the process by fostering a true European astronomical community. The 1963 European Astronomers Conference at Nijenrode Castle is an example. Organised by the Kapteyn Foundation for Research in Astronomy, the meeting brought together a number of upcoming scientists from across Europe. It was seen at least by some as an ESO conference, occurring as it did just one year after the signing of the Convention. At the meeting, Heckmann introduced the young audience to "Instruments for the ESO Project", André Muller to the site-testing in South Africa and Chile. Seen in the photo are: (1) Luc Braes, (2) Waltraut Seitter, (3) Trientjes Stuit, (4) unidentified, (5) unidentified, (6) Whitney Shane, (7) Ernst Raimond, (8) Kristen Rohlfs, (9) Adriaan Blaauw, (10) Jan Oort, (11) Otto Heckmann, (12) Jørgen Otzen Petersen, (13) Richard West, (14) Jörg Pfleiderer, (15) Peter Wild, (16) unidentified, (17) Ulrich Schwarz, (18) Erik Høg, (19) unidentified, (20) Bruno van Albada, (21) James Lequeux, (22) Ulrich Haug, (23) Jan Borgman, (24) Theodor Schmidt-Kaler, (25) Wim Rougoor, (26) unidentified, (27) Kees Zwaan, (28) Max Kuperus, (29) Tibor Herczeg, (30) Marcel Bonneau, (31) unidentified, (32) Raphael Steinitz, (33) Pierre Charvin, (34) Jørn Berentsen, (35) unidentified, and (36) Mart de Groot. (Courtesy: J. Merkelijn-Katgert)

On horseback: Frederik de Vlaming, André Muller and Otto Heckman at La Silla in Chile in 1964.

Chapter I-3

A Dramatic Twist

"When one is riding south along the Panamericana
from Vallenar to La Serena one sees the big massif,
well isolated from the higher mountain chains in the east,
towering over the surroundings.
We feel that the mountain is beautiful."

Gösta Funke, from a speech on the occasion
of the dedication of the road to La Silla, 24 March 1966.

The first two years of the new organisation entailed dramatic decisions of huge significance. For eight years, the founding fathers of ESO had worked hard to achieve their goal. It had certainly been an uphill struggle, requiring patience, diplomacy, political lobbying and a strong pioneering spirit, not least during the site tests. Now, with the Convention signed, they wanted to see practical results. The Convention, which was based on the CERN Convention, both provided the framework and set out specific tasks for the new organisation. It also described the governing structure, with a council comprised of Member State representatives[1] as the supreme organ of the organisation. The running of the organisation would be the responsibility of the Director General. An associated financial protocol established the Finance Committee as an advisory body to the Council.

A sentence from the preamble establishes the overall framework: *"The Governments of the States parties to this convention ... desirous of jointly creating an observatory equipped with powerful instruments in the southern hemisphere and accordingly promoting and organising co-operation in astronomical research... ."* ESO was not just a new observatory: the authors also clearly intended to give the new organisation a central role in European astronomy, although there is little evidence to suggest that they had a clear vision of what that role might be. It is also interesting that the Convention simply spoke about astronomy. Admittedly, this was before the opening up

[1] Two from each country, representing the national astronomical community and the government, respectively.

of the electromagnetic spectrum — the possibility of observing celestial objects in a variety of wavelength domains — had really occurred. But radio astronomy already existed. This notwithstanding, Article II specified that the purpose of the organisation was to *"build and fit out an astronomical observatory in the southern hemisphere."* The initial programme for the future observatory was also described in detail. The observatory was to comprise a *"telescope with an aperture of about 3 metres, a Schmidt telescope with an aperture of about 1.20 metres, not more than three telescopes with a maximum aperture of 1 metre, a meridian circle*[2]*,"* as well as the necessary auxiliary equipment and buildings. Importantly, the Convention incorporated the option of supplementary programmes — a provision that has enabled ESO to develop new activities over the course of the 50 years of its existence. Given the 1954 statement and the work undertaken in South Africa during the preparatory period, it was noteworthy that the Convention simply stated that the observatory must be located in the southern hemisphere. What had happened?

≈

Jürgen Stock was a German-born astronomer. He had spent part of his childhood in Mexico and later in his professional life he remained closely associated with Latin America. He did his PhD under Otto Heckmann (who was to become ESO's first Director General[3]) in Hamburg, but then moved to the US, according to Heckmann because during the long preparatory phase he had lost faith in the ESO project (Heckmann, *ibid.*). In any event, Stock was a true adventurer and became involved in site-testing activities in Chile[4], on behalf of his American employers[5]. During this work, he found Cerro Tololo, a 2200-metre-high mountain in Chile's Norte Chico (the Little North), a semi-desert area. Located in the Elqui valley area some 80 kilometres east of the Chilean town of La Serena, Tololo was to become the home of the first major observatory in Chile, the Cerro Tololo Inter-American Observatory (CTIO) — effectively the American Southern Observatory. However, Stock had also

---

[2] The meridian circle was never acquired. Instead, ESO installed an astrolabe at the National Observatory of Chile, located at Cerro Calán, just outside the capital, Santiago de Chile. For many years, the astrolabe also formed the nucleus of the scientific cooperation between ESO and the Chilean astronomical community.

[3] Originally, Heckmann's official title was that of Director. This was changed in 1967 following a reorganisation that, among other changes, saw the appointment of Adriaan Blaauw as Scientific Director.

[4] The fascinating story of Jürgen Stock's activities is provided in some detail in the book *Geheimnisvolles Universum* by Dirk Lorenzen.

[5] Both American and European astronomers had on earlier occasions been active in Chile. An article in *The ESO Messenger* (Dürbeck *et al.*, 1999) describes their work and even a brief site search in the Atacama Desert, conducted by Heber D. Curtis in 1909.

spent time at the Boyden Observatory in South Africa, maintaining contact with European astronomers and with Heckmann, his old mentor. With his experience of site testing, Stock quickly understood that Chile offered conditions far better than those found in South Africa and he informed Heckmann accordingly. By the spring of 1961, the ESO Committee, a predecessor of the ESO Council, became interested. Heckmann met Stock in California and Stock showed a number of photographs from northern Chile that left a deep impression on Heckmann. In June, Jan Oort, the chairman of the Committee, contacted his American colleague Donald Shane, who was involved in a US project to install a telescope in Chile, with a view to possible collaboration. The South African choice was no longer certain. It has later been a source of speculation whether the political situation in South Africa played a role in, or possibly was the cause of, ESO's sudden change of interest[6]. However, all the available evidence, for example, as provided in writing and orally by Danjon, Blaauw and Oort, suggests that political considerations played only an indirect role, in that they may have stimulated an interest in considering alternative sites. When the scientific findings became available, no further convincing was necessary: in terms of seeing, temperature variations during the night, and the number of clear nights, the Chilean data were significantly better. In fact the number of clear observing hours was more than 50% higher for the Chilean sites. Unsurprisingly, the Committee did not find it difficult to reach a unanimous decision.

In November 1962, a few weeks after the signing of the Convention, a two-man ESO team arrived in the Chile: André Muller, who had considerable experience from the site tests in South Africa and was already a long-time ESO stalwart, and Paul McSharry, a young South African geodesist who had also assisted in the site-testing effort in his home country. They joined Jürgen Stock, conducting tests at Tololo and a site further to the north. Their findings convinced ESO that the Chilean option was worth pursuing, so in June 1963, Jan Oort, Otto Heckmann, Charles Fehrenbach, Heinrich Siedentopf and André Muller went back to the country, where they met with their American colleagues, Frank K. Edmondson, Nicholas Mayall and Jürgen Stock, representing the (US) Association of Universities for Research in Astronomy (AURA), now in charge of the main US project.

Together, they undertook a visit to the area around Tololo, partly by helicopter, courtesy of the Chilean Air Force, partly on horseback — preserved for eternity by

[6] This speculation was not unreasonable given the growing worldwide concern over the situation in South Africa, dramatically enhanced by the Sharpeville massacre in March 1960.

The photo to the left was taken at the meeting in June 1963 between ESO representatives and their US counterparts. From left to right: Siedentopf, Oort, Frank Edmondson (AURA President), a tour guide, Muller, Heckmann, Fehrenbach, two unidentified participants and a Sr Marchetti (an architect). The picture was taken at Cerro Morado, a mountain near Cerro Tololo, on 10 June. The photo to the right is the famous picture of the group exploring the terrain. (Courtesy: Frank Edmondson)

the famous photograph. Early discussions conducted by Heckmann with Chilean authorities and universities also confirmed that ESO would be welcome. From this time, it seems that Chile had *de facto* been selected as the host country for Europe's observatory. In November of that year, the ESO Committee took the formal decision, albeit subject to confirmation by the ESO Council, which would be set up upon ratification of the Convention. But the question of exactly where the observatory should be built remained open. Despite most cordial relations, the initial idea of placing it together with the American observatory proved not to be feasible. As an international organisation at intergovernmental level, ESO possessed a legal status that was difficult to reconcile with that of AURA, itself an association of national universities. This became a serious issue in connection with the question of land ownership and although the discussions between the Americans and the Europeans remained inconclusive, ESO began to consider alternatives. This was painful for some, not least for Jan Oort, who favoured as close a collaboration with the American astronomers as possible. Common scientific interests and close personal friendships certainly played a strong role. However, the feeling, especially among the government representatives behind ESO, was that an independent observatory would better serve Europe's interests in the long run. So, in October 1963, Heckmann returned to Chile. He had been authorised to approach the Chilean authorities with a view to understanding the conditions under which the Europeans could establish their observatory, but things went not just very smoothly but also very quickly: faster than ESO had anticipated. Heckmann went the whole way: within a couple of weeks, he completed and signed the formal agreement between ESO and Chile, called the *Convenio*, setting the legal framework for ESO's activities in the Republic of Chile. It was a daring act to put it mildly, since it had neither been presented to the ESO Committee in advance, nor had the Convention been ratified. The reasons for Heckmann's swift, unilateral move, however, were understandable: his conversations had been helped significantly by the German immigrant community in Chile. One name that recurs is that of Father Bernhard Starischka, who headed the German School, the Liceo Alemán, in Santiago.

Photo left: Having obtained a PhD in astronomy in Bonn, Father Starischka was no stranger to European astronomy. Here he is seen with Waltraut Seitter and Hans Schäfer at the Observatorium Hoher List. (Photo: The Seitter Estate (Hilmar Duerbeck), courtesy: AIP Emilio Segre Visual Archives). Photo right: Heckmann (to the left) at La Silla.

With a common nationality, faith and professional background, Heckmann found it easy to develop a good rapport with Father Starischka, who he had already met in Germany in 1959. Father Starischka was quite influential in Chile and had connections to people in high positions in Chilean society, including ministers in President Jorge Allesandri's administration. Elections were due and it was not clear what changes they might bring. Heckmann saw a unique but narrow window of opportunity. It was not to be lost and he acted accordingly. Grudgingly, because of the circumstances under which it had been concluded, the ESO Committee went along with the deal and it was finally approved at the first formal meeting of the ESO Council on 5 February 1964. On 17 April, the Chilean government also endorsed the *Convenio*: ESO's future now lay in the Chilean desert. Hardly more than a month later, the ESO Council selected the location for its observatory, a 2403-metre -high mountain, some 100 kilometres northeast of La Serena and 90 kilometres north of Tololo. At the time the mountain was known as Cinchado Norte, but today it is called La Silla — Spanish for "the saddle" — because of its saddle-shaped top. The final selection had been based on desert plots that were government properties, where water could be found and mining activity was limited. Given the extensive search in the preceding years and the resources that had been allocated to finding the best possible site, it is almost unbelievable that no real site-testing was carried out at La Silla. Yet so strong was the conviction that *"almost any site in the La Serena area at the level of 2000–3000 metres, well isolated from surrounding mountain peaks, should be adequate,"* as Blaauw wrote. Heckmann certainly had no doubts. He described it as a stroke of luck (*"ein Glücksfall"*). Time would show that he was right.

It had been a long journey. Ten years had passed since the Leiden Statement. But within the last two and a half years or so, things had changed completely: the Convention had been agreed, signed and ratified, the basic institutional settings defined — and a location for the new observatory had been settled, albeit in a completely different place than originally expected. Even so, for a while La Silla — or, as it became colloquially known, "the mountain" — would remain exactly that: a lone mountain in the middle of the desert.

Following the site decision, ESO could begin to build its observatory — and its future. On-site conditions were tough, but so were the people. Seen here are André Muller and Hans-Emil Schuster (right) during a relaxed moment. This picture was taken after the completion of the road, possibly as late as 1967. (Photo: Jan Doornenbal)

Chile would, however, develop into one of the most important countries for ground-based astronomy in the world, with observatories at Cerro Tololo (AURA, US), La Silla (ESO, Europe) and Las Campanas (also known as CARSO, owned by the Carnegie Institution for Science, US), in addition to the Chilean observatory at Cerro Calán. The Soviet Academy of Sciences was also installing an observatory not so far from Santiago, but withdrew after the political events in 1973. Later, other sites would be opened, including Cerro Paranal, Llano de Chajnantor and Cerro Armazones.

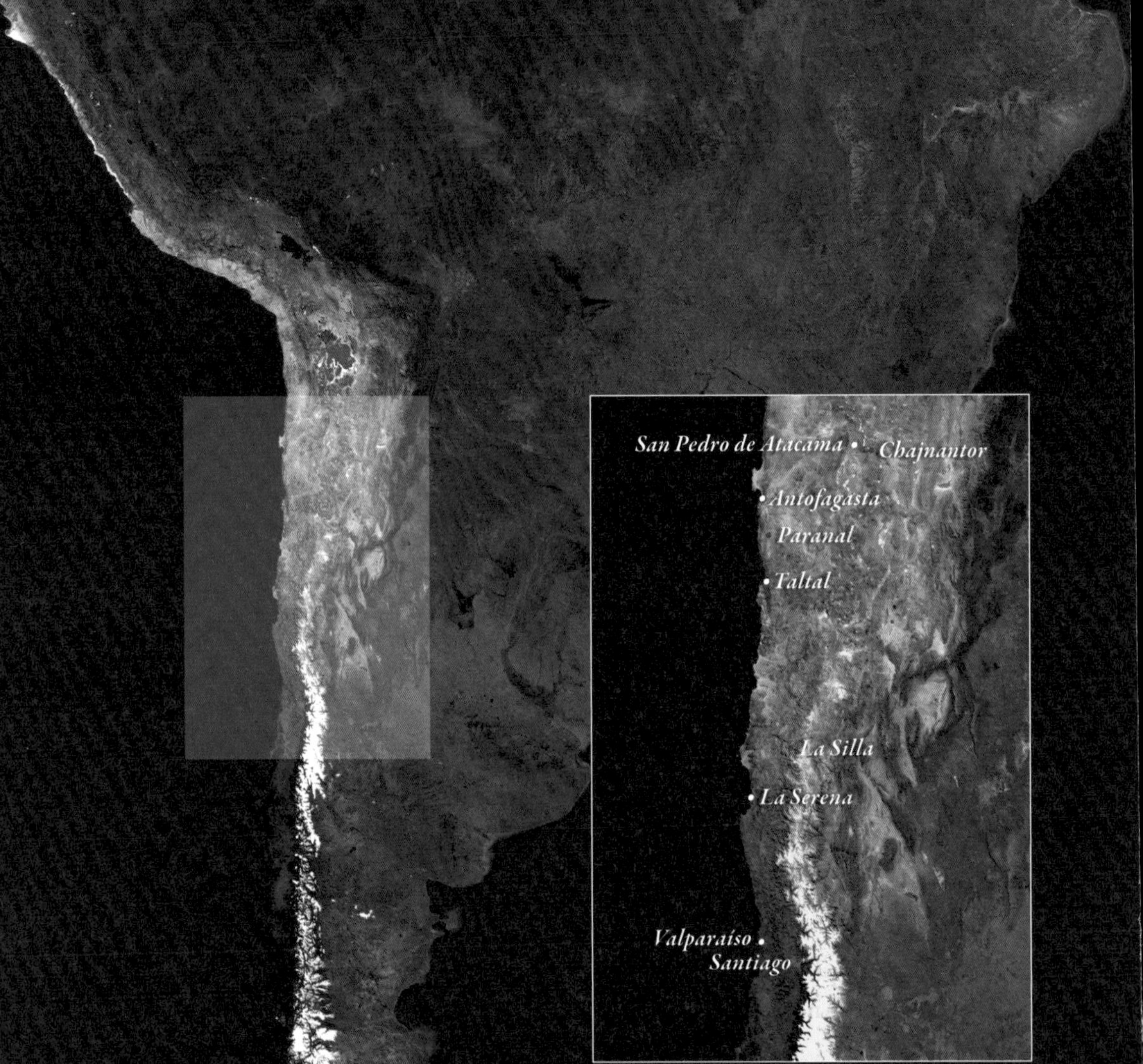

San Pedro de Atacama
Chajnantor
Antofagasta
Paranal
Taltal
La Silla
La Serena
Valparaíso
Santiago

Chapter I-4

In the Most Remote Place God Could Find

"En vez de dejar que estas maravillas se perdieran,
Dios las dispuso todas en el lugar más remoto de la Tierra."

Leyenda de la Creación de Chile.

According to a legend, much treasured in Chile, when God had created the wonders of the world, He had a bit of everything left over: rivers and valleys, glaciers and deserts, mountains, forests, meadows and hills. Instead of letting these wonders go to waste, God put them all in the most remote place on Earth. This was how Chile was created. Despite the charming story, scientifically minded people subscribe to a different explanation, but they too would undoubtedly agree that Chile offers spectacularly beautiful scenery and extraordinary landscapes.

Chile became fully independent from Spain in 1818 — in the aftermath of the Napoleonic wars — and in the period 1879–83 it fought a war with Peru and Bolivia (the War of the Pacific), which gave the country control over large parts of the Atacama Desert with its vast mineral riches. The heroes of the war, at least from a Chilean perspective, are still revered. Their names grace public squares, parks, streets and avenues in many places in the country.

Located on the western rim of South America, and stretching roughly from the 15th parallel to the southern tip of the continent, Chile is almost 4300 kilometres long, with all climate zones except the tropical. The northern part is mostly desert, thanks to the particular conditions found here: the Pacific anticyclone shields the area from the cold air of Antarctica, while the cold Humboldt current running along the coast creates an inversion layer — with a semi-permanent cloud layer — at altitudes of about 1000 metres. Precipitation occurs over the ocean or on the coast itself, but a low mountain range running north to south along the coast is enough to prevent clouds from penetrating further inland. On the other side of the country, the Cordillera de los Andes effectively blocks humid continental air from the East.

The mountain, which became known as La Silla, is located in the IV Region, with La Serena as its capital. From La Serena, *Ruta 5* — known elsewhere as the Pan-American Highway — winds through sparse scrubland and the low coastal mountain range. Shortly after the pass of La Cuesta de Pajonales, at La Frontera — a modest place where truck drivers check in for a meal and some rest — travellers can spot a mountain ridge with white shining domes to the east: the La Silla Observatory. Back in 1964 the area was largely empty and the Chilean government ceded to ESO a property of 627 square kilometres around the mountain to ensure proper protection from mining and other disruptive activities. The price was 8000 US dollars, to which should be added another 6000 US dollars, paid to a family subsequently claiming to hold a title to a part of the land. Corrected for inflation, the total price would amount to about 100 000 US dollars today. It might appear that for such a large stretch of desert the price was quite favourable, but it did reflect the market value at the time. The odd fact that more than one party held a title to a particular piece of land gave rise to a smile or two rather than to serious concern. The story was, however, to repeat itself at a later point in ESO's history — this time on a very different scale.

≈

With the acquisition of La Silla, Europe's astronomers could begin to realise their new observatory. The first telescopes to be built for La Silla were mid-sized telescopes and we note an interesting, if not surprising, correlation between traditional national areas of expertise (in the Member States) and the projects they worked on. The Dutch took a lead in developing a 1-metre photometric telescope following their scientific interests and expertise, particularly at the Kapteyn Laboratory in Groningen. The French built a 1.5-metre spectroscopic telescope, essentially a copy of the one at the French national observatory at Haute-Provence (Observatoire de Haute-Provence, OHP) as well as the double astrograph (GPO) already mentioned, while in Germany work focussed on the Schmidt telescope (a speciality of the Hamburger Sternwarte) — and the 3.6-metre telescope to which we shall return. The telescopes were all very different, of course, but in terms of detectors, they reflected the practice of the day: they either relied on photographic plates or on direct photoelectric measurements of the light coming from the object of observation, i.e. measuring the intensity of the light in different, distinct colours.

Meanwhile, in Chile, the overall observatory infrastructure had to be created. A base camp was established at the foot of the mountain, and a 20-kilometre-long road built to enable the transport of the heavy and bulky, yet delicate, instrumentation from the camp to the top of the mountain. A number of general facilities such as workshops, dormitories, a power station etc. were also needed to support the observatory.

The base camp at Pelícano ca. *1967. (Photo: Jan Doornenbal)*

The base camp was established where the road from the Pan-American Highway reaches the foot of the hills that lead up to La Silla. The name of the camp was Campo Pelícano. Blaauw's well-chosen quote from the ESO Annual Report of 1964 lets us catch a glimpse not only of how far (or how little) the ESO project had evolved, but also the conditions the small group of ESO employees, truly pioneers, were facing. At the end of 1964, the La Silla operation comprised:

"a) Office in La Serena, functioning with five persons active,

b) Camp Pelícano, with two old houses and four new ones installed, a carpenter's workshop in use, fifteen persons active, animals' camp installed and functioning with five horses and six mules, two wells ready with one pump installed.

c) Road project [i.e. planning and lay-out of the road], ready from camp Pelícano to the top of La Silla."

The camp manager, a Chilean, had organised life according to Chilean traditions, where the staff ate separately according to rank and status. The freshly arrived

Europeans displayed a more egalitarian view and soon changed this, so that from then on, all staff would frequent the same quarters and eat together.

In 1966, the road to La Silla was completed. It was now possible to bring equipment up to the top of the mountain, and soon after, the first scientific work at La Silla began. Later in that year, the 1-metre photometric telescope was the first telescope at La Silla to become operational. This was certainly a milestone, but the 1966 Annual Report dryly noted that, *"the astronomical observing activity was started late in November, after the 1-metre Photometric Telescope had been installed in a provisional building."* And in December, the first visiting astronomer, Jean Pierre Brunet from the Observatoire de Marseille, arrived at La Silla, *"to initiate a photoelectric programme of observations of stars in the Magellanic Clouds."* However, it took until mid-1968 before the system that we know today, with six-month observing periods, was put in place, following a decision by the Scientific Programmes Committee (see also the following chapter). The first call — for what became P2[1] in ESO jargon — announced in the *ESO Bulletin No. 4*, illustrates nicely the atmosphere of ESO at the time:

"Applications for the use of the ESO 1-metre Photometric Telescope for the period March 1 — September 1, 1969.

Applications for the use of the above mentioned telescope within the above period may now be submitted to the Directorate of the European Southern Observatory, 131 Bergedorfer Straße, 205 Hamburg 80, W. Germany.

The applications should be received by the Directorate not later than September 1, 1968. Applicants may expect to be informed by November 1, 1968 whether, and how much, time will be granted.... The application should normally be endorsed by the Director of the applicant's Institute. It will be reviewed by the ESO Scientific Programmes Committee....

The ESO Budget provides for travel funds and for fixed allowances for lodging and food to such an extent that, as a rule, it will not be necessary for the applicants to whom observing time is granted (or for their Institute) to contribute financially. Defrayal of

[1] The numbering system was referred to in the ESO Annual Report for 1972, with the period 1 November 1969–1 May 1969 as P1.

travel expenses of accompanying wives is foreseen to a limited extent and only in case the observers will have to stay in Chile for a period of at least 6 months.

Hamburg-Bergedorf, June 1968."

So long observation runs were considered — and the remark about accompanying wives clearly suggests that observers were assumed to be male.

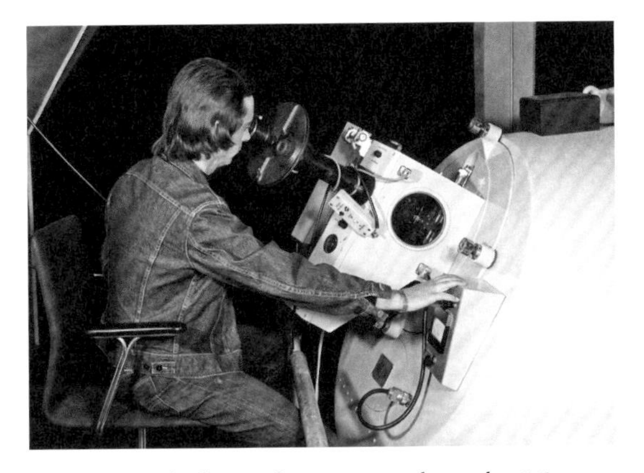

Working with the coudé spectrograph at the 1.5-metre telescope.

In 1968, the 1-metre telescope was also followed by the GPO and the 1.5-metre telescope, which became "the king of the hill" — at least as regards its size. It was originally fitted with a Cassegrain spectrograph, which became known as the "Chilicass"[2], and soon after with a coudé spectrograph. Observational astronomy was very different then from today, as is illustrated by a small anecdote told by the Belgian astronomer Jean-Pierre Swings, the son of Pol Swings, one of the founding fathers of ESO. *"My first experience at La Silla was in January 1972 when I was a Carnegie fellow in Pasadena, doing research on peculiar emission-line stars with infrared excess. I used the 1.5-metre telescope coudé spectrograph. One of the objects was the star HD 45677, a peculiar Be star that my father had already worked on in the 1940s. It has a tremendous infrared excess due to circumstellar dust in an equatorial ring. I decided to make a very long exposure — of three nights. In the middle of the second night, there was an earthquake — so we closed the shutter and resumed observations afterwards. Afterwards we developed the plate. Nothing had moved, so my spectrum was perfect!"*

[2] Richard West offers a fitting description of what it was like to observe with the 1.5-metre telescope in the early years. Observing a 13th magnitude nova in the Large Magellanic Cloud (LMC), *"I spent three nights at the Chilicass spectrograph ... exposing continuously for 4, 5, and 7.5 hours, respectively. To do the visual guiding properly.... I had to balance most of the time in total darkness, high up on a ladder at the edge of the floor platform. It was indeed a rewarding feeling when I finally saw a useable spectrum on the small plate in the dim darkroom light at the end of the night."* (West, 2002)

Whilst scientific life had begun at La Silla, more mundane issues had to be addressed as well, such as securing the water supply. This picture was taken in October 1971, when ESO was searching for additional water sources. The search was led by Jacques Rouel (sitting with his back to the camera). To his right is Emile Leroy, to his left Raul Villena with his hand on his cheek. The man to the left is unidentified. (Photo: Svend Laustsen)

In the following year (1969), according to the ESO Annual Report, the first scientific publications were registered by ESO: four by ESO staff (Arne Ardeberg, Eric Maurice, James Rickard and Bengt Westerlund) and one by a visiting astronomer, François Noël. Since then, ESO has kept track of scientific publications based on observations with telescopes at ESO, and this record provides an important key to understanding the impact that ESO has exerted on European astronomy over the years[3].

Among the early observers was a Czech astronomer, Luboš Kohoutek, who, after the political turmoil in his homeland in 1968, had found a position at the Hamburger Sternwarte. His main scientific interest was the study of planetary nebulae, objects that are stars in their final evolutionary phase rather than planets in formation as the name seems to suggest. His first observing run, in January 1974, however included a somewhat different object: Comet 1973f, which he had discovered during the previous year and which, at the time, was expected to become the brightest comet of the

[3] The current record year for publications was in 2011, with 783 refereed papers.

century. It did not[4], but bright comets have always aroused public interest, so a press conference was organised at the hotel at La Silla. Bengt Westerlund, as ESO Director for Chile, chaired the session, but the "star attraction" was the comet discoverer himself. Such events were rare at La Silla, and not really part of the La Silla ethos at the time. Nonetheless, it provided the public with a glimpse of the work of the observatory and the fascination of astronomy. And at that time, La Silla had truly become a working observatory — a factory, as Kohoutek later described it. The 1974 ESO Annual Report lists 66 publications (albeit including a few articles published in popular magazines). Unsurprisingly, most of the research activities focussed on the classical domains of European astronomy — the study of the Milky Way and its components — although some extragalactic work had begun.

≈

"La Silla 19xx?" — Under this heading the March 1978 ESO Messenger *cartoon addressed the question of the proliferation of telescopes at La Silla. (Drawing by Karen Humby)*

The sharp increase in scientific activity and thus in publications was not only due to the implementation of the initial telescope programme, as foreseen in the Convention. In 1967, the installation of a telescope not foreseen in the Convention had already begun: a 0.6-metre telescope owned by the Ruhr-Universität Bochum. This was to become the first of a number of national telescopes that with time added to the growing telescope park at La Silla. Unlike the ESO telescopes that were essentially purpose-built, the Bochum telescope was a standard off-the-shelf instrument, supplied by Boller & Chivens

[4] The comet reached magnitude 0, making it the 10th brightest comet of the century. By contrast, Comet West (C/1975 V1), discovered a couple of years later by ESO astronomer Richard West and not expected to become very bright, reached magnitude –3 and, in the 20th century, only superceded by Comet Ikeya-Seki (C/1965 S1).

Joachim Dachs preparing for a night's observations with the Bochum photoelectric reflector in 1982. (Photo: by the author)

(B & C) in the US. Initially installed in a provisional dome (the one used for the 1-metre telescope and later to be used by the Dutch 0.9-metre telescope), it was equipped with a photometer built at the Georg-August-Universität Göttingen. Built *"to supplement in an optimal way the instrumental plans of ESO"* (Schmidt Kaler & Dachs, 1968), it became a highly productive and reliable workhorse at La Silla for many years. Bochum observers also became the equivalent of what Mir cosmonauts became to spaceflight — known for their long-duration observing runs, often lasting for months[5].

The Bochum telescope was soon to be followed by a 0.5-metre photoelectric telescope equipped with a four-channel photometer for Strömgren narrowband photometry. The telescope belonged to the University of Copenhagen (Københavns Universitet) and was admitted to La Silla following the accession of Denmark to the ESO Convention in 1967. This telescope began its career in the provisional dome, until it was placed in its own building in 1971. The telescope had been built in Denmark and in 1971 ESO acquired a copy of it, partly to serve as a testbed for technologies to be applied to the 3.6-metre telescope[6]. The idea of using a telescope that was in service

[5] The long observing runs gave rise to the expression "the good-weather catastrophe". Astronomers obviously need good weather, but after many weeks or months of clear skies — and thus uninterrupted work, even the most dedicated scientists may have longed for the occasional break under the excuse of an overcast night.

[6] The case of what became the ESO 0.5-metre telescope is also interesting from other points of view. According to the Annual Report of 1968, *"The Copenhagen Observatory generously agreed to build this telescope at their own cost. The instrument is to be provided with a digitally controlled presetting system. It should not be considered as an additional telescope in the sense of the ESO Convention... ."* Over the years, again and again, Member States have made additional voluntary donations to ESO, ranging from instruments and various other pieces of equipment to complete telescopes. The reference to the ESO Convention shows that there was an ongoing discussion, and perhaps some unease in certain corners, about how freely the Convention could actually be interpreted.

in this way was repeated twenty years later when the NTT played an important role in the preparation programme for the Very Large Telescope. Further national telescopes to follow included the Dutch 0.9-metre telescope, the Danish 1.54-metre telescope, and three Swiss telescopes, first a 0.4-metre telescope (installed under a special arrangement since Switzerland was not yet a member of ESO), later a 0.7-metre and, finally, the 1.2-metre Leonhard Euler Telescope, used now for exoplanet hunting — a topic to which we shall return.

For a while La Silla would be populated with about as many national telescopes as telescopes belonging to ESO. The organisation provided logistical support and maintenance in return for part of the available telescope time. This significantly increased the scientific life at the observatory and brought a whole generation of young astronomers into contact with ESO (and with each other) at an early stage in their careers, but it also drained resources and it was not liked by all ESO Member States. Blaauw alludes to discussions in Council about the proliferation of telescopes, with concerns about available space and status of the national telescopes. And in his book *Europe's Quest for the Universe* — and on a more critical note — Woltjer states that *"[the] multiplicity of efforts may well have been detrimental for the progress of the main project, the 3.6-metre telescope."* We shall return to the 3.6-metre project shortly, but it is worth noting that the question of prioritisation has remained with ESO and has resurfaced on several later occasions.

≈

The ESO office in Vitacura in December 1968. The main building to the left contained offices for the staff, whereas the building to the right of the centre was the first Astroworkshop, providing mechanical services for the observatory. The photo was taken from a small tower at the adjacent United Nations compound. (Photo: Eric Maurice)

In parallel to the development of the La Silla site, ESO had established an office complex on land donated by the Chilean government in Santiago's Vitacura district, adjacent to the United Nations Economic Commission for Latin America (CEPAL). Furthermore, already in 1964, ESO had acquired a guesthouse, a charming colonial-style house in the agreeable nearby district of Las Condes. The decision to create an office in Santiago was preceded by discussions both about the location and the role of this office. AURA and CARSO, the two US observatories, had set up their central offices in La Serena[7], but ESO preferred to locate their offices in Santiago to maintain closer contacts with the universities, the government authorities and also out of consideration for the needs of its staff — schools, medical services, cultural life and, in general, the amenities that a capital city can offer. The close scientific interaction with the American astronomers, however, fell somewhat by the wayside. To understand this, it should be mentioned that the question of whether ESO should have its own staff of scientists — or not — was still undecided. ESO's Santiago Centre with offices, library, photographic service and a technical workshop was finished in March 1969, at the same time as the inauguration of La Silla.

The inauguration ceremony at La Silla. From left to right: Gabriel Valdés Subercaseaux, Minister of Foreign Affairs of Chile; Olof Palme, Minister of Education of Sweden; the President of Chile, Eduardo Frei Montalva and Henk Bannier, President of the ESO Council.

[7] ESO also established an administrative office in La Serena in 1964.

Chapter I-5

ESO — *Quo Vadis?*

"Bei ESO lag nichts vor als die Pariser Konvention und ihr Finanzprotokoll."

Otto Heckmann, 1976.

The establishment of La Silla, even if the main telescope was not yet there, meant that a dream had come true for Europe's astronomers: they could begin to do science. Astronomers are driven by their unrelenting thirst for knowledge and the need to understand of some of the greatest riddles the Universe has to offer. To address these questions, they need telescopes, ever more advanced and (most of the time) ever bigger. Given the magnitude of this overall task it is perhaps not surprising that their focus is precisely on that: getting access to telescopes. This was their purpose in creating ESO in the first place. But what did that mean in practical terms? What kind of organisation would ESO be? How should it be run? How would it fit into the overall scientific landscape? And how would it adapt to changes, whether scientific, technical or political? There were many different opinions about these issues, but very little is set out in the Convention, which — when it came to defining the initial telescope programme — is otherwise remarkably specific and precise. Developing the institutional framework for ESO's activities thus became the task for the Director General together with the various committees, and in particular the ESO Council.

Otto Heckmann had been appointed as Director (General) of ESO in 1962. He was highly respected as a scientist and had been Director of the Hamburger Sternwarte[1], a post he assumed in 1941. During the war, he had managed to keep many young scientists on his staff on the pretext that they were vital for the war effort, a rather unusual label for astronomers, but one that may have saved the lives of young scientists. In 1942, he also published the book *Theorien der Kosmologie*, and, as Alfred Behr later noted in Heckmann's obituary in *The ESO Messenger* (Behr, 1983), *"whoever witnessed the period of intellectual suppression in the late thirties in Germany, can*

[1] Hamburger Sternwarte is located in Bergedorf, on the south-eastern fringes of Hamburg, some 15 kilometres from the city centre. The present site was opened in 1912, in what was then relatively tranquil countryside.

imagine the hazardous enterprise of such a publication at that time."[2] But Heckmann was also a man of the old school — disciplined and autocratic. Even so, he was highly respected among his senior collaborators, as someone who would listen to advice and support his people. Heckmann felt comfortable in Hamburg and he used the services of the observatory there to help him during the early years. Here he surrounded himself by a small team, with Jöran Ramberg (of Sweden) as Technical Director and Johan Bloemkolk, a Dutchman, as Head of Administration. He also used his private resources: his wife, Joahanna, acted as an unpaid secretary during his tenure. There was little reason, he thought, to look for an alternative place for ESO's Headquarters, despite the statement in the Convention that the Headquarters *"shall be provisionally established in Brussels"*[3].

Aerial photo of the Hamburger Sternwarte in 1960. (Courtesy: Hamburger Sternwarte)

---

[2] Behr's account is not undisputed. Interested readers are referred to a detailed discussion by Jochen Schramm (Schramm, 1996). Whatever facts, Heckmann's appointment was endorsed by prominent scientists including Oort and Blaauw, who, given the proximity of the war, were unlikely to accept any German candidate solely on the basis of scientific merit.

[3] With the appointment of Blaauw as his successor, it is therefore no wonder that there was some speculation that ESO might move to Amsterdam, but this was never seriously considered.

≈

The first specialised ESO Committee, established even before the organisation formally came into being, was the Instrumentation Committee. This was created in 1961 and continued until 1974, chaired by Charles Fehrenbach until 1971, with the exception of a short period in 1961/62, when Otto Heckmann led the proceedings. From 1971, Jan Borgman chaired the meetings. As the predecessor to the Council, we have already met the ESO Committee, which was created in 1953 and ceased in 1963, chaired for most of the period by Jan Oort[4]. In 1967, a Scientific Programmes Committee was established, led by Bengt Strömgren. This committee initially looked after both the allocation of observing time and scientific policy, but split in 1971 into an Observing Programmes Committee and a Scientific Policy Committee. The Scientific Policy Committee, in turn, absorbed the Instrumentation Committee to become the Scientific Technical Committee, which still plays a key role today in ESO as the principal forum for discussing the scientific orientation of the organisation. What were the issues confronted by these committees? The main questions were how an operational model for the allocation of telescope time could be developed, now that the first telescopes had become available, basic issues of how much work should be done within the organisation and how much could be carried out by external contractors and the question of scientific staff and the rules for employment. Let's take each of these issues, but in reverse order.

≈

One of the fundamental questions was whether ESO should be an *observatoire de mission* — a service provider operating observational facilities on behalf the Member States and their scientific communities — or become a scientific centre in its own right, with its own scientific staff. This issue has been a source of discussion and tension that has surfaced on several occasions during ESO's history, with the pendulum taking a few swings, although the general idea that ESO needed a group of active scientists became widely accepted in the mid-1970s (albeit following dramatic discussions, which both Blaauw and Woltjer have described in their books). We shall return to this later, since it had implications that went far beyond the question of hiring staff. One aspect, however, should also be mentioned at this stage — the idea that staff would be hired for limited periods of time, or, as it was expressed in those days, as semi-permanent staff (originally thought to mean for three years). Astronomers

[4] Bertil Lindblad chaired two meetings, Otto Heckmann and Paul Bourgeois one each.

joining ESO were thus expected to have positions in their home countries to which they could return after their stint at ESO. Although this was seldom the case, the discussion cast a long shadow forward to the wider mobility question — still unresolved — within European science today in the context of the European Research Area.

The question of what should be done by the organisation and what could be done outside it was no less of a source of diverging opinions. Today we would term this outsourcing, but this would fail to describe the situation in the early 1960s accurately since ESO in fact had a minute staff. The question, therefore, was rather what insourcing might be required and how that could be achieved. This turned out to be far more serious than many had thought and the failure to solve this question in a timely fashion undoubtedly was a major contributor to what emerged as ESO's biggest problem at the time: the snail-paced progress of the 3.6-metre project. With the benefit of hindsight, the answer today would have been that ESO both needed to possess in-house competencies and work closely with industry. But hindsight was not available in the early 1960s. On the contrary: *"... even if you put together the best European astronomers and their instrumentally minded colleagues, they had no experience at all in realising a project of that size!"* as Adriaan Blaauw expressed it in an conversation with the author many years later.

On the question of allocation of telescope time, the original thinking was that applications would first be considered by the country from which the application came and subsequently by ESO, reflecting a traditional view of European cooperative projects. However, in the end, the national pre-selection was never carried out and ESO simply collected and assessed all the applications through the appropriate committee. One reason why this could happen may have been the small number of applications at the beginning — quite natural given the limited number of telescopes. In 1967, ESO organised an information meeting in Hamburg for applicants. Five astronomers attended. This would soon change and today, ESO receives about 2000 applications each year. Nonetheless, this single-stage European-level evaluation of proposals clearly preceded practices in other fields of science by decades and it is interesting to note the many similarities between the selection procedure at ESO and the one used by the European Research Council (ERC), established 40 years later and widely praised for its work and its new approach.

≈

That the European view on matters prevailed within ESO (despite the small number of Member States) is also evident in its support for the creation of the journal *Astronomy and Astrophysics (A&A)* in 1968. *A&A* was the result of a merger of five (later six) national astronomical journals in Belgium, France, Germany, the Netherlands and Scandinavia, initially at the initiative of Stuart Pottasch, Jean-Louis Steinberg, and Anders Reiz (Blaauw, 2004). Since the new journal had no legal personality of its own, ESO made its good offices available for the purpose of concluding contracts on behalf of the journal. In the usual reserved style of the day, the 1968 Annual Report stated that *"ESO lent its administrative and legal advice to the foundation of a new international astronomical journal"* Behind this factual statement lies the fact that *A&A* contributed to instilling a sense of European identity to its own astronomical community. Meanwhile, *A&A* has become one of the leading journals for professional astronomers and ESO has maintained its links with it and its board of directors ever since.

ESO on television — While ESO had begun to address some of its institutional issues, work proceeded on the telescopes to be installed at La Silla. Dutch television showed the construction of the 1-metre telescope at Rademakers in Rotterdam. This photo, taken from a television screen, shows (from left to right) the head engineer at Rademakers, Blaauw, Jan Doornenbal (at the time mechanical engineer at the Kapteyn Laboratory in Groningen) and Martin de Vries, astronomer at Kapteyn. The person seen from behind is Ben Hooghout, who designed the telescope. The programme was transmitted in March 1965. (Courtesy: Jan Doornenbal)

The 3.6-metre telescope and Coudé Auxiliary Telescope (CAT) buildings under construction in early 1975. The crates contain material for the dome.

Chapter I-6

Towards the 3.6-metre Telescope

"The design of a large telescope, though not possible without the help of astronomers, is essentially an engineering problem. It may well be that the organisation of a close co-operation between engineers and astronomers is one of the really important steps towards a modern large telescope."

Klaus Bahner, *ESO Bulletin* 5, 1968.

On 25 March 1969, La Silla was officially inaugurated. The formal ceremony took place in the (empty) Schmidt telescope dome, with high-level visitors from Europe and Chile, including the President of Chile, the Chilean Minister of Foreign Affairs and the Swedish Minister of Education of the time, Olof Palme. Yet despite the festive occasion and the joy of having passed an important milestone towards the realisation of the initial programme, ESO still had a long way to go. Work on the main project, the 3.6-metre telescope, as well as on the Schmidt survey telescope progressed in Germany, albeit slowly. The initial idea, harking back to the first discussions in 1953/54 had been that ESO's main telescope should be a copy of the 3-metre telescope at the Lick Observatory in the US. This observatory dates back to the late 19th century. Placed on a mountaintop near San José in California it illustrates well how American astronomy progressed through the opening of observational sites under superior conditions. The 3-metre telescope (120-inch), known today as the C. Donald Shane Telescope, was the largest at Lick and the second largest reflector in the world when it was commissioned in 1959. Today we would call it a classical telescope, meaning one with a massive, monolithic primary mirror and a telescope structure on an equatorial mount, contained in a huge dome. Yet at the time, the Lick telescope was a modern telescope, just entering into service as ESO was being formed. However, it soon became clear that what one today might call a copy-and-paste solution would not be the right approach for ESO. Firstly, large telescopes usually represent one-off solutions, developed to satisfy the particular scientific requirements of the time. Secondly, the chosen location has a significant influence on the particular design, as have a number of perhaps more mundane, but very practical considerations such as access to the site. It is therefore not surprising that the history of the development of the astronomical telescope is one of constant refinement, improvement and adaptation. Historically, this has been a fairly smooth progression and, it could be argued, with only a few paradigm changes: the switch from the refractor to the reflector, the introduction of multi-mirror (array) telescopes and, in the late 20th century, the development of thin-mirror and segmented-mirror telescopes, to which we shall return later. At this stage it suffices to understand that ESO's decision not simply to build a copy of the Lick telescope was fully in line with this tradition of continued development and ultimately gave European astronomers a better (and larger) telescope. There was another advantage, perhaps not fully appreciated at the time: the well-proven value of learning by doing. By the end of the 3.6-metre telescope project, ESO had built up the skills, and the self-confidence, enabling it not simply to think of the next generation of astronomical telescopes but to lead their development, at least as regards some of the crucial technologies needed. What a learning curve! But let us return to the situation in the late 1960s.

≈

Originally, the task of designing the main mechanical structures for the telescopes had been given to an engineering company in Hamburg, run by Walter Strewinski. Strewinski possessed an ingenious, but complex mind. He clearly exerted a strong influence on the design of 3.6-metre telescope but, due to his erratic ways, he also caused much frustration and in any case, his small design firm was clearly inadequate to carry out a task of this magnitude. By 1969, the lack of tangible progress with the design of the mechanical structure caused ESO's Council to lose patience and ESO began to take the 3.6-metre project into its own hands. While Council and the Instrumentation Committee had long been uneasy about the situation, this new development had started in the previous year. A small group was set up at the Hamburg Office[1] and charged with addressing automation aspects[2] of the telescope. The group was led by a Danish astronomer, Svend Laustsen, and initially included another Dane, Mogens Blichfeldt, a mechanical engineer. In Chapter I-4, we saw how the 1-metre telescope carried a strong Dutch signature and the 1.5-metre a clear French signature. From now on, the 3.6-metre project would acquire a Danish character, even if the magnitude of the project obviously meant the involvement of experts

[1] The group was not physically located at the observatory, but occupied rented rooms, first at the Iduna-Haus, in the centre of the town of Bergedorf.

[2] Computer control of telescopes was in its infancy in those days.

from many countries, with Bernth Malm (from Sweden) and Jean Weber (a Frenchman) among the first recruits. The group formed the nucleus of what was to become the ESO Telescope Project Division, created in 1970. Building a major astronomical telescope is not a simple task: the technical challenges are numerous. New and innovative solutions, going hand-in-hand with technological progress, mean that, normally, no two telescopes are alike. Finding people who can master telescope design and construction is a challenge in itself. Reaching the critical mass of competent scientists and engineers necessary to realise a major project, such as the 3.6-metre, is no less of a challenge. The small group in Hamburg quickly got its teeth into the project, but it, like Strewinski's group, was nowhere near sufficient to deliver a modern telescope of this size[3].

Before we turn the next page of the tale of the 3.6-metre, let us briefly stop to look at the telescope as it was planned. It was a quasi-Ritchey-Chrétien telescope[4] with a primary mirror 3.5 metres in diameter. There were three foci: an *f*/3 prime focus at the top of the telescope with a camera including a Gascoigne plate corrector (later to be replaced by a triplet corrector offering a wider field of view), an *f*/8 Cassegrain focus below the primary mirror and an *f*/30 coudé focus located below the observing floor. The increase in size of the primary mirror over that of the Lick telescope was chosen to compensate better for the obscuration of the prime focus cage, which had to accommodate the observer. As it happened, upon delivery of the primary mirror, it turned out that the useable area could be extended to about 3.6 metres, increasing the light-collecting area and giving the telescope its name. The telescope was to have an equatorial mount with a combined horseshoe and fork structure, one of the modifications that Strewinski had proposed to the Lick design. Strewinski had also suggested the use of interchangeable top ends, with those not in use being stored outside the telescope rather than being carried within the telescope's mechanical structure at all times.

The primary mirror, provided by Corning Glassworks in the US (and paid for with the grant from the Ford Foundation) was made of fused silica and had a mass of 11 tonnes. The order was placed in January 1965 and the blank was delivered in February 1967 and shipped to REOSC, the French optical company, for polishing.

---

[3] From the outset, designing the entire telescope was not part of the task of the group.

[4] A Ritchey-Chrétien optical design, named after its inventors Henri Chrétien and George Ritchey, combines a concave primary with a convex secondary mirror. To improve the image quality over a wider field of view than the classical Cassegrain telescope, both mirrors are hyperbolic.

The 3.6-metre primary mirror at REOSC in 1968.

Due to flaws in the surface, it had to be returned to Corning, but — with a new top layer — was back at REOSC in 1969 and finally accepted by ESO in 1972. Despite the delay caused by the defects in the original blank, when the mirror arrived in Europe there was no mechanical structure in sight. Heckmann's faith in Strewinski and his company had been in vain. At the same time, Heckmann's term in office was coming to an end after having been extended in 1967. Council chose Adriaan Blaauw, who had been associated with ESO since day one, to succeed him. They also made it clear that the most important task for the incoming Director General was to complete the 3.6-metre telescope[5]. Blaauw assumed office officially on 1 January 1970, but by the autumn of 1969, he had already started to look for solutions to what was now a major problem for ESO. He first contacted CERN in Geneva, asking for help. Soliciting help from CERN was natural: CERN had in many ways served as a model for ESO, and delegates of three of ESO's Member States also represented their countries on the CERN Council. Most importantly, CERN also possessed the necessary skills and capacities needed to develop advanced scientific instrumentation on a scale that made it possible to engage in telescope design, even if the field was new to the CERN engineers. Even better: at the time, CERN was about to complete what was the world's first proton–proton collider, the Intersecting Storage Rings (ISR), while the next project, the Super Proton Synchrotron (SPS) was still a while away. At precisely the right time, CERN had free capacity in terms of engineering work.

In parallel, Blaauw also made contact with ESRO, the European Space Research Organisation[6], whose Director General Hermann Bondi — himself an accomplished cosmologist — also viewed potential cooperation favourably. Blaauw has provided

[5] The pressure was on, as can be seen by a letter of 4 June 1970 from the Chairman of the ESO Science Policy Committee, Bengt Strömgren, to the ESO Council: *"... any postponement regarding the 3.6-metre Telescope would endanger the future of ESO."*

[6] Together with the European Launcher Development Organisation (ELDO), ESRO was transformed into the European Space Agency (ESA) in 1975.

The process of establishing institutional links between ESO and CERN was greatly eased by personal relations and friendships. Seen in this picture, obtained at CERN in December 1966, are Henk Bannier (left) and Gösta Funke. Bannier was President of the CERN Council in 1964–66, followed by Funke (1967–69). Funke presided over the ESO Council in 1966–1968, Bannier in 1969–1971. (Photo: CERN)

the details of the discussions both with CERN and ESRO, so that we shall restrict ourselves to recalling that, after due consideration by the ESO Council and subsequent successful negotiations with CERN, an agreement was signed on 16 September 1970, allowing ESO to establish its Telescope Project (TP) Division on the CERN premises in Meyrin on the outskirts of Geneva, straddling the Franco-Swiss border. The agreement foresaw that ESO would set up its operations with its own personnel, but with help from CERN's scientific and technical divisions. Furthermore, CERN would provide space for the erection of ESO offices as well as administrative and logistical support, while ESO would appoint a head of the division, with the same authorisation and powers as a CERN division head. We shall return to the political aspect of the CERN–ESO collaboration in Chapter IV-9. At this moment, however, our focus will remain on the 3.6-metre telescope project.

At the first Committee of Council[7] meeting, held in Hamburg on 6 May 1970, Blaauw presented a thorough document — *"the first one to describe in comprehensive way the various aspects of the [3.6-metre telescope] project, their interrelations, the time schedule and the personnel planning... ."* The main author of the report was Svend Laustsen, assisted by the small technical group as well as Jöran Ramberg. Three paragraphs in the introduction defined the new path, an almost complete relaunch of the project:

"A. ESO must form its own group of astronomers, engineers, etc. which group shall be able to conduct the project through all its phases including the first period of operation of the instrument in Chile... .

B. The group must at any time have all parts of the project under firm control. But in order to keep the group within reasonable bounds, a major part of the design work and all construction work will be done by consulting and manufacturing firms... .

C. The group should be offered the best possible working conditions. For its tasks in Europe it should be located in a scientific and technological milieu and be offered good service facilities."

The initial plan foresaw the staff complement growing from 5.5 full-time equivalent positions in July 1970 to 26 by 1 January 1972.

Svend Laustsen was appointed as head of the new ESO division with Alfred Behr and Anders Reiz as close collaborators. On 1 October 1970, Laustsen's group moved from Hamburg to Geneva. During the subsequent years he oversaw the building up of a highly competent group of engineers, some on ESO contracts and some seconded from CERN. One of the key recruitments from CERN[8] was Wolfgang Richter, who became chief engineer of the 3.6-metre telescope. Other recruits of key importance involved staff from CERN's Services and Buildings Division, including Jacques Rouel and Emile Leroy. But for some skill sets ESO needed to look elsewhere. On the occasion of ESO's 40th anniversary Svend Laustsen wrote in *The ESO Messenger*: *"In the optical field, however, ... we had not succeeded ... in attracting optical technicians.*

---

[7] The Committee of Council was intended to allow Council Members to meet under less formal circumstances and, among other things, discuss sensitive issues ahead of the regular Council Meetings. The Committee of Council was instituted in December 1969.

[8] Several key people remained on CERN contracts though they worked full-time for ESO.

Finally Alfred Behr and I agreed to ask Ray Wilson at the Zeiss Works, whether he knew of any young man he could recommend to us. He replied: 'No, I do not know of any technician for that job, but I can offer myself to ESO as an optician... .' Shortly after taking up his duties he presented plans for an optics group, and according to this Francis Franza, Maurice Le Luyer, Daniel Enard ... were engaged."[9] (Laustsen, 2002) All of them were key recruitments, but as we shall see, the hiring of Ray Wilson and Daniel Enard would turn out to be of especially decisive importance for the organisation.

As an initial task, the TP Division drew up a revised schedule and presented it to the ESO Council in February 1971. The plan foresaw the completion of the 3.6-metre building by December 1975, followed by the telescope assembly and culminating with first light in 1976.

≈

The way forward had now been defined, but it was less straightforward than many would have liked. Telescope design had not stood still in the intervening years and the long delay meant that Strewinski's design was no longer up to date, particularly in terms of exploiting the benefits offered by the new electronic control systems. Since the telescope optics were ready, the new telescope team was forced to accept a number of technical compromises. Both time and financial pressure made this unavoidable. But sometimes this pressure turned out to be beneficial. Thus the next step was the crucial decision to discard the existing plans for the telescope building and begin again from scratch. The original design, developed by the Hamburg-based firm of Lenz Architekten & Ingenieure, foresaw a huge, rectangular coudé floor, protruding from the circular telescope building and involving very high cost. Too high, the Finance Committee thought[10]. In the end, a much smaller and undoubtedly better solution was found. The most important changes involved the reduction of the coudé floor by more than 50% and a dramatic reduction in the area to be heated[11], paving the way for the design of the clean, cylindrical building that has since graced the highest point of La Silla.

---

[9] The Optical Group was established in March 1973.

[10] Chapter II-3 will look at the question of the coudé focus from an instrumentation point of view.

[11] Over the decades, an important trend in telescope design has been to remove as many heat sources as possible to avoid thermal effects within the telescope enclosure that could degrade the seeing. At ESO, the fruits of this were reaped with the 3.5-metre NTT and later with the VLT, but these were early days and the complexities of dome seeing were poorly understood.

By 1973, construction contracts had been awarded to industry for the building (to the Dutch company Interbeton), the 350-tonne steel dome (Krupp, Germany) and the main mechanical structure (Creusot-Loire of France). Further contracts included the main gears for the telescope (MAAG, Switzerland) and the provision of the diesel-driven power turbines (Motoren Werke Mannheim)[12]. Major industrial contracts are coveted by industry, and therefore often involve difficulties as Member States vie for them. It appears, however, that the awarding of the main contracts for the 3.6-metre telescope presented fewer problems than some of the large contracts for later telescope projects. It is also possible that, in this situation, time pressure had a healthy effect on the decision making. With major national projects on their way in both France and Germany, ESO had to deliver on its promise to European astronomy, if it were not to lose its relevance before it had really begun in earnest.

At La Silla, construction started in June 1973 and 18 months later, after 3000 cubic metres of concrete with 350 tonnes of reinforcing material had been poured, this work was completed (Leroy, 1975). In Europe, the mechanical structure was assembled for test purposes at a factory in Saint Chamond near Lyons in the autumn of 1974. The tests were completed by November 1975. Before it was shipped out from the Port-St-Louis-du-Rhône near Marseille, the ESO Council had a chance to inspect the test assembly.

In 1973, the first ESO TP staff were transferred to Chile with a gradual build-up in numbers as the 3.6-metre project progressed. For many TP staff the prospect of moving to Chile involved interesting tasks in an unusual environment, and it was not difficult to convince people to leave Europe for a period of time. However exciting the task may have been for the individual staff members, though, there was a price to pay for their families, most of whom would settle in Santiago, 600 kilometres away, and accept that while they might live a materially comfortable life there, neither Chile nor Santiago could, at that time, offer medical and educational standards comparable to those in Europe. Spouses — almost always husbands — spent long and exhausting

[12] Along with these contracts, ESO also awarded a contract for the building to house the planned Danish 1.54-metre telescope. This illustrates that the new TP Division almost instantly had additional tasks on its hands, beyond delivering the 3.6-metre telescope. According to the ESO Annual Report of 1974, it also included tasks previously carried out by the Technical Department of ESO-Chile, including *"construction of support facilities such as new stores and workshops, living quarters for the local staff, the improvement of the water, heating and electrical distribution networks as well as the road improvement programme."* Clearly this was far from negligible: Laustsen later estimated that these tasks may have taken up to 50% of the capacity of the TP Division. Perhaps we also see the seeds to the 1987 suggestion that the VLT could be built with only a small addition to the staff, an idea that we today regard as totally unrealistic.

The main mechanical structure of the 3.6-metre telescope at the factory site of Saint Chamond.

duty tours at La Silla, and La Silla became the burial ground for many a marriage. *"ESO is a disaster for families,"* as one staff member expressed it.

≈

But for one employee, simply transferring to Chile was not good enough[13]. As project leader, Svend Laustsen felt that he had to be reachable on the site all the time. Married and with four school-age children, he took over some abandoned drawing offices that had been erected next to the construction area (close to where the Swiss 1.2-metre Leonhard Euler Telescope is now located) as his private quarters and moved there with his wife and two younger children. His wife looked after their school education, often served as a charming host for many visitors from Europe and created an atmosphere of normality in this otherwise rather unusual place — a combination of a remote mountain observatory and a large construction site. The *Casa Laustsen* became a popular place on the mountain, often referred to as "the Culture House".

[13] There are a few examples of staff living for extended periods at La Silla, with or without family. Mart de Groot is one and Hans-Emil Schuster another.

A rare example of family life at La Silla — Kirsten Laustsen with daughter Susanne (photo left) and Susanne having fun with one of the desert foxes that thrive on waste food at La Silla. (Photo: Svend Laustsen)

≈

On 7 April 1976, the Spanish vessel *MV Riviera* arrived at the Pacific seaport of Coquimbo, 150 kilometres southwest of La Silla, with the mechanical parts for the telescope in its hold. Twenty-three trucks were waiting, forming a 500-metre-long convoy, to transport 450 tonnes of material to La Silla (Plathner, 1976). It took three days, but everything arrived safely and the assembly of the main mechanical structure could begin.

Snow at La Silla in 1975. The building to the left is the Danish 1.54-metre telescope. Behind, the 3.6-metre building, still with an uncovered dome, is seen. (Photo: Svend Laustsen)

On 26 May, La Silla was hit by a severe storm. *"There was ice all over as we drove up to the dome. When we came out again, the wind had pushed the cars quite a bit from where we'd left them. The wind speed at the top of the mountain reached at least 200 km/h, then the anemometer was blown down,"* Laustsen later recalled. The result was severe damage to the inner insulating shell of the dome.

Soon after, on 2 June 1976, the telescope optics left the factory site at Ballainviliers near Paris. They travelled first to Dunkirk and then onboard the *m/v Anjou* to Coquimbo. CERN's press office did not miss the opportunity: its press release carried the headline: *"Precious Cargo — Don't Drop!"* They didn't. But if Laustsen had thought that all the troubles had now been put behind them, he was wrong. On 26 August, as the installation work at La Silla was running at full speed, lightning struck the dome of the building. *"I was in the dome at that moment together with some colleagues ... suddenly there was an enormous bang. It did not seem to come from any specific direction; we were right in the middle of it. We later found a hole in the dome. The lightning destroyed a lot of the electrical equipment, giving us a month-long delay."* In earlier times, some might have interpreted these events as wrath of the heavens, but 20th century scientists had a more rational approach to the problems. They simply carried out the repairs.

By mid-September, the optical alignment and testing of the telescope optics began, carried out by Ray Wilson, Francis Franza, Maurice Le Luyer and André Bayle from REOSC, with tracking tests by Charles Fehrenbach. On 8 November 1976, the moment that everybody had been waiting for finally arrived: the slit of the huge dome opened under the velvety dark sky and the 3.6-metre telescope mirror caught its first photons from the star-speckled firmament. First light at last[14]! Laustsen made his first entry in the observation log for the new telescope: a small field in the beautiful globular cluster Omega Centauri[15]. Selecting the targets for the first objects to observe was his prerogative, though several astronomers made suggestions and soon saw impressive new images of their pet objects. It had been a long journey for ESO and some intensive years for Laustsen and his team. Never a man for theatrical gestures or bold statements, he was deeply satisfied by reaching this goal. In a brief report in the *The ESO Messenger*, dated 12 November, he wrote: *"During the first five nights we have*

[14] As is normal, the 3.6-metre had seen technical first light a few days before. At that time, however, the primary mirror had not been finally adjusted and no astronomical images could be obtained.

[15] At the suggestion of Ray Wilson, who was interested in a field with many stars to assess the optical quality of the telescope.

taken about 30 plates and the image quality looks very good. We still have to determine the limiting magnitude, but it should be at least 24m[16]*, possibly even fainter."*[17] One of the first observations was of a 16th magnitude irregular dwarf galaxy in the constellation of Sculptor. This object had been discovered a year earlier by Richard West on a plate obtained with the ESO 1-metre Schmidt telescope. The combination of a wide-field survey telescope and a large telescope, which had proved its value on Mount Palomar, was now also available to Europe's astronomers.

The 3.6-metre telescope and the 1.4-metre Coudé Auxiliary Telescope. (Photo: John Launois/Black-Star)

[16] As a technical term magnitude is an expression of the brightness of an astronomical object. The brightest star in the sky, Sirius (α Canis Majoris) has a magnitude of –1.5. The faintest stars that can be seen with the naked eye are around 6. The difference in brightness between two objects whose magnitudes differ by one is roughly a factor of 2.5.

[17] Laustsen's prime focus photos clearly demonstrate the optical quality and success of the 3.6-metre telescope (Wilson, 2010), but it would take a very long time before the telescope was able to reach its full performance, with the dome seeing being the main culprit. At the time of first light, this was not clear. As Ray Wilson later wrote: *"The terrible 'chimney effects' of dome air were (fortunately) mitigated by the Gascoigne plate prime-focus corrector which sealed off the tube as well as correcting the spherical aberration and field coma"* (Wilson, *ibid.*). In Chapter III-8 we return to the 3.6-metre upgrade, which was undertaken in the 1990s.

The Cassegrain focus was used for the first time on 19 April 1977, in photographic mode since the photometer was still a few months away from installation. At that time, about 700 plates had been obtained at the prime focus, with *"many ... under evaluation by astronomers in the ESO countries,"* as Laustsen reported in the *The ESO Messenger* (Laustsen, 1977). Nonetheless, the use of the telescope remained restricted for a while. This is not unusual: the commissioning phase of any new telescope is a period of fine-tuning and testing as much as it is a time for initial science to be done.

Finally, with the observing period commencing in October 1977, the 3.6-metre telescope was opened for visiting astronomers. A total of 49 applicants submitted 54 proposals for the first round, with 26 being accepted. The first observing period coincided with ESO's 15th anniversary. A fine birthday present, even if many wished that it had come earlier.

Among the first users was a young Italian astronomer, Massimo Tarenghi, who came to observe clusters of galaxies. The 3.6-metre still had teething problems, and Tarenghi quickly developed an interest in the technical aspects, becoming what today would be called an instrument scientist. He developed an automatic plate changer for the 3.6-metre prime focus, so that the astronomer would no longer need to ride in the cage during the observations[18].

≈

The 3.6-metre building is a towering building. It stands 45 metres tall, overlooking the mountain ridge with its retinue of telescope domes. Next to it, a much narrower cylindrical tower, connected to the main building through a horizontal tube and a suspended walkway can be seen: this building that houses the Coudé Auxiliary Telescope. The building was originally supposed to host a 1.4-metre siderostat, feeding light into the coudé focus, but in 1973, it was realised that installing a telescope instead would greatly enhance the scientific capacity. So the mirror, which

[18] Partly because of the delay, the ESO 3.6-metre telescope came right at the time when technological advances enabled major changes in observing practices. The tension between what was and what could be is nicely illustrated by an anecdote, told later by Daniel Enard: *"Around the Cassegrain adapter and instruments arose a fierce debate, in particular as to whether the astronomer should sit in the Cassegrain cage to guide the telescope through an eyepiece or whether it was at all thinkable to trust a television camera and perform the control from the control room! Although an eyepiece was included, the 'modern' school eventually won, but not before making three prototypes of the 'Cassegrain chair', an improbable object somewhere between a medieval torture device and a dentist's chair for cosmonauts."* (Enard, 2002). Illustrations of these remarkable contraptions appeared in *The ESO Messenger*, presented with the dubious comment: *"these three chairs represent the latest in European modernity!"* (Richter, 1976).

Photomontage showing the planned siderostat at the 3.6-metre telescope. (Photo: OHP)

was already at hand, was used for the new telescope. The CAT marked ESO's first step away from classical telescope design. To accommodate the telescope within the dome, which had already been designed, Wolfgang Richter had proposed an unusual alt-alt mount for the telescope, instead of the classical equatorial mount, which had become common. However, because of the pressure to finish the 3.6-metre telescope, he soon left the CAT project to a newly-arrived young engineer, Torben Andersen. Andersen's design contained several novelties for ESO. The telescope had no fewer than four secondary mirrors, mounted in a turret. Each mirror had a different coating, optimised for a particular wavelength domain. A tertiary mirror rotated to keep the output light beam pointing in a fixed direction. Finally, Andersen introduced the idea of direct drives, which have since become the norm in telescope design. The mechanical structure was produced by MAN in Germany and the optical elements were delivered by Grubb Parsons (UK). In 1979 the telescope was assembled at CERN for testing before being shipped to Chile. The CAT became a very popular telescope and ensured that the coudé spectrograph in the 3.6-metre building could be used independently of the main telescope. In the mid-1990s, it was one of three telescopes that, for a while, were used in remote-control mode from Garching, but at the end of September 1998 it had its last observing run. In 2006, the telescope returned to Europe, having been bought by Lund Observatory, now also Andersen's home institute, and mounted in a former water tower, retrofitted with a proper dome.

Torben Andersen in front of the CAT at the assembly hall at CERN in Geneva. (Photo: Roy Saxby)

The ESO 1-metre Schmidt telescope at La Silla. (Photo: Bernard Pillet)

Chapter I-7

Sky Mapper

"The stars with deep amaze
Stand fixed in steadfast gaze, ..."

John Milton, 1645.

In addition to the 3.6-metre telescope and the other smaller telescopes, the initial programme also foresaw the installation of a Schmidt telescope at La Silla. At Mount Palomar, the combination of a major telescope and a survey telescope had proved extremely successful, with the survey telescope finding interesting objects that could subsequently be studied with the bigger telescope with its special instrumentation. A Schmidt telescope is a giant photographic camera with a large field of view and excellent aberration-free image quality, thanks to the combination of a spherical primary mirror and an elaborately shaped corrector plate — a huge aspheric lens — at the top end of the telescope, at the centre of curvature of the primary mirror. The detector — in those days large photographic glass plates — is located inside the telescope. Since the focal plane is curved, the glass plates had to be bent accordingly during exposure, requiring the plates to be fairly thin, so as not to break in the plateholder.

Schmidt telescopes are associated with the Hamburger Sternwarte. It was here that Bernhard Schmidt, a somewhat introverted optician of Swedish–Estonian origin, developed the original concept. With ESO being so closely linked to this observatory, it was natural that ESO's Schmidt telescope would be built in Hamburg. The task was given to Walter Strewinski, the engineer who we have met in connection with the 3.6-metre telescope. Strewinski had already built a Schmidt telescope for the Hamburger Sternwarte with a 1.2-metre mirror and a corrector plate of 80 centimetres. This telescope had been in service since 1954. The ESO Schmidt was originally expected to be a copy of the telescope at Mount Palomar[1,2], but instead it became a bigger sister of the Hamburg one, with a primary mirror of 1.6-metre and a 1-metre corrector plate. One of the reasons for this choice was that ESO wanted to be able to fit an objective prism and the smaller aperture (than the Palomar one) would help to save weight and thus potentially provide for greater stiffness of the tube. But, importantly, the telescope was designed to have the same focal length — and hence image scale as that of the Palomar[3] telescope, 1 millimetre = 67.6 arcseconds[4].

Strewinski demonstrated his skill in the mechanical design. As Bruce Rule (of the Palomar Observatory) put it: *"Most of the new ideas and the new mounting proposals involve rearrangements of the equatorial mounts to combinations closer to symmetrical designs."* (Rule & Sisson, 1965). This was also what Strewinski had in mind. He designed the telescope so that if an imaginary sphere surrounded all the movable parts, the centre of gravity of these parts would coincide with the centre of a sphere around which the telescope rotated, so that a minimum of force was required to move the telescope. He had already tested this solution with success at the Hamburg Schmidt.

The primary mirror was made of Schott Duran, a borosilicate glass with minimal thermal expansion, while the single corrector plate was made of ultraviolet-transmitting UK50 glass and the objective prism of UBK 7 with a resolution of about 450 Å/millimetre in blue light (near what astronomers call the Hγ line at 4340 Å). The primary mirror and the additional optics were figured by Carl Zeiss in Oberkochen, Germany. The telescope offered an unvignetted image field of 5.4 × 5.4 degrees on the sky, to be recorded on photographic plates, each 30 × 30 centimetres in size.

In terms of detection capacity, to use modern-day terminology, each Schmidt plate would contain the equivalent of something like 800 megapixels or more — depending on the particular kind of photographic emulsion. But, seen from today's perspective, there would be serious drawbacks as well. Firstly, the nonlinear transformation

---

1 Now known as the Samuel Oschin Telescope.

2 We will recall the association of Walter Baade with the Hamburger Sternwarte. Coming from Hamburg, Baade was familiar with the work of Bernhard Schmidt and his telescope design and he naturally became involved in the Palomar Schmidt Telescope. It is illustrative of the mutually enriching relationship between the two continents after the war that the Europeans now thought of copying the US telescope.

3 The UK Schmidt Telescope (UKST), which would be added to the southern hemisphere telescopes at the Anglo-Australian Observatory in Siding Spring, New South Wales, somewhat later also featured that focal length.

4 Angular measure, subdivison of 1 degree (= 60 arcminutes = 3600 arcseconds).

of intensity into density, depending both on the intrinsic characteristics of the type of emulsion, the production batch and the actual processing, made photometry a relatively complicated and uncertain affair. Secondly, the data would be analogue. To exploit them scientifically — to extract the information in a digital form — it was necessary to scan the plates, a cumbersome and time-consuming exercise. Thirdly — and perhaps the worst problem — the general sensitivity, known as detective quantum efficiency, was low, and at low intensity levels the inherent sensitivity dropped even further, an effect called the Schwarzschild effect or linear reciprocity failure. To counter the latter problem, the plates were subjected to a hypersensitising treatment, typically by baking them in an oven filled with nitrogen or hydrogen shortly before exposure. This increased the sensitivity markedly, but occasionally caused inhomogeneities across the plates. Often, working with these plates was more of an art than anything else, but at the time, there was no alternative and photographic plates, therefore, remained the primary detector in astronomy for a century.

≈

Heckmann taking a stroll outside the Schmidt telescope. (Photo: Svend Lauststen)

The ESO 1-metre Schmidt telescope was delivered to La Silla in late 1971, two years late. The delay was partly the result of Strewinski's simultaneous involvement with the 3.6-metre telescope. The installation was overseen by Heckmann, who had kept his involvement with ESO for a while after his retirement as Director General, a fairly unusual occurrence in the history of ESO. By 1972, the telescope began operations but, over the following couple of years, it would undergo a series of adjustments and major improvements, notably in the drive and control systems, optical collimation and mirror handling, an effort led by André Muller. Muller had been involved with ESO in Chile since the beginning and also in South Africa. During the early Chile years, he had the function of superintendent, which meant being the chief trouble-shooter. In 1969, he went back to ESO's Headquarters in Bergedorf to take charge of the organisation of observing programmes, while Bengt Westerlund took over in Chile, now with the title of Director for ESO Chile. As problems continued to plague the Schmidt telescope, Blaauw decided to send Muller back to La Silla. At La Silla, he worked with the Dutch engineers Jan van der Ven and Jan van der Lans, as well as Hans-Emil Schuster, a German astronomer, who had previously been involved with the Hamburg Schmidt telescope. Schuster had joined ESO in early 1964 and arrived in Chile in October 1964 as assistant to Muller, so the two knew each other well, and together they formed an excellent team.

During the 1950s, the Palomar Schmidt telescope had been used to produce the National Geographic Society/Palomar Observatory Sky Survey[5], which had covered all of the northern sky as well as the part of the southern sky that could be reached from the site near San Diego in California (down to –30 degrees). This survey, carried out in two colours (using blue-sensitive and red-sensitive photographic plates), had strongly influenced contemporary astronomical research. It provided the first major inventory of objects in the sky and went much deeper than any previous survey. Targets found on the images could be chosen for study with larger telescopes, and the photographs themselves could be used to study variable objects (by comparing plates obtained at different times) and moving objects — such as comets and asteroids. The plates could also be used for astrometric purposes, the positional determination of objects in the sky, as well as for photometry.

The 1954 Leiden Statement made explicit reference to the Palomar Observatory and also called for a Schmidt telescope of a similar size (the latter being repeated in the

[5] The National Geographic Society/Palomar Observatory Sky Survey was led by Rudolf Minkowski, a friend of Walter Baade from his Hamburg days and a pre-war émigrée from Germany.

1962 Convention). Though not specifically stated, it seemed an obvious idea to carry out a survey of the southern sky similar to the Palomar survey and ESO intended to do exactly that. A major argument for placing the observatory in the southern hemisphere was that this part of the sky had not been as well explored as the northern part. When the United Kingdom began to consider installing its own Schmidt telescope (the UKST) at the future Anglo-Australian Observatory in 1967, it was with the expressed objective of extending the Palomar survey to cover all of the southern sky. This, for ESO somewhat surprising development, led Blaauw to contact Vincent Reddish, who was in charge of the British plans, to see if it was possible for ESO to join forces with the Science Research Council of the UK and eventually this resulted in an agreement to carry out a joint survey. The agreement was signed in January 1974.

Guido Pizzaro during an observing run with the Schmidt telescope. (Photo by the author)

When the Schmidt telescope at La Silla was commissioned, Schuster became responsible for its operation, and thus for the implementation of the survey, and remained so until 1991[6]. At the Schmidt, Schuster had two Chilean assistants, Oscar and Guido Pizzaro. Together they formed a very dedicated and capable team, albeit much

[6] From 1984 to 1987 he assumed the post of acting director at La Silla and from 1987 he oversaw the site testing in the north of Chile until 1991.

smaller than that of the other large southern hemisphere Schmidt — the UKST. To deal with the Schmidt plates ESO established its Sky Atlas Laboratory in Geneva in 1972. The laboratory was headed by Richard West, assisted by French photographers, Bernard Dumoulin — who joined from CERN — and Bernard Pillet. The task of the laboratory was to produce both glass and film copies of the plates from both Schmidt telescopes[7]. It evolved to become the larger of the world's two specialist laboratories capable of dealing with Schmidt plates, combining industrial production methods with the extraordinarily high demands in terms of quality and precision that were inherent to this kind of scientific work[8].

In the *ESO Bulletin No. 10*, West introduced the new project *"at the request of the Director General of ESO in order to inform in detail the ESO Council and its Committees as well as other interested parties of the purpose and scope of the ESO Sky Survey Project."* (West, 1974). In the article, he pointed out that rather than simply copying the Palomar approach and using it for a southern sky survey, new methods in astronomy, as well as progress in the field of astronomical photography, warranted a new approach. The ESO/SRC survey would set completely new standards. It is interesting to note that while Palomar and its owners, the California Institute of Technology (Caltech), provided valuable help to ESO to get the project going, when in 1987 it was decided to undertake a second Palomar Survey, ESO returned the favour and worked closely with the Americans in the realisation of what eventually became the Palomar Observatory/ESO Sky Survey.

The Southern Sky Survey, now carried out jointly by ESO and SRC, comprised 606 fields and covered the area from the celestial pole up to –20 degrees (securing an overlap with the Palomar Survey, quite prudent since the southernmost Palomar plates did not have the same quality as the other plates, due to the larger intervening airmass[9]). The survey was carried out in two spectral bands, the *J*-band (plates obtained with the UK Schmidt telescope) and the *R*-band (with the ESO Schmidt). Kodak IIIa-J and IIIa-F emulsions were chosen for the plates and both kinds were

[7] Unlike for the Palomar Atlas, it was decided not to offer copies on photographic paper. This was because paper prints would not do justice to the quality — and thus scientific usefulness — now possible thanks to advances in the copying process.

[8] By 1990, the ESO Sky Atlas Laboratory had produced more than 300 000 high-fidelity copies of Schmidt plates for the Atlas (ESO Press Release, 26 January 1990).

[9] Observing celestial objects from the ground implies looking through the terrestrial atmosphere. Since the atmosphere introduces various negative effects, such as scattering and absorption of light as well as wavefront deformation (see Chapter III-3), astronomers prefer to observe objects at zenith, where the amount of intervening air is smallest.

hypersensitised. For the red survey (covering 6300–6900 Å, thus including the Hα line at 6563 Å), the aim was to reach a limiting magnitude of 22.5. This would require an exposure time of 120 minutes (90 minutes for the *J*-plates), placing extremely tough demands on the quality of the telescope guiding and drive systems. Given that observations could not be carried out in moonlight, the demands on the performance of the telescope and the intricacies of hypersensitisation, particularly for the red plates[10], it was clear that this project would last quite some time. In fact it turned out to be a herculean task. It was decided, however, that ESO would initially produce what became known as the Quick Blue Survey (QBS), to *"accelerate the [optical] identification of thousands of radio and X-ray sources ... in parts of the sky not covered by the Palomar Atlas."* (Lauberts, 1982). The Quick Blue Survey would use coarse-grain unsensitised IIa-O plates exposed for 60 minutes to reach a limiting magnitude of 21.5 and covering the spectral range of 3850–5000 Å. The Quick Blue Survey was completed in 1978, after which ESO concentrated fully on the red survey[11].

High-fidelity copying of the large Schmidt plates was both a science and an art. Here, Bernard Dumoulin of the Sky Atlas Laboratory is seen processing an atlas copy plate. (Photo: Bernard Pillet)

≈

[10] The IIIa-F emulsions used for the red part were considerably more difficult to sensitise.

[11] The QBS coverage was later extended from –20 degrees to the celestial equator.

An early research project linked to the survey involved the University of Uppsala (Uppsala Universitet). The ESO Annual Report of 1973 explains that *"it was ... felt that the expertise (and necessary time) was not available within ESO for a large-scale searching of the Schmidt plates. In order to avoid haphazard first exploitation of this field and to put the searching for new objects into a system that would be of advantage to all parties involved, the idea arose that ESO should collaborate with an observatory with experience in this field."*[12] Andris Lauberts of Uppsala University carried out the measurements using specially designed survey machines built at the university. They were to cover all New General Catalogue (NGC) and Index Catalogue (IC) galaxies, all galaxies with an angular extension of more than 1 arcminute (1 millimetre on the original plates) as well as clusters and planetary nebulae. This led to the ESO/Uppsala Survey, published in early 1982, with positional coordinates, angular extension, B-magnitudes, classification and, in some cases, radial velocities. The catalogue listed more than 18 000 objects, of which *"about 60% [were recorded] for the first time"* (Lauberts, *ibid.*). This work was followed up by a systematic scan of the galaxies found on the Schmidt plates, this time using a PDS scanner at ESO Headquarters. The new survey also included the deep red plates. It was carried out by Andris Lauberts and Edwin Valentijn and completed with the publication of *The Surface Photometry Catalogue of the ESO-Uppsala Galaxies* in 1989.

The Optronics scanner was originally installed at ESO in Geneva. With the move to Garching in 1980, it was transferred, together with the entire Sky Atlas Laboratory equipment.

The main survey (also called the Deep Survey) was finished in 1990 when the last copies were produced and distributed. At that time, however, plans were under way

[12] Such an institute was the Uppsala Observatory, whose director, Erik Holmberg, had worked at Mount Palomar. Under Holmberg, Peter Nilson in 1973 had published the *Uppsala General Catalogue of Galaxies*, based on the Palomar Sky Survey, and listing almost 13 000 galaxies visible from the northern hemisphere (Nilson, 1971).

to repeat the Palomar Survey, and Caltech contacted ESO for advice. Although ESO was prepared to help Caltech re-establish its own production facility, the final outcome of the discussions, concluded in January 1990, was that ESO took over the copying process of the new survey, extending the life of the Sky Atlas Laboratory until 2002, when it finally closed.

As a scientific project of its own, in terms of its duration and the resources invested, it was arguably the largest project that ESO has ever carried out. Further to the main programme, the Schmidt also carried out a smaller wide-field survey in the near-infrared (which was not subject to copying) and also produced many plates with the objective prism. Its legacy is several thousand wide-field photograph plates, each a record of a part of the sky at a particular epoch. The project also constituted a concrete link between ESO and UK astronomers in a period before the UK joined the organisation.

≈

With the completion of the ESO survey, there was free capacity for other, user-proposed projects such as a quasar survey by Lutz Wisotzki and Dieter Reimers. The search, known as the Hamburg/ESO Survey (HES), began in 1993. It was an objective-prism survey covering the southern extragalactic sky (about 9000 square degrees, partitioned into 373 ESO Schmidt fields). The purpose was to record the largest possible number of bright quasars[13], both for statistical studies and for detailed follow-up studies with other telescopes (Reimers & Wisotzki, 1997) — in other words the classical function of a sky survey. With the scanning of the plates, the survey was completed towards the end of the decade. Another project was an asteroid project led by Claes-Ingvar Lagerkvist, together with a group from Deutsche Luft- und Raumfahrtforschung (DLR) Berlin, to study asteroids near Jupiter in orbits similar to the planet, but located 60 degrees ahead of it in the Jovian L4 Lagrangian point. These objects are often referred to as Trojans, because they carry names from Greek mythology related to the Trojan War. The telescope searched about 900 square degrees around the Jovian L4 point twice, with follow-up observations of selected objects by the Bochum telescope, as well as by the NTT.

---

[13] Quasars — short for quasi-stellar radio sources — are extremely luminous, seemingly star-like objects at large (cosmological) distances. Today, quasars are considered to be galaxies with very massive black holes at their centres. Aside from the intrinsic interest in understanding quasars, their brightness makes them useful for the study of the intervening space. The discovery of quasars, which goes back to the late 1950s and early1960s, is an interesting story, demonstrating the interplay between radio and optical astronomy.

But slowly, twilight fell on the Schmidt telescope and in December 1998, it was closed down[14]. Other telescopes with wide-field capacity, not comparable to the Schmidt as regards the field of view, but with much more sensitive detection systems[15], began to take over. For ESO, the closing of the Schmidt led to an interruption in survey programmes that has only been convincingly ended in December 2009 when the 4.1-metre Visible and Infrared Survey Telescope for Astronomy (VISTA) telescope began operations at Paranal. As VISTA entered into full operation, a research group under the name of LS-Quest from Yale University began using the ESO Schmidt telescope again, now equipped with a mosaic of 112 thinned, back-illuminated CCDs (each with 600 × 2400 13μ × 13μ pixels), covering almost the full field of view of the telescope (approximately 3.6 × 4.6 degrees), and built at Yale University and Indiana University. Thanks to the giant leap in sensitivity, compared to the photographic plates of yesteryear, it is possible to cover 1200 square degrees of the sky, *twice*, in just one night. The plan is to return to the same field with a four-day rhythm for a continuing search of 4800 square degrees to find low-redshift Type Ia supernovae, RR Lyrae stars and trans-Neptunian objects. The camera was previously used at the Samuel Oschin Schmidt telescope at the Palomar Observatory and installed in early 2009, with first light in April 2009.

[14] The UK Schmidt Telescope, from the outset better resourced than the ESO operation, continued its classical operations with photographic plates, until 2005, although the transition to other observing modes had begun long before.

[15] CCD mosaics.

Chapter I-8

Of Heaven and Hell, the Vatican and the Mission

"There is general unrest in the country, but for the time being it does not affect the ESO establishments."

ESO Council meeting minutes, June 1971.

In the past chapters, we have looked at the establishment of the La Silla Observatory and the build-up of the telescope park, more or less as stipulated in the Convention. We have also seen how additional telescopes were added. The desire to exploit the new site was clearly strong, perhaps insatiable. La Silla was becoming a Heaven for Europe's astronomers. Or perhaps more appropriately expressed, the best place next to Heaven. This view was undoubtedly shared by the new Director General, Adriaan Blaauw. He was fully occupied by the work involved in reviving the moribund 3.6-metre project, but Blaauw was also an astronomer himself. Staying at La Silla gave him the occasional opportunity to observe. In those early days the atmosphere at ESO was a curious blend of a pragmatic "let's-do-it", hands-on ethos and at the same time a rather elevated and reserved atmosphere. Blaauw's wish to observe might well have been fully understood and appreciated by his scientist colleagues[1], but some administrators found it hard to understand[2] that a Director General would take time to *"look at the stars"*, as they put it. But, Blaauw's wish to observe was not only the desire of a scientist, but also a way for him to see for himself how ESO really functioned for visiting astronomers: the core business of the observatory.

[1] Staying active in science whilst also holding high office is obviously not easy. Blaauw therefore hired Richard West as a scientific assistant to carry out some of his observations. West was also charged with building up a scientific library before he became head of the ESO Sky Atlas Laboratory.

[2] One of the challenges of science organisations like ESO is to turn what could easily become a destructive tension between an administrative backbone, acting within strict legal and fiscal rules and regulations and, not to forget, traditions and a free-wheeling academic world into a fruitful symbiosis. Over the years, however, I believe that ESO has been remarkably successful in this respect.

Blaauw was an unpretentious and congenial Director General, well liked by his staff. Christiaan Sterken, who at the time served as a junior astronomer[3] recalls Blauuw's straightforward manner: *"In 1971 (mid-April or early May) Adriaan Blaauw spent considerable time in the guesthouse, and before returning to Hamburg he went on a shopping expedition with Hilde Fritsch (in charge of the Guesthouse), and he bought a typical Chilean carpet. He asked for a hammer and nails, climbed a chair, and started hammering the thing on to the living room wall (between the window doors to the garden). I had to hand him the nails, one by one, till the DG's [Director General's] handiwork was done."* Another example serves to illustrate his management style. Life in Chile was not easy in the early 1970s, either for the local population or for the foreign ESO staff. Travel between the observatory and Santiago became more difficult, and staff had to accept long and time-consuming coach rides. For the occasional visiting astronomer, this may have been quite acceptable, but staff who regularly had to commute found it hard. When Blaauw asked a young staff member about his wellbeing at ESO and he seized the opportunity to express unhappiness about this continued nuisance, Blaauw's reaction was prompt. *"Give me your bus ticket,"* he said. *"You can have my plane ticket instead!"* But Blaauw's generous manners could not compensate for the fact that the situation in the country was worsening. Strikes, powercuts, shortages of all kinds of everyday supplies, from spare parts and fuel to food, and a burgeoning blackmarket, were sure signs of Chile's downward spiral in the early 1970s. Uncertainty and fear became widespread and this affected the families of the ESO staff, however privileged, as well as every citizen of the country. The blackmarket economy undoubtedly created temptations and in a few rare cases some ESO staff could not resist, creating difficulties for the organisation. To prevent such problems, Westerlund imposed various regulations on his staff.

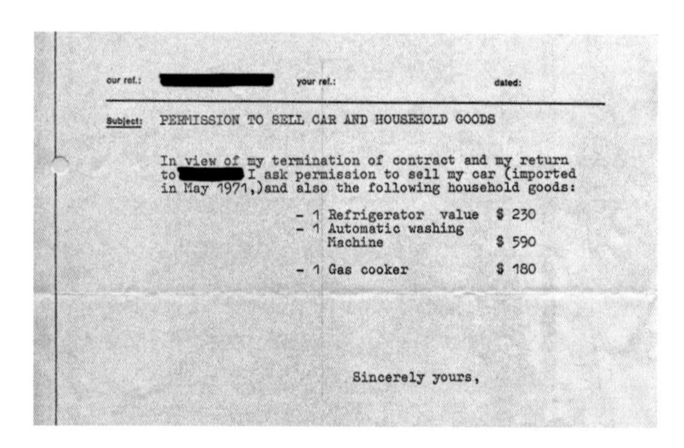

our ref.: ■■■ your ref.: dated:

Subject: PERMISSION TO SELL CAR AND HOUSEHOLD GOODS

In view of my termination of contract and my return to ■■■ I ask permission to sell my car (imported in May 1971,)and also the following household goods:

- 1 Refrigerator value $ 230
- 1 Automatic washing Machine $ 590
- 1 Gas cooker $ 180

Sincerely yours,

ESO's international staff were able to import goods from Europe duty free. To sell goods, however, they were required to obtain permission from ESO's administration. In this letter, a departing staff member requests permission to sell his car, refrigerator, automatic washing machine and gas cooker.

In November 1972, the ESO Council paid a visit to Chile. Council meetings in Chile have

[3] The equivalent of a student in today's ESO terminology.

been most helpful in fostering a deeper understanding of the operational issues on the ground — far away and quite remote from the perspective of ministry offices in Europe. On the fourth day of the meeting — 24 November 1972 — a Council delegation was received by the President of the Republic, Salvador Allende. Allende showed great interest in the ESO project. He also expressed an interest in cooperation between Chilean universities and ESO, *"not only for the education of young astronomers and their preparation for fellowships in Europe, but also in the technological sector"*[4]. A few years earlier, at the inauguration of La Silla, Allende's predecessor had, almost poetically, declared that *"... for us [the ESO Observatory] not only means the construction of material buildings for a centre of scientific progress; we hope ... that it will benefit our country to have among us men who silently explore the heavens from this vantage point. May they reveal to us the harmony and greatness of the Universe and the humility which their profession and their dedication teach them."* Allende's comments could certainly be seen as a move towards a relationship with more concrete benefits for Chile. In the aftermath of the audience, Council discussed the issue of closer cooperation with Chile, and in general, there was a readiness, within the limits of the organisation, to see what could be done. But unknown to Council were the dramatic events that lay ahead — the general further deterioration of the overall conditions in the country[5] and the abrupt and gruesome end of Allende's presidency, less than ten months later.

≈

On 10 September 1973, Hans-Emil Schuster returned to Santiago after a visit to Europe. *"It was strange,"* he said in an interview with the author. *"Everyone seemed to leave the plane in Buenos Aires, where we made a ground stop. In the end, there were only three passengers continuing to Chile. In Santiago, we found the airport almost empty. There was no immigration check and no custom's officers. Outside, all the taxis had disappeared. I called the ESO Guesthouse in Las Condes and they sent a car to pick me up. We drove through deserted streets and arrived at the Guesthouse. On the next morning, a bus came to pick us up and take us to La Silla. We saw planes in the sky and at the edge of the city we were stopped by the police at a roadblock. 'No-one can*

[4] No direct record of the discussion has been available to the author. The quote is based on the recollection of the ESO Council President, Augustin Alline, who took part in the encounter and is contained in the minutes of the Council debate that followed immediately afterwards.

[5] In a report to the ESO Council, Bengt Westerlund noted that *"from early 1973, the ESO staff were really living in a 'special hazard' situation. A civil war seemed imminent, travel on the Pan-American Highway became risky..."* and of course, the overall supplies of goods, including food, had dwindled.

travel to the North,' we were told, so we returned to the Guesthouse. Bengt Westerlund told me to stay and 'guard' the Guesthouse. He would remain at the ESO Office in Vitacura. We had a visitor ... [from an Eastern European country] ... who told us that all foreigners were requested to report to the nearest police station or military patrol. He insisted on following this order, so I locked him in his room. That may have saved his life. During the night we heard shots. For three days, we did not know how things were at La Silla." On that day, Monday 11 September 1973, the military staged a bloody coup d'état, bringing an end to the government of Salvador Allende and the beginning of a 16-year long dictatorship. *"Finally, we obtained* laissez-passer *documents and came to La Silla."* At La Silla, the Belgian administrator, Georges Anciaux, had been at a police station to help with the release of a close relative of an ESO employee, who had been detained. Other staff members off duty had also tried to reach the observatory, but few had succeeded in passing the checkpoints and only under caution that no-one could guarantee their safety. Whilst communication with La Silla was interrupted, the Vitacura office managed to establish shortwave radio contact with the German embassy in Buenos Aires and thus families in Europe could be notified about the well-being of their loved ones.

The events in Chile sent shockwaves through the ESO Member States and the ESO Council. Nonetheless, neither ESO nor the Americans considered withdrawing. ESO's mission was clear and unambiguous — to serve the scientific communities of its Member States, and that essentially meant to serve science. ESO found itself in a classical dilemma — seen in many other cases where fundamental political issues become entangled with human activities such as sports and the arts as well as science. Could serving science, unquestionably a noble goal, stand above politics? The Council had no doubts and did not waver. The task would therefore be to ensure that ESO's mission was not jeopardised, despite the difficult times. It would be for ESO management to manoeuvre with extreme care in the years to come to avoid any sign of support or collusion with the military regime, yet to protect the organisation and its ability to operate as necessary and in accordance with its remit. Contacts with the regime were kept to a necessary minimum and at as low a level as possible. ESO would be correct and polite, as befits conduct in international affairs, but it would refrain from large public manifestations in the country. Thus there never was a festive inauguration of the 3.6-metre telescope and the NTT inauguration in 1990, during the final days of Pinochet's rule, took place in Europe, with a relatively low-level parallel event in Chile. The ESO Council did not pay any further visits to the country until a new democratic government took over.

≈

The turbulent situation in Chile, both before the *coup d'état* and afterwards fed into an ongoing discussion about ESO's presence, especially in Santiago. These deliberations had already come up in 1972 in connection with the TP Division, which had begun to become active in Chile. A plan for expanding the facilities in Santiago was not approved by Council. Many argued that the distance between the observatory and the ESO centre in Santiago had an adverse impact on the efficiency of the organisation. Furthermore, the office in La Serena could play an increased role, when La Silla was fully developed, meaning at the time the when 3.6-metre project was completed. Conversely, it was felt that perhaps some of the functions carried out at the Santiago centre could perhaps be transferred to Europe. This scheme may appear rather confusing, but it illustrates a more basic issue that has remained with ESO, in sense because it has been "built into the system": the constant dialogue, and occasional tension, about the proper balance between the main research facilities in Chile, for which the organisation exists, and the Headquarters in Europe, without which it cannot function. There have been, and are, many aspects of this dialogue and many operational decisions have hinged upon the contemporary state of it. It is therefore understandable that the ESO Council discussed these matters, although often in an inconclusive way. The visit to Santiago in 1972 had convinced some members that a move might be unwise. Even so it was thought that the La Serena office could play a bigger role. ESO also acquired land to build a compound for its international staff there. But discussions about the individual merits of the various solutions ceased when the incoming Director General, Lodewijk Woltjer, presented a much more radical approach — to concentrate activities at the observatory proper — at La Silla. In his proposal to Council, submitted for the December 1974 meeting, he argued cogently why, both from a point of view of cost and scientific efficiency, the existing solution needed to change. In his arguments we also see a return to the subject of a scientific centre in Europe (see also the next chapter). Since such a centre was lacking, Santiago had taken over some of these necessary functions, he argued, an altogether unsatisfactory situation in his view. On the other hand Woltjer also underlined that ESO staff should be *"free to live where they prefer"*, and continuing to live in Santiago, an option that most preferred, could be made possible by the introduction of regular flights between Santiago and Pelícano, the old base camp at the foot of La Silla, where an airstrip had just opened. A further element of his restructuring plans was to reduce the number of international staff and increase the hiring of local staff from Chile. And so, it was decided.

The decision to relocate almost all ESO staff in Chile to La Silla made sense at the time. It improved the overall efficiency of the operation and it, literally, distanced, and shielded the organisation and its staff from the grim political situation in the country. Woltjer defends the decision in his book, adding that *"... in any case, it was not ESO's role to finance an astronomical centre [in Santiago]"*. But it came at a price. In the short term the cost was simply the additional expenses incurred to keep the full staff on a desert site, with additional financial compensation and much more travel. In the longer run it may have contributed to isolating ESO from Chilean society. La Silla was sometimes compared to a monastery, also in the European media. And it may have contributed to creating the impression that ESO did not care about the host country and its inhabitants. This was hardly justified, but — twenty years on — this isolationist image would come back to haunt ESO.

≈

Before we conclude this chapter, we shall briefly return to the issue of internal communication. With facilities in several locations, some of which lay 12 000 kilometres apart, communication constituted a major challenge. Before the advent of modern information technology this was even more daunting, but necessary for the proper functioning of the observatory[6] and for the organisation as a whole. Working in relative isolation, the risk was high that each facility would develop its own solutions and, in an understandable attempt to achieve the best results, acquire ever more functions locally, sometimes leading to rivalry and what might appear as wasteful duplication. When these changes were denied by the management, disappointment and frustration might occur. Keeping everyone informed was therefore of huge importance. When solid information was lacking, it left ample space for stereotypical expectations. Such problems were often countered with humour, albeit mixed with a degree of sarcasm that was hard to overlook. Garching — meaning the ESO Headquarters — became known in Chile as "the Vatican", while La Silla enjoyed the doubtful honour of being called "the Mission" in Europe. The importance of communication was recognised by Adriaan Blaauw, who instigated the creation of an ESO newsletter, *The ESO Messenger*. In May 1974, the first issue of the new in-house magazine appeared. The primary aim of this new publication (which succeeded the *ESO Bulletin*) was to strengthen internal communication, but also *"to give the outside world some impression of what happens inside ESO."* (Blaauw, 1974). The first issue was rather modest,

[6] As late as the early 1970s, the only means of telecommunications at La Silla was a shortwave radio, though telephone and telex connections followed later in the decade.

with just six pages, written in English, but with selected texts also reproduced in Spanish. It provided various titbits of information about new staff, technical projects, construction progress and general issues, but it soon also included articles about astronomy. This was originally meant to motivate the non-astronomical staff at ESO. However, it soon became clear that the magazine could provide a window onto ESO for the outside world. The editor therefore sought a balance between light and easy-to-read stories of general interest and more technical articles. It occasionally even published letters to the editor, such as the following in the December 1979 issue, allegedly from a certain 'H.D.': *"Since* The Messenger *is evolving in the direction of serious journals, one should consider the problem of quoting articles in lists of references. The other day I found the reference: 'ESO Mess' — which is perhaps not the best compliment to the otherwise fine organisation… ."* A particularly amusing article, written by the editor, discussed the transformation of observing, from the classical style with

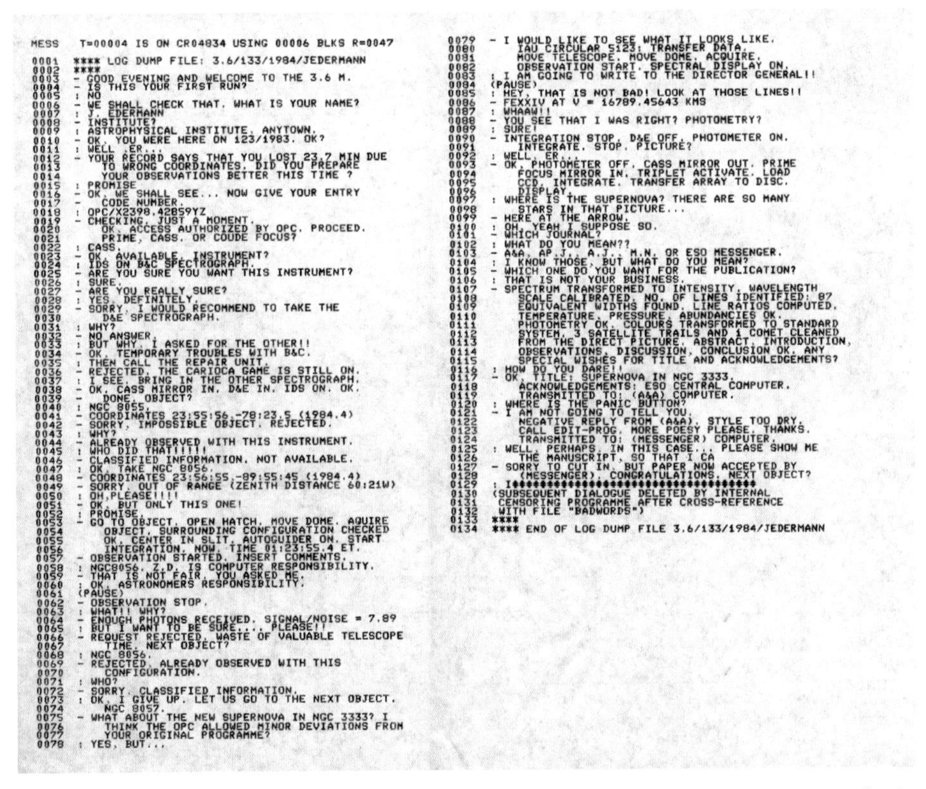

```
MESS   T=00004 IS ON CR04834 USING 00006 BLKS R=0047
0001   **** LOG DUMP FILE: 3.6/133/1984/JEDERMANN
0002   ****
0003   - GOOD EVENING AND WELCOME TO THE 3.6 M.
0004   - IS THIS YOUR FIRST RUN?
0005   : NO
0006   - WE SHALL CHECK THAT. WHAT IS YOUR NAME?
0007   : J. EDERMANN
0008   - INSTITUTE?
0009   : ASTROPHYSICAL INSTITUTE, ANYTOWN.
0010   - OK, YOU WERE HERE ON 123/1983. OK?
0011   : WELL .ER...
0012   - YOUR RECORD SAYS THAT YOU LOST 23.7 MIN DUE
0013       TO WRONG COORDINATES. DID YOU PREPARE
0014       YOUR OBSERVATIONS BETTER THIS TIME ?
0015   : PROMISE
0016   - OK, WE SHALL SEE... NOW GIVE YOUR ENTRY
0017   -   CODE NUMBER.
0018   : OPC/X2398.42859YZ
0019   - CHECKING, JUST A MOMENT.
0020       OK, ACCESS AUTHORIZED BY OPC. PROCEED.
0021       PRIME, CASS. OR COUDE FOCUS?
0022   : CASS.
0023   - OK, AVAILABLE. INSTRUMENT?
0024   : IDS ON B&C SPECTROGRAPH.
0025   - ARE YOU SURE YOU WANT THIS INSTRUMENT?
0026   : SURE.
0027   - ARE YOU REALLY SURE?
0028   : YES, DEFINITELY.
0029   - SORRY, I WOULD RECOMMEND TO TAKE THE
0030       D&E SPECTROGRAPH.
0031   : WHY?
0032   - NO ANSWER.
0033   : BUT WHY, I ASKED FOR THE OTHER!!
0034   - OK, TEMPORARY TROUBLES WITH B&C.
0035   : THEN CALL THE REPAIR UNIT.
0036   - REJECTED. THE CARIOCA GAME IS STILL ON.
0037   : I SEE. BRING IN THE OTHER SPECTROGRAPH.
0038   - OK. CASS MIRROR IN. D&E IN. IDS ON. OK.
0039   -   DONE. OBJECT?
0040   : NGC 8055.
0041   - COORDINATES 23:55:56,-78:23.5 (1984.4)
0042   - SORRY, IMPOSSIBLE OBJECT. REJECTED.
0043   : WHY?
0044   - ALREADY OBSERVED WITH THIS INSTRUMENT.
0045   : WHO DID THAT!!!!!
0046   - CLASSIFIED INFORMATION. NOT AVAILABLE.
0047   : OK, TAKE NGC 8056.
0048   - COORDINATES 23:56:55,-89:55:45 (1984.4)
0049   - SORRY, OUT OF RANGE (ZENITH DISTANCE 60:21W)
0050   : OH,PLEASE!!!!
0051   - OK, BUT ONLY THIS ONE!
0052   : PROMISE.
0053   - GO TO OBJECT. OPEN HATCH. MOVE DOME. AQUIRE
0054       OBJECT. SURROUNDING CONFIGURATION CHECKED
0055       OK. CENTER IN SLIT. AUTOGUIDER ON. START
0056       INTEGRATION. NOW. TIME 01:23:55.4 ET.
0057   - OBSERVATION STARTED. INSERT COMMENTS.
0058   : NGC8056. Z.D. IS COMPUTER RESPONSIBILITY.
0059   - THAT IS NOT FAIR. YOU ASKED ME.
0060   : OK. ASTRONOMERS RESPONSIBILITY.
0061   (PAUSE)
0062   - OBSERVATION STOP.
0063   : WHAT!! WHY?
0064   - ENOUGH PHOTONS RECEIVED. SIGNAL/NOISE = 7.89
0065   : BUT I WANT TO BE SURE.... PLEASE!!
0066   - REQUEST REJECTED. WASTE OF VALUABLE TELESCOPE
0067       TIME. NEXT OBJECT?
0068   : NGC 8056.
0069   - REJECTED. ALREADY OBSERVED WITH THIS
0070       CONFIGURATION.
0071   : WHO?
0072   - SORRY. CLASSIFIED INFORMATION.
0073   : OK, I GIVE UP. LET US GO TO THE NEXT OBJECT.
0074       NGC 8057.
0075   - WHAT ABOUT THE NEW SUPERNOVA IN NGC 3333? I
0076       THINK THE OPC ALLOWED MINOR DEVIATIONS FROM
0077       YOUR ORIGINAL PROGRAMME?
0078   : YES, BUT...
0079   - I WOULD LIKE TO SEE WHAT IT LOOKS LIKE.
0080       IAU CIRCULAR 5123: TRANSFER DATA.
0081       MOVE TELESCOPE. MOVE DOME. ACQUIRE.
0082       OBSERVATION START. SPECTRAL DISPLAY ON.
0083   : I AM GOING TO WRITE TO THE DIRECTOR GENERAL!!
0084   (PAUSE)
0085   : HEY, THAT IS NOT BAD! LOOK AT THOSE LINES!!
0086   - FEXXIV AT V = 16789.45643 KMS
0087   : WHAAW!!
0088   - YOU SEE THAT I WAS RIGHT? PHOTOMETRY?
0089   : SURE!
0090   - INTEGRATION STOP. D&E OFF. PHOTOMETER ON.
0091       INTEGRATE. STOP. PICTURE?
0092   : WELL, ER...
0093   - OK, PHOTOMETER OFF. CASS MIRROR OUT. PRIME
0094       FOCUS MIRROR IN. TRIPLET ACTIVATE. LOAD
0095       CCD. INTEGRATE. TRANSFER ARRAY TO DISC.
0096       DISPLAY.
0097   : WHERE IS THE SUPERNOVA? THERE ARE SO MANY
0098       STARS IN THAT PICTURE...
0099   - HERE AT THE ARROW.
0100   : OH, YEAH I SUPPOSE SO.
0101   - WHICH JOURNAL?
0102   : WHAT DO YOU MEAN??
0103   - A&A, AP.J., A.J., M.N. OR ESO MESSENGER.
0104   : I KNOW THOSE, BUT WHAT DO YOU MEAN?
0105   - WHICH ONE DO YOU WANT FOR THE PUBLICATION?
0106   : THAT IS NOT YOUR BUSINESS.
0107   - SPECTRUM TRANSFORMED TO INTENSITY, WAVELENGTH
0108       SCALE CALIBRATED. NO. OF LINES IDENTIFIED: 87
0109       EQUIVALENT WIDTHS FOUND. LINE RATIOS COMPUTED.
0110       TEMPERATURE, PRESSURE, ABUNDANCIES OK.
0111       PHOTOMETRY OK. COLOURS TRANSFORMED TO STANDARD
0112       SYSTEM. 3 SATELLITE TRAILS AND 1 COMET CLEANED
0113       FROM THE DIRECT PICTURE. ABSTRACT, INTRODUCTION,
0114       OBSERVATIONS, DISCUSSION, CONCLUSION OK. ANY
0115       SPECIAL WISHES FOR TITLE AND ACKNOWLEDGEMENTS?
0116   : HOW DO YOU DARE!!
0117   - OK, TITLE: SUPERNOVA IN NGC 3333.
0118       ACKNOWLEDGEMENTS: ESO CENTRAL COMPUTER.
0119       TRANSMITTED TO: (A&A) COMPUTER.
0120   : WHERE IS THE PANIC BUTTON?
0121   - I AM NOT GOING TO TELL YOU.
0122       NEGATIVE REPLY FROM (A&A). STYLE TOO DRY.
0123       CALL EDIT-PROG. MORE POESY PLEASE. THANKS.
0124       TRANSMITTED TO: (MESSENGER) COMPUTER.
0125   : WELL, PERHAPS, IN THIS CASE... PLEASE SHOW ME
0126       THE MANUSCRIPT, SO THAT I CA
0127   - SORRY TO CUT IN, BUT PAPER NOW ACCEPTED BY
0128       (MESSENGER). CONGRATULATIONS. NEXT OBJECT?
0129   : I****************************************
0130   (SUBSEQUENT DIALOGUE DELETED BY INTERNAL
0131    CENSORING PROGRAMME AFTER CROSS-REFERENCE
0132    WITH FILE "BADWORDS")
0133   ****
0134   **** END OF LOG DUMP FILE 3.6/133/1984/JEDERMANN
```

The article "Brave New World", published in the December 1979 issue of The ESO Messenger, *offered a dark-humoured preview of how computer automation might determine the way astronomers would work in the future.*

the astronomer glued to the eyepiece, patiently guiding his telescope for an hour or more, to the modern automated (as it is described) mode of observation. Nonetheless, from September 1976 (issue no. 6) onwards it increasingly became a forum for scientific users and instrument builders to describe their projects. With some variations in emphasis, depending on the respective editors in charge, this would continue up to the present. *The ESO Messenger* might even sometimes be used for fast presentation of new research results: though without a peer-review system it never aspired to become a formal journal. The Spanish summary was discontinued in March 1988, although two articles, in much abbreviated versions, appeared slightly later. The last one, with the title, 'Mi Visita a La Silla', described a recent visit to the observatory by André Muller, the ESO Superintendent in Chile of the early years. This article was published in December 1988. It is gratifying to note that *The ESO Messenger* today has a wide readership outside the organisation and thus evidently continues to fill a specific need, but it failed to become the internal communication tool that Blaauw had in mind.

Despite the continued need for internal communication, later attempts also failed. In 1988, Harry van der Laan encouraged the creation of the *ESO Echo*, but this ran to only two issues. Again in 2000 Catherine Cesarsky instigated the creation of an on-line newsletter, called *Open Skies*. Although it survived for several years it never really caught on as a major information and communication channel within the organisation.

Chapter I-9

The Changing of the Guard

"I was lucky to come to ESO when I did ... it [was] possible to develop ESO as a scientific organisation with gradually the astronomical community on the continent identifying itself with their organisation, something fostered with the VLT project and its policy of building instruments in the institutes of member countries."

Lodewijk Woltjer, communication with the author, November 2011.

The arrival of Lodewijk Woltjer to ESO in 1975 initiated a phase change within the organisation. Before coming to ESO, he had been chairman of the astronomy department at Columbia University. He was not primarily an observational astronomer, but a theoretician. He had a broad overview of the science, understood what European astronomy needed to do to evolve and become competitive, and his approach was not at all theoretical. As Johannes Andersen, one of the European astronomer who was heavily involved in the early instrumentation programme and known in the scientific community for his occasional incisive remarks, put it: *"Having a theoretician as Director General was of great help to the instrument builders, because he actually believed that the laws of physics also applied to instruments."*[1] Woltjer is a strong-willed person, in his own quiet way direct and blunt if necessary, but also a master of tactics and thus able to navigate the sometimes complex and sensitive landscape of science politics.

What brought Woltjer back to Europe was a strong desire to push European astronomy forward and he saw that ESO had the potential to achieve this goal. He possessed both the will and the intellectual strength to get his way, even if — this being

[1] It would be hard to argue otherwise, but Woltjer chose another no-nonsense explanation: *"Technical matters concerning large optics are generally better left to optical engineers than to scientists."* (Woltjer, *ibid.*). And he proceeded accordingly.

Europe — not everyone agreed with him from the outset. A key to his success was a much closer link between active scientists, engaged in frontline research, and the telescope and instrument builders at ESO. This could only be achieved by having active scientists working directly at ESO, carrying out their own science, but also interacting with the engineers — pushing them, but also working with and helping them to achieve excellence in all aspects of the ESO operation. As he said in an interview with the author: *"I wanted to have the opportunity to make ESO also a first-class scientific institution and not only a place where telescopes were built... ."* But that was precisely what some Member States did not want, sticking to the concept of the *observatoire de mission*. Blaauw had already pushed for a scientific group, but failed to convince the Council. At the same time, Council and its President, Bengt Strömgren, had wooed Woltjer for some time.

With Blaauw's term expiring, Council gave in. Blaauw quotes from the minutes of the Committee of Council meeting in November 1974, only eight weeks before his term of office as Director General came to an end: *"As to the creation of an astronomical Centre in Europe, which had been made a condition by Professor Woltjer for his acceptance of the position of Director General*[2] *... a course of action would seem to be acceptable ... to start recruiting a nucleus team... ."* But the battle was not over yet[3]. Woltjer recalls a stormy Council meeting in October 1975 in which this development was under threat of being undone. Woltjer did not yield. In the end, a showdown was avoided through firm handling by the otherwise convivial Council president of the time, Bengt Strömgren. The matter came up again at the December Council meeting, but Woltjer stood fast. Introducing his plan for *"The Scientific-Technical Centre of ESO in Europe"* he once more presented the case for having a dedicated scientific group. *"Some ... have expressed ... that it would be more efficient to have a small number of persons who would spend all their time on [supporting the 3.6-metre telescope and its instrumentation] and for the duration would forget their about their research*

[2] In his book, Blaauw mentions his own struggle to convince Council of the necessity of establishing a science group. He also writes that *"an astronomer of outstanding qualification had, in private, expressed to the Director General interest in [leading this group]. However, contrary to expectations raised at Committee of Council, the Council meeting of November 1972 held in Chile acted reluctantly. It authorised the Director General to approach the person concerned about the intended association with ESO, but rejected creation of the research-oriented group. As a result, interest on the part of the person concerned faded." "The person concerned"* was Woltjer, whom Council now tried to win for the post of Director General.

[3] In fact, the *observatoire de mission* discussion resulted in a proposal in Council to reduce the TP Division on the grounds, that the 3.6-metre telescope was now close to completion, after which ESO would no longer need telescope and instrument development capabilities on this scale. With the clarity and determination characteristic of him, Woltjer — not yet in office — made it clear that he did not intend to lead ESO in this condition. The discussion then subsided.

activities.... Could one really believe that the optimum use of the 3.6-metre telescope — or the European cooperation with that in mind — could be effectively promoted by second rate-scientists? Or does one have the illusion that one could get such scientists to an organisation where research is judged to be superfluous or in any case an impermissible luxury?" he wrote. Both Blaauw and Woltjer have dealt with this topic in their books. It illustrates how deep-rooted this discussion has been within ESO, and it has returned on later occasions, albeit in more benign forms. This is illustrated by the frequently changing status of the science group, sometimes forming a division, sometimes an office linked to the Office of the Director General, sometimes a directorate — sometimes headed by a deputy director for science, a division head or even something else[4]. It also shows why Woltjer was the right leader for ESO at the time. Only someone with a strong personality and a clear vision for the future could have brought ESO forward in a decisive way. Woltjer's vision — and victory — was a victory for ESO and for European astronomy. He soon began to unfold that vision. During his term of office, he carefully cultivated ESO's potential, exploited fresh ideas from the scientific community and wove them into the ESO programme, preparing the ground for the breakthrough that ESO achieved towards the end of the century.

≈

In ESO's newly established scientific group, Woltjer gathered a group of young and talented scientists from all over the world. The first three members were Jürgen Materne (Germany), Michel Dennefeld (France) and Phil Crane (USA). Even though Dennefeld was still a student, he already had some experience with ESO, having worked as a *co-opérant*[5] in Chile. Phil Crane had been involved in CCD detector development at Princeton. The group was soon strengthened with the arrival of John Danziger (Australia), who joined the group from Chile. With time it grew, adding short-term visitors, fellows and staff astronomers. Among others the group included Guido Chincarini and Massimo Tarenghi (Italy), Jorge Melnick and Hernán Quintana (Chile), Daniel Kunth, Danielle Alloin, Marie-Helène Ulrich Demoulin, Philippe and Mira Véron (France), Alan Moorwood (UK) and Peter Shaver (Canada). Their diverse national origins clearly reveal one of the decisive features and strengths of ESO: its commitment to scientific excellence and its ability to

[4] Currently, the science group is vibrant and strong, headed by a Director for Science. The latest change, which occurred in early 2008, followed a strong recommendation by an international visiting committee and was fully accepted by the Member States. For the time being, at least, this discussion has been consigned to its grave.

[5] The civilian equivalent of military service in France.

free itself of national constraints when it came to recruitment. Woltjer nurtured the group, even walking from office to office to pick up "his scientists" for lunch at the CERN canteen. The group soon began to exert its influence on the operations and the future development of the organisation. It was initially headed by Franco Pacini, who, in 1978, was succeeded by Per-Olof Lindblad, a mild-mannered Swedish scientist with a decidedly stately air and the son of Bertil Lindblad, one of the astronomers who had signed the 1954 declaration. In the years to come, the Science Division remained a highly dynamic group, with members leaving (often returning to their home countries) and others joining. Early newcomers to the group in Garching included Helène Sol, Joachim Krautter, Gösta Gahm, Ian Glass and many other, mostly young, scientists — many more than we can mention in this book.

Gerhard Bachmann (left) entertaining guests during the lunch offered by the German President on the occasion of the inauguration of the Headquarters building in Garching. (Photo by the author)

Whilst the incoming Director General was supported on the science side by the changing heads of the science group (later Science Division), the key person on the administrative side was Gerhard Bachmann, who was ESO's Head of Administration from the end of 1972. This post was generally considered to be the second highest staff position in the organisation. Bachmann, a German national, had worked in a German state government administration and in NATO before he joined ESO as

Head of Finance and subsequently took over as Head of Administration from Johan Bloemkolk. Bachmann remained in this position for almost 20 years. With his broad Berliner dialect, he could be jovial, but he ran a tight ship for his staff. His tasks were not confined to budgets, bookkeeping and personnel issues. A great admirer of French diplomacy in particular, he was often called upon to help solve delicate issues of a diplomatic nature over the years.

Lodewijk Woltjer, Robert Fischer, Bernard Dumoulin, Philippe Crane and Jürgen Eichler, project manager for the ESO Headquarters building on behalf of the Max-Planck-Gesellschaft. The person to the extreme left is thought to be Walther Nöbel, an assistant to Fehling and Gogel, the architects of the building.

The new Headquarters building in Garching, photographed in 1980. (Photo by the author)

Chapter I-10

Garching United

"The Federal Republic of Germany, together with the Max Planck Society for the Advancement of Science, offers ESO a site at Garhing near Munich for its European Headquarters."

Introduction, official proposal to the ESO Council 29 April 1975.

Setting up the TP Division also threw up the question of where ESO's long-term home should be. The choice of Hamburg had been largely coincidental, while the selection of Geneva constituted a very deliberate choice, driven mainly by the need to gain access to a rich and highly competent technical reservoir. However, there was also a small "ESO group" in Marseille, a relic of the early years, where everybody chipped in with whatever they had to offer[1]. With facilities at La Silla, in La Serena and in Santiago as well, some people muttered that ESO seemed to be an excuse for travelling. Reducing the geographical spread was necessary. The arrival of ESO at CERN, had stimulated a group of Swiss astronomers, notably Marcel Golay, the Director of the Observatoire de Genève, to consider Swiss membership of ESO. With the first technical departments now on Swiss territory and the recognised need for bringing ESO's European activities together in one place, the idea of suggesting that the new headquarters could be in Geneva was in the air[2]. At the

---

1 The Marseille group was not formally part of ESO.

2 In the end, nothing became of this. The Swiss decision to join ESO came too late for this idea to be seriously considered. Nonetheless, the disappointment in Switzerland was palpable: in a jab at the Germans, on 4 December 1975, the newspaper *Tribune de Genève* titled *"We want to keep what we have" ("Nous voulons garder ce que nous avons"),* but did not spare the Swiss of criticism either: *"The Swiss Science Council saw this too late. The Confederation risks one day becoming the sad figure of the Far West cowboy who closes the stable door after all the horses have bolted." ("Le Conseil suisse de la Science a vu juste trop tard. La Confederation risqué d'avoir un jour la triste figure du cow-boy du Far West qui ferme soigneusement l'écurie après que tous les chevaux son parties.").* This might appear to be a drastic comment, but it must be understood against the background that Geneva had also lost its bid to host the European Molecular Biology Laboratory (EMBL). In any event, on 1 March 1982, upon completion of the ratification formalities, the Swiss Confederation joined ESO as its seventh member.

same time, however, the German government was wary of losing ESO. For historical reasons few other international organisations had settled in Germany at that time and this had not gone unnoticed by the German Federal authorities. In June 1973, the German Council delegate read out a formal statement, in effect offering to host a new ESO Headquarters in Germany. Blaauw has described how this became intertwined with the still unresolved discussion about whether ESO should develop into a scientific centre in its own right or follow the idea of being an *observatoire de mission*. Irrespective of this, in December 1973, Council authorised the Director General to *"study the offer of the German government... ."* In the course of 1974 the offer became more specific, with Garching being the site of choice, leading to a formal proposal by the German government presented to the ESO Council at its meeting of 29–30 April 1975.

≈

Garching was a small rural village, located some 15 kilometres north of Munich, on the high plain that leads to the Alps. The quality of the soil is poor and did not allow prosperous farming and so land was cheap. In 1958, the German government chose this rather uninspiring place as the site for Germany's first experimental nuclear reactor, which, because of the shape of its building, became known as the "Atom Egg". Over time the site became the breeding ground for a dynamic research campus and a powerhouse for work in the physical sciences. By the late 1970s, the Max-Planck-Institut für Plasmaphysik and the Max-Planck-Institut für Extraterrestrische Physik had been joined at Garching by the Max-Planck-Institut für Astrophysik and the new Max-Planck-Institut für Quantenoptik was also expected to settle on the campus. It was clear that the stimulating environment here was precisely what ESO needed.

In December 1975 the Council accepted the German offer and half a year later, the administration left the Bergedorf premises for a new start in an interim location — an apartment building in Garching. The decision to bring together all ESO facilities in Europe also required a new Headquarters Agreement between ESO and the Federal Government. The Agreement formally confirmed the commitment by the government to provide land and a Headquarters building free for a period of 99 years. This agreement was signed on 31 January 1979. ESO could now look forward to bringing its European staff together under one roof, but — perhaps understandably — not all staff saw this positively. Many of the staff, especially those in Geneva, decided against a life in Munich and, in the wake of the transfer, ESO lost a sizable fraction of its very competent engineering and technical staff. While recovery was certainly possible, it slowed down a number of activities, such as instrumentation development. Construction of the new home for ESO began in October 1978 and the building was ready in September 1980.

The German government was clearly prepared to put an adequate headquarters building at the disposal of ESO. Through the Max-Planck-Gesellschaft, which oversaw the construction project, they appointed the upcoming German architects Hermann Fehling und Daniel Gogel to be in charge of the project. They had already designed the adjacent building for the Max-Planck-Institut für Astrophysik and had made a bit of a name for themselves. Yet it was the ESO Headquarters building that elevated them to the status of star architects (even if the expression may be ill-chosen in the context of an astronomy book). The result was a modernistic unusually-shaped concrete building with offices, an auditorium, a large library, laboratories and an assembly hall for instrumentation. All in all, 4200 square metres of space awaited the ESO staff as they began to arrive, either from their interim offices elsewhere in Garching, or from Geneva. At the time, there were a total of 40 staff members in Europe (Fischer & Walsh, 2009). With 120 offices, an auditorium, laboratories and storage rooms, this was certainly a generous space allocation, but it would not take long, before the growth of the organisation, led by an increasingly dynamic and ambitious programme, meant that an expansion became necessary[3].

The main idea behind the unusual design was to create open "communication spaces", such as the entrance area with its bridges and stairs and to enable a fast transition from one area of the complex building to other areas: at the price of creating what to many newcomers must appear as a bit of a maze. The German weekly, *Die Zeit*, tried to describe the challenge that the architects had seen and attempted to solve. In a somewhat free translation of the article, it said that *"organising the building was more complicated than [the Max-Planck-Institut für Astrophysik], because [at ESO], thinkers and engineers, software people and technicians work together, as closely as possible, yet*

[3] The first extension was built in 1986, adding a wing of 1300 square metres to the existing building. In addition, the workshop area was expanded by 500 square metres. A further expansion occurred in 1990, when an additional storey of 575 square metres was added. Alas, this expansion also only provided a temporary reprieve. In 1998, as a provisional measure, ESO began to install offices in containers and, as the need for offices continued to rise, ESO rented office space elsewhere on the Garching campus. By 2008, almost half the ESO staff in Garching were housed outside of the Headquarters building. In that year, Council agreed to add a completely new building complex to the existing Headquarters, based on the outcome of an architectural competition in the preceding year. Construction work began in early 2012.

clearly separated and without disturbing each other."[4] (Sack, 1980). Less epideictic and in his own unique way, then-editor of *The ESO Messenger* Philippe Véron offered his view of the new headquarters building: *"A short period of familiarisation was needed during which everybody lost their way during a few hours or a few days in what, at first sight, looks like a labyrinth of the kind used to test the intelligence of rats. But human beings are on average cleverer than rats and the problem is quickly solved; after a while it appears that everybody found the building to be very convenient and a pleasant place to work in."* (Véron, 1981). It is probably impossible to achieve consensus about modern architecture, and some technical staff who had to fit their equipment into curved rooms — in some cases even to cut large holes in massive concrete walls simply to get the equipment into these rooms — may not have seen the building in the same way as the journalist from *Die Zeit*. Yet the ESO Headquarters remains a landmark building, and the goal of fostering communication and interaction between various groups in ESO by means of the architecture has undoubtedly been achieved[5].

Lodewijk Woltjer welcoming the inauguration guests at the Garching Headquarters (photo left). President Karl Carstens and Woltjer during a relaxed moment. (Photos by the author)

On 5 May 1981, the new Headquarters building was inaugurated in the presence of the President of the Republic of Germany, Karl Carstens, the Prime Minister of Bavaria, Franz-Josef Strauss, diplomatic representatives of the Member States, the President of the Max-Planck-Gesellschaft, Reimar Lüst, and the ESO Council and numerous members of the scientific community. Because of the large number of

[4] *"Die Organisierung dieses Hauses war komplizierter als die des anderen, weil hier Denker und Ingenieure, Rechner, Mechaniker miteinander arbeiten, möglichst nahe beieinander, aber deutlich und störungsfrei voneinander getrennt."*

[5] For many years the ESO Headquarters building remained an object of study by groups of architectural students.

participants, the assembly hall, intended for the integration of large components (even telescopes) had been transformed into a festive hall for the day. The inauguration itself was followed by a formal lunch, offered by President Carstens, and an evening state reception at the Antiquarium, the stunningly beautiful 16th-century renaissance hall of the former Royal Palace in Munich. In between these events, and on the following day there was a scientific symposium entitled Evolution of the Universe. The symposium featured high-profile speakers, such as Hubert Curien, two Nobel Laureates (Hannes Alfvén and Manfred Eigen), a former Director General of CERN (Léon van Hove) and two astrophysicists of world renown, Dennis Sciama and Jan Oort, the spiritual father and long-time associate of ESO. The strong showing at the inauguration marked the arrival of ESO on the research campus and helped to put Garching on the world map of physics.

Jan Oort speaking at the scientific symposium in connection with the inauguration of the Garching Headquarters. (Photo by the author)

≈

With the move in September 1980, the reorganisation, as Woltjer had planned it, had been achieved. In its new Headquarters, ESO Garching was becoming an important focal point for optical astronomy in Europe. And a natural meeting place for astronomers from the Member States and beyond. The Visiting Astronomers office, led by Jacques Breysacher and ably assisted by Christa Euler — perhaps the most famous non-astronomer female in European astronomy — handled the applications for observing time and organised the meetings of the biannual Observing Programmes Committee (OPC), which tended to grow as the number of proposals kept increasing.

The many ESO conferences created important fora for the exchange of ideas and drew many scientists to Garching. Others came to use the measuring machines, the PDS and Optronics scanners with which they digitised their photographic plates, or to reduce their data with the help of computers at ESO.

The Users Room at Garching where visiting astronomers could reduce their observational data obtained at La Silla. (Photo by the author)

This was a time when microelectronics began to assume great prominence. It was not limited to control systems for telescopes and the development of new detectors, but also to the treatment of astronomical data — or "image processing" as it became known. At ESO, the first step in that direction was taken by a young Dutchman, Frank Middelburg. The son of the Dutch Ambassador to Chile, Frank Middelburg had followed the early epoch of ESO's presence in the country, and in 1967, he joined the organisation to work at La Silla. Middelburg developed an interest in the use of computers for image analysis. At the time ESO's Santiago office operated a Grant

Measuring Machine, a one-dimensional scanner used for spectra on photographic plates. In 1975, he was transferred to Geneva, where ESO acquired a two-dimensional S-3000 Optronics scanner (for the scanning of Schmidt plates) and from then on the system, which he almost single-handedly developed and named the Image Handling and Processing System (IHAP), became an official ESO project. In the years to come, IHAP would be expanded and it not only became the workhorse image processing software at ESO, but was also implemented at 15 major institutes in Europe. IHAP owed its success to the fact that it contained most of the functions of systems still to come. However, it was highly optimised for the HP-2100 mini-computers that ESO had selected for its facilities. Porting the system to other platforms, or even using it with operating system updates, required a major effort and posed considerable problems (Grosbøl & Biereichel, 2003). With new non-photographic detectors, notably the CCDs, emerging and in general, the move towards digitisation, it was clear that more powerful computer systems would be needed, and ESO moved to VAX 11/780 computers.

Even though IHAP was a well-developed and popular system by 1980, ESO had embarked on developing a more versatile software system, which received the name MIDAS — the Munich Image Data Analysis System. King Midas acquired his fame by turning everything he touched into gold[6]. ESO's MIDAS could hardly compete with such a performance, but it did become the dominant image processing software for astronomy in Europe for more than ten years, with yearly releases and substantial user support. More significantly, MIDAS was conceived as an open and modular system. The MIDAS development was led by Phil Crane, initially supported only by Klaus Banse, but later the group[7,8] grew in size.

≈

So where was ESO by the early 1980s? From the perspective of the Member States it would seem to have accomplished its task. It had completed the initial programme

6 A software project more likely to make gold was software with the same name, but developed for the financial sector, and at the time unknown to ESO. For this reason ESO had to change the formal name of its software product to ESO-MIDAS (Banse, 2003).

7 Others working on MIDAS were Preben Grosbøl, Charlie Ounnas, Daniel Ponz, François Ochsenbein, Michèle Péron and Rein Warmels.

8 The Image Processing Group was ultimately absorbed into the Data Management Division to which we shall return. In the end, given that a competing image processing system called IRAF had been developed in the USA, Riccardo Giacconi decided that ESO should concentrate its resources elsewhere.

foreseen in the Convention and it had established a highly efficient setup in the sense of extremely smooth operations and a high degree of user satisfaction. With all European activities finally concentrated in one place, the ESO "system" was complete. And the organisation had begun to prepare for the future, embracing new ideas and concepts. On the other hand, by that time, ESO had not really achieved anything that had not been done before by others. For all its merits a 3.6-metre telescope as the major observational facility was perhaps not so much to write home about after all. Some individual Member States had in parallel embarked on programmes that would develop national 4-metre-class telescopes of their own. On Hawaii, France was building the 3.6-metre Canada France Hawaii Telescope. In Spain, the German Max-Planck-Gesellschaft was erecting a 3.5-metre telescope. Among the countries that would later join ESO, the UK was already operating the 3.9-metre Anglo Australian Telescope (AAT) at a site in New South Wales, Australia and was planning to build the 4.2-metre William Herschel Telescope on La Palma in the Canary Islands. In addition Italy's 3.5-metre Telescopio Nazionale Galileo was being considered and would later also be built on La Palma. From a global perspective, there were nine telescopes in the 3.5–6-metre range, though only three of these were in the southern hemisphere. Also, in the field of detectors and instrumentation, crucial for a telescope's ability to collect scientific data and generate new knowledge, others — notably the AAT — had moved more astutely. Finally, telescope designers had begun to develop and test new ideas, such as implementing alt-azimuth mounts and dreaming about arrays of interconnected telescopes. Ideas for much larger telescopes existed as well. If ESO had caught up, reaching the goal also meant realising that this goal was itself moving fast.

In the world of science, ESO's reputation was still mixed. To many astronomers, perhaps especially in non-member states, ESO was the target of scorn, perhaps also of envy. Some American astronomers assailed ESO for its luxurious habits, such as flying its staff and visitors in and out of La Silla instead of having them take the bus. Woltjer coolly dismissed the criticism, simply referring to the higher efficiency and the matter was put to rest. But the fact remains that in the global community, scepticism prevailed. It is almost as if ESO was the ugly duckling of world astronomy. But ugly ducklings may become beautiful swans, and ESO's Member States understood this.

Part II:
Years of Experimentation

Chapter II-1

Upping the Ante

"It was no longer a question of catching up with the rest of the world, but of taking the lead."

Lodewijk Woltjer, in his book *Europe's Quest for the Universe.*

Optical astronomy is fundamentally a question of photon-hunting. If we disregard the Sun, we cannot travel even to the nearest star to carry out detailed *in situ* measurements. Hence astronomy is really an example of remote sensing, catching a little of the electromagnetic radiation that reaches us from objects in the sky and analysing it. But that radiation is weak and some therefore say that astronomy is photon-starved. The remedy is to set up traps — telescopes with large collecting areas — the larger the better. In December 1977, scientists gathered at CERN in Geneva to discuss exactly how to do that. The occasion was an ESO Conference entitled Optical Telescopes of the Future.

In his introduction to the meeting, Woltjer mentioned the long lead times for new telescope projects, twenty years or more. He also stressed that, in spite of the merits of the Hubble Space Telescope, whose launch was then expected in 1983, in many areas of astronomical research large ground-based telescopes would offer better performance. The science case that he outlined covered studies of the chemical composition of main sequence stars in globular clusters in the Milky Way and of similar objects in nearby galaxies, detailed spectroscopic observations in the field of cosmology, studies of faint optical counterparts to the mysterious radio- and X-ray sources, and also *"high time-resolution photometry of X-ray sources, pulsars and other objects."* A man of great clarity, he did not hide his ambitions: *"The scientific case for one or more very large future telescopes is strong. The investment costs ... would not much exceed the sum ... of the present ESO, German and French large telescope projects. As such, the construction of a very large telescope of the future would be a fitting project for European cooperation."* (Woltjer, 1977).

Of the more than 240 participants, 19 came from the US. Not surprisingly, though, with 15 out of a total of 46 presentations, they played a prominent role. There were also many presentations by scientists from the UK and Canada. In the part dedicated to incoherent arrays and multi-mirror telescopes — arguably the core part of the conference — US astronomers gave seven out of the twelve talks.

It was evident that new ways had to be found to overcome the existing barriers in telescope design, whether technical or financial. Topics such as active and adaptive optics were mentioned as theoretical possibilities. Interferometry was mentioned as well, although with widespread scepticism. Nonetheless the options presented by Don Hall from Kitt Peak for what was then called the KPNO Next Generation Telescope — a 25-metre equivalent aperture optical and infrared telescope — more or less described the general situation. They were:

1) An "Arecibo Bowl" solution, i.e. a stationary, spherical primary composed of passive segments. The celestial object would be followed on the sky by pivoting the secondary mirror;
2) A "Rotating Shoe" solution, reminiscent of the Arecibo Bowl, but enabling the entire structure to rotate in azimuth. (This solution came to be realised in 1997 with the 9.2-metre effective aperture Hobby-Eberly telescope at the McDonald Observatory in Texas and its sister, the Southern African Large Telescope [SALT] in Sutherland, South Africa);
3) A giant steerable dish, with suggested 25-metre diameter (again for the KPNO project);
4) A Multi-Mirror Telescope, i.e. several mirrors on a single mount. This represented a scaled-up version of the University of Arizona/Smithsonian MMT[1] on Mount Hopkins, which would soon go into operation;
5) An array of individual telescopes in the 8–10-metre-class, the solution finally selected by ESO; or
6) Arrays of smaller telescopes, in some proposals up to one hundred or more 2-metre-class telescopes.

[1] The Multiple Mirror Telescope (MMT) constituted a radically new approach to telescope design. It served as an important source of inspiration also for the Europeans in the years to come.

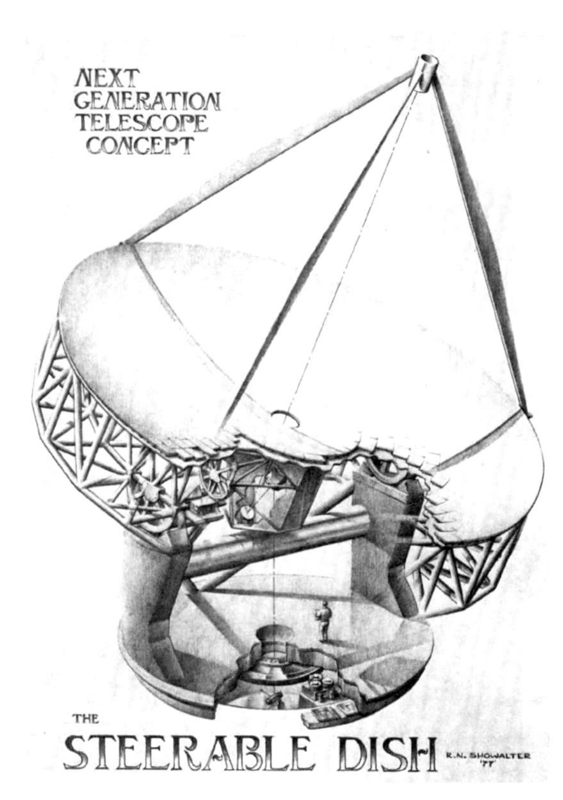

The 25-metre steerable dish — one of the telescope concepts presented by Don Hall at the 1977 ESO conference at CERN.

From ESO, both Ray Wilson and Wolfgang Richter gave presentations, but they elected to describe generic issues rather than a particular project. Nonetheless, the presentation by Richter contained a drawing of a 16-metre *f*/8 Cassegrain telescope with a segmented mirror. The telescope was housed in an enclosure, but outside the enclosure, Richter had placed a 60-metre windmill. This was intended to reduce the wind load on the telescope, and simultaneously produce energy for the facility (Richter, 1978). The windmill, not surprisingly, joined the not-so-small collection of curious, but long forgotten ideas, but the notion of the 16-metre telescope remained.

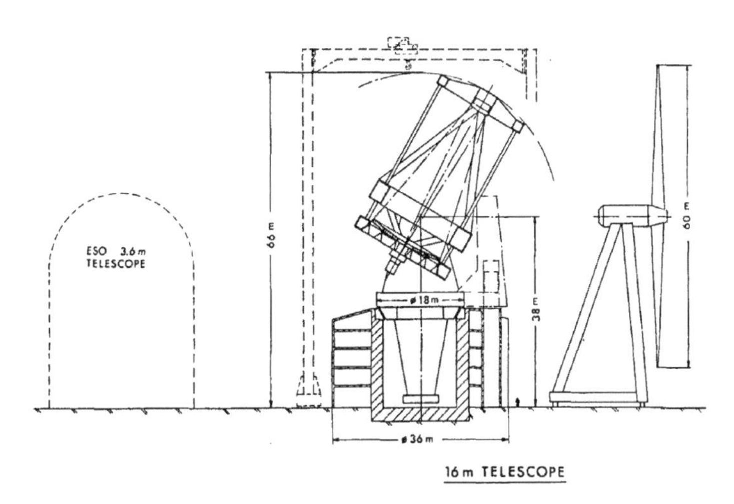

Wolfgang Richter's proposal for a 16-metre telescope, to be operated in the open air, but with a giant windmill "to diminish the influence of the wind".

≈

While much of the conference focussed on the technical challenges and merits of the individual ideas, lurking in the background was also a deeper issue, about how astronomy should be conducted in the future. More precisely, where should the balance lie between investments in large telescopes, expected to be used only by a few elite scientists, and smaller telescopes, of which there could be plenty and thus also be available for a much larger number of astronomers. This issue came to the fore in the final panel discussion, chaired by Jesse Greenstein of Caltech/Palomar Observatories. Harry van der Laan expressed the concern of many vividly: *"The astronomical community does not consist of one mastermind and* n *robots. The creativity and enthusiasm that goes with individuality and rivalry are aspects that must have great weight when considering the next generation of facilities."* In spite of the fairly sharp debate, a consensus seemed to emerge. The technology was not ripe enough to think in terms of really large single-dish telescopes. But without increased light-collecting power, many of the science goals could not be achieved. Therefore the solution lay in arrays rather than in single-dish extremely large telescopes. On the other hand, the individual telescopes of such arrays ought to be bigger than the already existing large telescopes to allow real progress.

≈

Despite the relatively low profile of ESO [2] at the meeting — at least as regards the number of presentations — the Geneva conference was pivotal for the development of a European Very Large Telescope project. It was the first major conference in this process, which would span two decades. Further milestone conferences, still to come, were the ESO conference on the Scientific Importance of High Angular Resolution at Infrared and Optical Wavelengths (1981), which ensured that adaptive optics and interferometry remained on the agenda, the VLT workshops in Cargèse (1983) and in Venice (1986), among many other meetings.

The Geneva conference opened up interesting perspectives for the future and Woltjer was clearly determined that ESO should play a leading role in that future. So, after the conference, Woltjer asked Wolfgang Richter to set up a study group at ESO. Not surprisingly, the group looked at three options: a single 16-metre telescope, four

[2] There were no presentations of an ESO large telescope project *per se*, but in his introduction Woltjer *"for illustrative purposes, ... [considered] some aspects of a 16-metre aperture equivalent telescope" (!)*

8-metre telescopes and sixteen 4-metre telescopes. At the same time Woltjer started to ponder a name for ESO's new project. *"Why don't we call it what it is?"* was Ray Wilson's straightforward answer. *"Why not simply call it the Very Large Telescope?"* And so, under the heading "Ten Nights at the Very Large Telescope (VLT)!", in 1978 *The ESO Messenger* opened its pages to contributions from members of the scientific community, describing the science they would wish to conduct with the new telescope. The fire had been lit.

The work of the study group, however, was disrupted by the move to Garching and the ensuing reorganisation of ESO's technical departments. In addition, two new telescope projects, to which we shall return below, meant that resources, always scarce, had to be reallocated to those other tasks. But in January 1981 a new study group was established, initially chaired by Ray Wilson and, from April 1981, by Jean-Pierre Swings. In the course of 1982, it became clear that the limited array concept (of four 8-metre telescopes) was the preferred option[3], although a number of technical questions remained to be clarified and, indeed, solved. One of the main issues was the feasibility of interferometry on the scale required for such a facility. The idea behind the array was the light beams of each telescope could be combined, either incoherently or, preferably, coherently. In the first case, the array would act as a "light bucket" with a light grasp equivalent to a 16-metre telescope. In the second case, the array could be used as a giant interferometer, yielding an angular resolution comparable to that of a much larger telescope. We shall look at this later. The outcome of the work by the study group was presented and discussed at a meeting of about 50 scientists in Cargèse, Corsica, in May 1983. At the meeting there was support for the basic idea that the VLT should consist of an array, rather than a single-dish telescope. Following the workshop, ESO's Scientific Technical Committee — and subsequently Council — decided that ESO should set up a dedicated project group within the organisation. Daniel Enard was appointed to lead this group. Before joining ESO in February 1975, Enard — an outstandingly gifted, yet modest engineer — had worked at the Optical Division of MATRA, the French automotive and defence company. At ESO, he had made his mark mainly by developing scientific instrumentation, an area in which ESO had been trailing badly. Since 1980, he had been in charge of the instrumentation group and as leader of the VLT project group he would play a crucial role in developing the project to the stage where it could be formally submitted to the ESO Council for approval.

---

[3] The 16-metre option was found to be too risky from a technical point of view and the 4-metre was viewed as not ambitious enough and therefore politically risky (Wilson, 2012).

The project group would interact with an advisory committee, comprised of members of the scientific community and led by Jean-Pierre Swings. The committee would, in turn, be assisted by dedicated working groups dealing with critical issues such as instrumentation (imaging and spectroscopy), site selection and interferometry. To ensure consistency and cohesion, the chairs of the working groups were also members of the advisory committee.

≈

The new VLT project group again reviewed the three main options, this time also from a technical/industrial point of view, and increasingly the array solution emerged as a winner. In an article in *The ESO Messenger*, Daniel Enard provides the main arguments for the final choice of the 4 × 8-metre telescope array. But, as he also acknowledged, this choice had its own challenges: *"Indeed, the array concept presented a number of problems which had to be matched by adequate solutions. There were three of them: the feasibility of the primary mirror, an efficient way to recombine the beams and a building concept combining a low-cost, a minimal degradation of seeing, the best use of the site's topography, and an optimal arrangement for interferometry."* (Enard, 1987). ESO therefore carried out a number of detailed feasibility studies, mostly contracted out to European industry, leading to a proposal that could be discussed again with the broad scientific community. For this purpose, a conference was organised at the Cini Foundation in Venice in October 1986.

The participants in the landmark VLT conference in Venice in October 1986.

Building on the outcome of this meeting, ESO then formulated its formal proposal to the Member States for the construction of the VLT. We shall return to this proposal in Chapter II-6, but for now, we need to look at a few parallel developments within ESO, some of which had a direct bearing on the VLT project and on ESO as a whole. At this stage, we shall just note that *"the VLT was very much in the air"*, to quote Jean-Pierre Swings[4].

Nonetheless, ESO would need a few interim steps, partly because they were highly prudent from a development point of view, partly to deal with the growing short-term needs of ESO's users, not the least due to the entrance of Italy and Switzerland, which had become a reality around 1980 (even if the formal membership process with its political and legal steps had not been completed).

≈

The addition of two new Member States meant dealing with a larger user community, adding to the pressure on the ESO facilities. To alleviate this pressure somewhat, it was decided that the entrance fee of the two newcomers should be used to construct a new large telescope. At the same time this new facility, which became the 3.5-metre New Technology Telescope, could be used to test the many new ideas about telescope design that were emerging in these years and without which the much larger telescopes of the future could never be built. The construction of the new telescope fitted nicely with the proposal by Italy to pay part of its entrance fee in-kind by providing the primary mirror[5].

Around the same time, another option came up that would expand the La Silla telescope park further. The Max-Planck-Institut für Astronomie, located in Heidelberg, was undertaking a substantial effort on its own, which included the construction of a 3.5-metre telescope, to be placed at Calar Alto in southern Spain, as well as two 2.2-metre telescopes, one for Spain and one to be erected in Namibia. The Namibia telescope had been built but, for political reasons, was never installed in its foreseen

[4] Interview with the author for the ESO 1992 promotional film *The VLT Tale*.

[5] In1968 Italian astronomers had acquired a 3.5-metre blank for a planned Italian National Telescope. As was custom at the time, this was cast with an aspect ratio (thickness *vs.* diameter) of 1:5 for a traditional, passive telescope. With the new ideas about thin telescope mirrors emerging, the idea was to slice the Italian mirror into two, but it turned out that only the top part could be used. This was then thought to become the primary mirror of the Italian National Telescope Galileo on La Palma, a modified copy of the ESO NTT. The Italian in-kind contribution to ESO was converted into orders to Italian industry.

location and instead had been stored in boxes for several years. Following an agreement between the Director of the Max-Planck-Institut, Hans Elsässer (who had actually himself taken part in the early ESO site tests in South Africa) and ESO's Director General, the 2.2-metre telescope was put at ESO's disposal on a 25-year loan from the Max-Planck-Gesellschaft, on the condition that ESO installed the telescope at its own cost[6]. Woltjer, in turn, had not only committed to doing this in record time, but also allocated a mere five million deutschmarks for the job. Undoubtedly, Woltjer was keen to demonstrate ESO's prowess, but there was a scientific need for speed, too: the 2.2-metre was a fine telescope and should be available to scientists sooner rather than later. He asked Massimo Tarenghi to become project scientist, but soon after changed his status to project manager. Tarenghi acted without delay and in a way that, for ESO, was unusual at the time. He brought two accomplished Chilean staff members, Gerardo Ihle and Manuel Cartes, up to Europe and put them in charge of the installation of the telescope. He also solicited help from Carl Zeiss, the company that had produced the telescope. Finally, at the suggestion of Robert Fischer[7], he decided not to develop a new dome, although initial plans had been to construct a scaled-down version of the 3.6-metre dome. Instead, he procured a commercially available dome from Observa Dome in the US, which was installed in a space of just six weeks in October/November 1982 (Bauersachs, 1982). Erection of the telescope began on 15 February 1983, and first light was achieved by 22 June of the same year. This must be close to a world record for such an undertaking! The initial instrumentation was a photographic camera and the Boller & Chivens spectrograph with a CCD camera. Later instrumentation included a Wide-Field Imager (WFI)[8] and FEROS, the Fibrefed Extended Range Optical Spectrograph, which was transferred from the ESO 1.52-metre telescope.

The telescope would serve as a most useful addition to the arsenal at La Silla and it also played an important role in a series of experiments that ESO undertook regarding possible remote control of telescopes. For a while three telescopes were being operated remotely from Garching, the CAT, the NTT and the 2.2-metre telescope. The rationale for remote control was partly to reduce travel costs and time, and to allow for a much more efficient operation of the telescope. Satellite communication,

---

[6] In addition the Max-Planck-Gesellschaft would receive 25% of the available observing time.

[7] ESO's Head of Contracts and Procurement.

[8] In the VLT era, much deeper surveys were needed than the ESO/SRC Survey and thus the requirements exceeded the capability of the Schmidt telescope. The WFI was a joint project of ESO and the Max-Planck-Institut für Astronomie to provide ESO with a wide-field capability, covering 0.5 × 0.5 square degrees with a 67-megapixel camera. We shall deal with the issue of survey telescopes again in Chapter IV-3.

which had become a readily available option, opened up new opportunities. However, the first attempts began much more humbly and without involving satellites in geostationary orbits. The first test was carried out over a distance of a few metres in June 1984 by Tarenghi and Gianni Raffi, a software engineer based at Garching. They tried the idea from a console a few steps away from the telescope, causing a few smiles at La Silla. After that, they moved to La Serena, to continue the experiment from the ESO office there. Since the 2.2-metre telescope was not really designed for remote control, a telescope operator (then called a night assistant) stayed behind to check that everything was working properly and also to carry out the few manual functions that could not be steered remotely. The computers in the control room and at the La Serena office (two HP-1000 systems) were connected by a telephone line. The observers thus had the choice of observing remotely or talking to the night assistant, when they needed his help. With a smile, Tarenghi later recalled: *"At a certain moment, the operation went so smoothly that we decided to continue observing without interruption. When we were finished — in the morning hours — we called the night assistant and told him that he could now close the dome. 'OK', he replied, 'but before I do so, could I please go to the toilet, because you have gone on for hours without talking to me.'"* (Tarenghi, 2010).

With the completion of the 2.2-metre, Tarenghi became involved in the NTT project, first as a consultant, but soon after as project manager. The NTT was a springboard to the VLT and we shall therefore deal with this telescope in some detail. The next chapter examines some of the underlying ideas that made this telescope so revolutionary.

Chapter II-2

Inventing a Game Changer

"The development of telescope optics is a fascinating story.... Over all its history, the optical development of the telescope has also depended on technical inventions, above all in mirror materials, glasses, support systems and means of achieving high reflectivity."

Ray Wilson, 1996.

The task of an optical astronomical telescope is simply to collect and focus the faint light from objects in the sky. It usually does so by means of one or more mirrors and the bigger the main (primary) mirror, the more light it can collect. Since a mirror reflects light, what really matters is the reflecting top layer of what is normally a glass substrate. For almost all telescopes, this layer is comprised of aluminium and, remarkably, only a few grams of it will suffice to create a thin, yet highly reflective coating, even for the largest mirrors. However, to keep this reflective layer in its place one needs a support — the glass substrate already mentioned — properly shaped, and fixed in a mount that can turn to point the telescope towards its target and to counter the rotation of the Earth during prolonged observations. One also needs equipment to detect the light as it arrives at the focus and thus the whole structure needed to keep these few grams of aluminium correctly positioned, and to exploit the light scientifically, becomes a giant machine weighing many tonnes, in fact sometimes hundreds of tonnes.

In the case of ESO's 3.6-metre telescope, to maintain the shape of the primary mirror accurately, irrespective of the direction in which it points, the glass mirror weighs 11 tonnes. It is supported by a mirror cell and a polar axis of 119 tonnes and the mechanical structure needed to keep the secondary mirror in its right place adds an additional 70 tonnes to the mass budget. Including the non-moving pedestal, the mass of this telescope is of the order of 245 tonnes. Clearly, any increase in the diameter of the primary mirror to improve the "photon catch" would involve a string of additional problems related to the overall mass growth and, even if these were solvable, would surely lead to a dramatic increase in cost. For traditional telescope designs, the relationship between cost and size seems to follow an empirical law, according to which the cost rises by roughly the 2.6th power of the diameter of the primary mirror. For these reasons, the practical upper limit in terms of diameter of the primary mirror was considered to be around 5 metres, the size of the famous Hale Telescope at Mount Palomar[1].

If astronomers wanted to build telescopes with larger light-collecting surfaces, they would have to fundamentally rethink the way telescopes were designed. Among the optical scientists who had thought about this was Ray Wilson, at the time working at Zeiss, where he was conducting a study for the optics of the ESO 3.6-metre telescope in 1966. Originally, it was neither the weight issue nor the cost that caused him to think about other solutions, but rather the inherent difficulties of maintaining a proper alignment of the main optical elements of a telescope — between the primary and the secondary mirror — as the telescope changed its orientation during tracking. Poor alignment resulted in an aberration that opticians call decentring coma. He dubbed the problem "Cassegrainitis", because it applied to all classical Cassegrain telescopes. His great insight was the realisation that if it were possible to measure the adverse effect on the image, then it would be possible by means of an active support system to correct the problem. It would also be possible to correct other kinds of aberrations such as astigmatism. At the time, however, Zeiss showed little interest in these thoughts and moving to ESO (in 1972) gave Wilson the opportunity to pursue the idea further. In 1977, after the completion of the 3.6-metre telescope, he published his ideas, followed by a more elaborate paper in 1982, in which he also introduced the term "active optics". The idea was stunningly simple. The telescope mirrors would be controlled by an active support system that would apply the necessary force to the mirrors to maintain them aligned with each other (in spite of tube flexure[2]) and correct for gravitationally induced deformation as the telescope changed its orientation. The correcting forces would be applied on the basis of analysis of an actual star image during the observations, so that the telescope would, in effect, be self-adjusting and always produce the optimal result[3]. But simple ideas do not always convince everybody. In a comprehensive article about the history of active optics in *The ESO Messenger*, Wilson recalls the reaction of an American scientist, when he explained

[1] The Russians had built a 6-metre telescope at the Special Astrophysical Observatory, which revealed the huge problems associated with conventional telescopes of that size.

[2] Deformation of the telescope mechanical structure under the influence of gravity.

[3] Alternative — but less elegant — solutions, pursued elsewhere, involved correcting the telescope according to pre-calculated flexure, rather than measuring the actual image deterioration.

the system. *"Well, my feeling is that such a system will never be realisable in practice,"* said the American (Wilson, 2003). Feelings or not, he was wrong.

But why was this idea so important? Firstly, significantly improving the image quality meant a dramatic improvement in the resolution of the images that could be obtained with the telescope. But higher resolution also means that more (light) energy is concentrated in a smaller area on the detector. This in turn makes it possible to detect fainter objects. However, the real advantage was that with an active support system, the primary mirror in particular could be made much thinner, breaking out of the straightjacket that led to a disproportionate weight growth when increasing the diameter of the mirrors. With lower mass, thermal effects would be minimised and the whole telescope structure could be slimmed down. And last, but not least, stringent specifications as regards the optical quality of the mirrors could be relaxed, because much could be corrected by the support system. Basically, the invention of active optics held the potential to push the door wide open to much larger telescopes, both technically and financially. If only it would work.

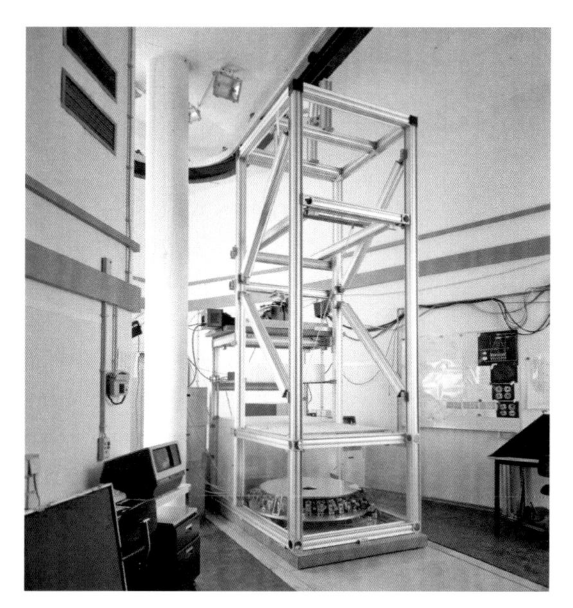

The active optics experiment at the ESO optical laboratory in Garching. (Photo: Hans Hermann Heyer)

To demonstrate the feasibility of active optics, an experiment was set up in the optical laboratories at Garching with a thin 1-metre test mirror resting on an active support system with 75 actuators, produced by Oberto Citterio of the Laboratorio CNR Fisica Cosmica e Tecnologie Relative[4] in Milan. In addition to Ray Wilson, the experiment mainly involved Paul Giordano and Lothar Noethe, the latter joining ESO in 1983 from Siemens. The setup was not just used to demonstrate that the active

[4] The mirror was later donated to the Deutsches Museum, the huge technical museum in Munich, where it was prominently displayed in the permanent astronomy exhibition.

support system worked but also the image analyser, based on a Shack–Hartmann system of lenslets and a CCD detector allowing for real-time analysis. The latter was of essential importance for the success of the entire system. The tests went on for about one year. Although the team had expected complications along the way, at least one problem was completely unexpected. After a successful start, the test results began to raise eyebrows. Results could not be repeated, and the system seemed to behave in a completely random way. Evidently, something was wrong, but what? In the end, the team decided to remove the mirror and check the actuator system below. This was not an entirely trivial exercise, but it — literally — brought the cause of the problems into the light of day. The ESO Headquarters building is located on a green-field research campus north of the small town of Garching near Munich. Looking out of their office windows staff could enjoy the charming view of fields being tended to by local farmers. Less pleasant, though, was that by autumn, the Headquarters building seemed to have been selected as cosy winter quarters for harvest mice. Some found the warmth and darkness underneath the 1-metre mirror experiment an almost ideal place to build a nest, with food right on their doorstep — since mice seem to like to eat the protective layer of plastic around electrical cables (and, if need be, the copper wire as well). The affair caused a great deal of amusement and discussions about the solution — some suggested simply getting a cat, but being a high-tech place, ESO instead acquired an ultrasound system which made the mice migrate, presumably to more welcoming locations.

Wilson's active optics became the main novel feature of the New Technology Telescope. But when it went into operation it combined a host of technical solutions that, in combination, made it one of the world's finest optical telescopes, perhaps the finest. We shall look at the NTT in more detail later, but at this stage it is important to realise that the NTT constituted a bold attempt to exploit a wide range of new ideas about telescope design. Woltjer's decision to go ahead, backed by the ESO Council, allowed ESO to cut free of its existing programmatic paradigm, to search for and to develop new pathways. In a sense, it was the birth of a new ESO, but, just like stars that form inside dense clouds of interstellar dust and gas, and are thus shrouded from our sight, it would take a while for the new ESO to become visible.

Whilst the first light of the NTT that we describe in Chapter II-9 undoubtedly marked the professional high point of Ray Wilson's career, he subsequently received several high honours, including the prestigious Medaille Lallemand of the French Académie des Sciences in 2005, and, in 2010, both the Tycho Brahe

Prize of the European Astronomical Society, and the Kavli Prize[5], awarded by the Norwegian Academy of Science and Letters (Det Norske Videnskaps-Akademi), the Kavli Foundation and the Norwegian Ministry of Education and Research (Kunnskapsdepartementet), for his seminal work.

[5] Shared with Roger Angel and Jerry Nelson of the US — other great telescope builders.

Chapter II-3

EMMI, SUSI, SOFI and the other Darlings

"The long time ... has led me into reconsidering whether the coudé spectrograph really deserves the high priority it has been given, and I am convinced that it does not."

Johannes Andersen (letter of 9 October 1974 to Ray Wilson).

Astronomy is as close to fundamental science as it gets. Yet progress in astronomy has always depended strongly on technology. Sometimes the necessary technologies existed already, sometimes they had to be invented and, in such cases, astronomers often gave a hand in the technical development. Galileo, the first person known to have used a telescope for astronomical observations, did not invent it. But he put it to use as soon as he could lay his hands on one. Conversely, John Herschel, the astronomer, played a major role in the development of photography. Astrophotography, it has been said, was the step that *"turned observational astronomy into a true science"* (Schilling & Christensen, 2009). Astronomers were no longer dependent on the human eye and the subjective and fallible perception caused by the signal processing in the human brain. Instead they had objective records that could be measured and compared. And they could collect light over long periods of time, adding photons as they came in. The introduction of solid-state detectors came as a godsend (if that term can be used in connection with astronomy)[1]. The increase in quantum efficiency was dramatic, data came in digital form and thus could be exploited scientifically much more easily. Cumbersome darkroom processing, often at the end of a long night and a tiring observing run, became a memory of a distant past. But CCDs were still far from perfect and once more, astronomers played an important

[1] In the course of the 1970s, the photographic plate was superceded as the detector of choice, in a sense simply because the emerging world of electronics provided exactly that — a choice! Initially image intensifiers were used in front of photographic film, and also electronographic cameras, pioneered by André Lallemand, providing a badly needed increase in sensitivity. Yet they were not very practical to use, and the next step — the Spectracon and McMullan cameras — was a short one, as subsequently, charge-coupled devices became available. Readers interested in electronic cameras will find an overview in a paper by Dennis McMullan and Ralph Powell (McMullan & Powell, 1977).

role, together with industry, in optimising the performance, with respect to spectral sensitivity, even better quantum efficiency, lower noise and more efficient read-out. We have already discussed why astronomers are keen to have bigger telescopes, but we see here that there is more to it than large mirrors. The auxiliary instruments and their detectors, the equipment with which astronomers actually record and analyse the light from their targets, are at least as important. But whilst astronomers have expended great efforts on developing their telescopes, for a long time, their focus has been rather less on what we today call instruments. In fact, in the past, the term was used almost arbitrarily, as can be seen in the ESO Technical Report No. 1, in which Svend Laustsen describes the 3.6-metre *telescope* as ESO's principal *instrument* (Laustsen, 1974)[2]. Fortunately, all of this has changed, enabled by technological progress and by changes in telescope design. This, in turn, has led to new operational concepts and to an explosion in the amount of scientific information — i.e. data — that we collect. The development of astronomical instruments has arguably been one of the most dynamic and significant areas for the evolution of the science. But we are jumping ahead. The roots of ESO's instrument development programme were laid in connection with the 3.6-metre project, and the early phase was anything but auspicious. It is indicative that Blaauw's book on the early history of ESO pays only scant attention to the question of instrumentation for Europe's new flagship telescope.

The 3.6-metre telescope had three foci: firstly a prime focus at the top of the telescope used for direct photography. Behind the main mirror was the Cassegrain focus, with the light being bounced from a secondary mirror through a central hole in the main mirror. And finally, a coudé focus was located a floor down in the telescope building, with the light being brought there by means of a series of mirrors. Much attention and effort, and many hopes, were directed towards the coudé focus of the telescope. In conventional telescopes, the coudé focus offers the only place where very heavy and bulky instrumentation can be placed. Unlike the other conventional foci (prime focus and Cassegrain), the instrumentation is not mounted on the moving structure and so the coudé was the location of necessity for the large instruments that were needed to obtain high spectral resolution[3]. We have already encountered

[2] To be fair it should be noted that the terminology was different in those days. What we today call an instrument was then called an auxiliary instrument. However, even this term betrays the somewhat subordinate status of this component in the minds of many telescope designers.

[3] The preoccupation with the coudé spectrograph, to which Johannes Andersen alludes at the beginning of this chapter, was not simply a technical question. Rather it was rooted in the fact that while it was a key instrument for stellar observations, it was less useful for the observation of faint galaxies. Yet at least some astronomers saw the 3.6-metre telescope as crucial for opening up extragalactic studies for European researchers.

Gerhard Schnur riding in the prime focus cage at the top of the 3.6-metre telescope. From this position, the observer guided the telescope and exchanged the photographic plates. (Photo: John Launois/Black-Star)

the term spectrograph many times and will continue to do so. The reason is simply that a spectrograph is the most important tool in the hands of an astronomer. In a typical spectrograph, a diffraction grating spreads (disperses) the light from the astronomical objects into its individual colours, or as the scientists would express it, wavelengths. These spectra provide a wealth of information about the target object. With an echelle grating it is possible to obtain high dispersion spectra, i.e. of high resolution, and a wide wavelength coverage at the same time. This is ideal for detailed studies of the chemical composition of stars, for example. The first instrumentation plan for the 3.6-metre telescope — i.e. for the instruments to be mounted at the three foci — was modest and at the same time reflected traditional thinking. Woltjer put it less diplomatically: when he arrived at ESO he found that *"... instrumentation developments were in a catastrophic state."* (Woltjer, *ibid.*).

A joint ESO/SRC/CERN conference on Research Programmes for the New Large Telescopes, held at CERN in May 1974, provided a forum for thinking about the kind of instrumentation needed to tackle some of the big research questions. One of the participants was Johannes Andersen, a young, dynamic astronomer from Denmark, who had recently gathered experience in spectroscopy first at the Observatoire de Marseille and subsequently at the Dominion Astrophysical Observatory in Victoria, Canada. Not a timid person, in his characteristic bold way he quickly developed a good rapport with Woltjer, who was poised to become the next Director General of ESO (though at that time he had not yet been formally appointed). Andersen brought fresh ideas with him and he became an important player — together with Ray Wilson[4] and Daniel Enard — in fundamentally revising the plans for the coudé instrumentation for the 3.6-metre telescope. They developed the design for a powerful, up-to-date scientific tool: the Coudé Echelle Spectrometer (CES), which would offer a resolving power of 100 000[5]. That, however, would take considerable time, and in the end, it was only put to use after the Coudé Auxiliary Telescope was installed in November/December 1980.

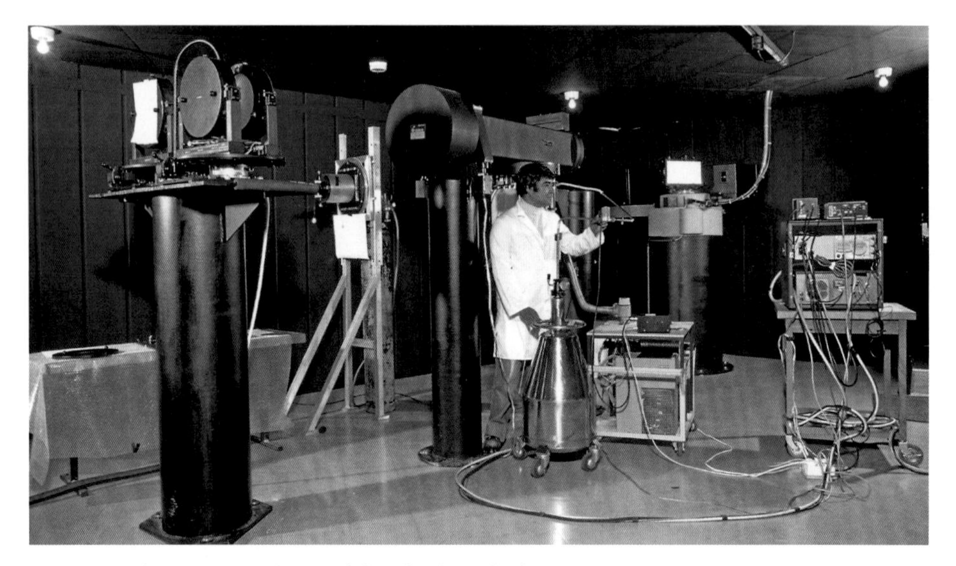

The CES at the 3.6-metre telescope. (Photo by the author)

So, at the start of operations, the 3.6-metre instrument complement was sparse. Aside from the prime focus camera and a photometer, the main instrument was a Boller & Chivens spectrograph, another off-the-shelf instrument bought in the US. The decision to go for a readily available commercial instrument was not undisputed, but it

[4] The working relationship between Wilson and Andersen is nicely illustrated by an anecdote, told years later by Andersen on the occasion of the retirement of Ray Wilson: *"Returning from my first observing run at La Silla in March 1973, I came on a fine morning to the newly-established ESO TP Division in Geneva to visit a Danish colleague. Treated to my then standard sermon on spectrograph design, he quickly introduced me to the new Head (and only member) of the Optics Group with a comment that we might have common interests to discuss. Indeed we had: after what seemed like five minutes, the cleaner politely suggested that we get out of his way; everybody else had gone home long ago!"* (Andersen, 1993).

[5] The resolving power of a spectrograph is normally given by a number that expresses its ability to resolve features in the electromagnetic spectrum in a given wavelength domain.

was taken for reasons of time and was, in any case, necessary for extragalactic studies. Yet it is also an indication of the lower importance given in the early years to planning and developing instruments for frontline science relative to the efforts of building the telescope itself.

≈

Aside from the CES, the first steps to develop dedicated instrumentation at ESO began with a spectrograph for the 3.6-metre Cassegrain focus, an instrument called CASPEC. CASPEC was an echelle spectrograph enabling a resolving power of 20 000 (and later up to 50 000). Such a resolution was previously the domain of coudé focus instruments, but since CASPEC did not need an elaborate optical train of mirrors — each resulting in additional light loss — it could be used to study much fainter objects (down to the 15th magnitude). This opened up new possibilities for studying stars in the Milky Way and nearby extragalactic objects. The development of CASPEC was led by Maurice Le Luyer and began in the TP Division in Geneva in the mid-1970s. The instrument saw first light on the 3.6-metre telescope in June 1983 (D'Odorico *et al.*, 1983). An SEC Vidicon tube had initially been foreseen as detector, but this was changed in favour of a 512 × 320 pixel, thinned, back-illuminated RCA CCD detector[6] — the first such detector at ESO[7] to come into operation in a spectrograph.

This instrument was followed by the ESO Faint Object Spectroscopic Camera, EFOSC, conceived by Daniel Enard. EFOSC was a focal reducer and spectrograph capable both of imaging and faint object spectroscopy and in this sense the first multi-purpose, or multimode, instrument at ESO. Focal reducers provide for a reasonable field of view even with a small detector, while better concentrating the light and thus ensure a better signal-to-noise ratio (Dekker, 2009). The optical design of EFOSC, which became the mother of many other FOSCs produced elsewhere, was due to Bernard Delabre, an ingenious young optical engineer from France and one of the unsung heroes at ESO[8]. It was first installed at the 3.6-metre telescope

[6] A CCD that has been thinned has higher sensitivity, especially to shorter wavelengths, and also has a higher quantum efficiency than front-illuminated CCDs, because the light can pass through the back layers instead of through the CCD's gate structure at the front which blocks some of the light.

[7] The first CCD detector at La Silla was used at the Danish 1.54-metre telescope for direct imaging (Pedersen & Cullum, 1982). Both systems had been procured in the US. Although in the days of the Cold War, such advanced equipment was not easy to get, thanks to the efforts of Phil Crane, ESO succeeded.

[8] Delabre has played a key role in developing the optical design for most of ESO's instruments.

in June 1984 and offered to the user community in April of the following year. The detector in this case was also a thinned, back-illuminated RCA CCD.

The move towards multi-purpose instruments was based on the experience that constant instrument changes, necessary on the 3.6-metre telescope to accommodate many different observational needs, were time-consuming and achieving a perfect optical alignment reliably could be problematic, quite apart from the wear on connectors that had to be constantly connected and disconnected with every change. Perhaps more importantly, there was no flexibility during an observing run. The reasoning was, therefore, that if the telescope had fixed instruments that could perform different types of observations, this would bring about a major improvement in efficiency.

≈

Multi-purpose instruments tend to be much more complex than their single-purpose brethren. This almost inevitably has consequences for their size and weight. And since Cassegrain instruments are mounted underneath the primary mirror — and thus riding on the structure as the telescope tracks the stars — there are limits to what can be done this way. The solution came not primarily from changes in instrument design, but from electronics. Or rather from the new possibilities created by the introduction of computer control for telescopes. In the earlier chapters we have discussed the equatorial mount used by most classical telescopes. A simpler solution is the altitude-azimuth (alt-az) mount, where the telescope rotates about two mutually perpendicular axes, one of which is vertical, leading to a cleaner, simpler and more symmetric design and load distribution. This was first introduced for large telescopes by the Russians with their 6-metre Bolshoi Teleskop Alt-azimutalnyi (BTA) telescope[9] at the Zelenchuk Special Astrophysical Observatory in the Northern Caucasus region. Tracking the stars with an alt-az mount is more complex, because the telescope has to move around two axes with continuously changing speeds. Furthermore, the field — the image, so to speak — rotates, so that the detector must rotate as well to keep it steady. However, with the advent of modern computer control, all of this is possible. The advantages are that alt-az mounts are lighter and cheaper to make. They can also carry heavier, i.e. larger, telescopes. More importantly in the context of instrumentation, however, is that they enable the use of another type of focus — the Nasmyth focus[10]. In this scheme a flat tertiary mirror diverts the light coming from the

[9] In operation since 1976, i.e. the year of first light for the ESO 3.6-metre telescope.

[10] In classical telescopes, the Nasmyth focus can be used only to relay the light, e.g. to a coudé instrument.

secondary mirror towards a location on the side of the telescope's main mechanical structure, which can easily accommodate heavy and complex instrumentation. And by rotating the flat mirror through 180 degrees, instruments can be placed on either side of the telescope. This was the solution adopted for the New Technology Telescope, which for a while became the main field of occupation for the ESO Instrumentation Departments[11], which had been established within the TP Division. The NTT meant a major change at ESO in terms of instrumentation development, not only because the technological advances opened up new perspectives for sophisticated design solutions, but also because instrumentation was included from the outset in the telescope project in a way that had not been the case before. Seeing instrumentation as integral to the entire project and as a vital means of optimising the scientific use of the telescope, in other words adopting a holistic approach to the development of a new "science machine", has been the hallmark of ESO ever since and is the foundation of the success of both the NTT and the VLT.

This photo of EMMI, taken in 1989 during the final testing at Garching, provides a good impression of the complexity of the new generation of instruments. (Photo: Hans Hermann Heyer)

The original instrument complement for the NTT comprised EMMI, the ESO Multi-Mode Instrument, and the infrared spectrometer (IRSPEC, which was transferred from the 3.6-metre telescope). The EMMI two-channel instrument (ultraviolet [UV]-blue and Visual-red) was specifically developed for the NTT and was a good initial all-round instrument for both imaging and medium- and high-resolution spectroscopy. It was efficient, flexible and had wide wavelength coverage[12]. However, its designers had not

[11] The reason for two departments was to separate work on optical and infrared instruments. These evolved at a different pace and in any case had somewhat different requirements.

[12] Years later, to illustrate the versatility of EMMI, Hans Dekker made an analogy with the mythical Bavarian animal, the "egg-laying woolmilkpig" (Dekker, 2009).

anticipated the spectacular first-light images produced by the NTT (see also Chapter II-9). These images provided a stark illustration of what could be achieved by controlling the dome temperature and ventilation to optimise the local seeing. But they also demonstrated the quality of the site — and it was realised that the instruments would not be able to take full advantage of superb atmospheric conditions when they occurred. Since it became clear that this actually happened *"for a relevant fraction of the observing time"* (D'Odorico, 1991)[13], ESO set out to quickly develop a CCD camera that, with a pixel size corresponding to 0.13 arcseconds, would be able to exploit these opportunities — the SUperb Seeing Imager, or SUSI, for short[14]. And quick it was: SUSI saw first light in April 1991, sitting on the so-called Nasmyth B platform of the telescope, together with IRSPEC.

≈

Let us briefly turn to the subject of observations in the infrared. Infrared observations complement observations at visual wavelengths. They provide us with additional information about the nature of the objects that we study and, in some cases, they allow us to study objects that cannot be seen at all in visible light. This is, for example, the case for stars in their early formation phase, which occurs inside dense clouds of gas and dust, impenetrable to visible light. Likewise, the central part of the Milky Way is heavily obscured by dust while it shines brightly in the infrared. At certain wavelengths, infrared observations can be carried out from the ground, but at others space-borne telescopes are needed. ESO, of course, focuses on the ground-based work.

If ground-based observational astronomy is normally a struggle *for* photons, it could be said that infrared observations are as much a struggle *against* photons. This is because of the high degree of infrared radiation in the local environment compared to that coming from distant celestial sources. Infrared radiation is basically heat and heat is of course generated almost everywhere. Even the telescope structure itself radiates heat and so does the large quantity of electronics associated with any modern telescope. Add to this the fact that the atmosphere is partly opaque to infrared radiation from the Universe (while itself being a strong infrared emitter) and the challenge

[13] According to the 1990 ESO Annual Report, about 10% of the observing time at La Silla offered a seeing better than 0.5 arcseconds (full width at half maximum [FWHM]).

[14] In 1995, ESO in collaboration with the Osservatorio Astronomico di Roma upgraded the instrument with a new imager with a much larger field and of a better camera as regards read-out speed and sensitivity.

facing infrared astronomers becomes clear. As ESO astronomer Hans-Ulrich Käufl expressed it rather graphically: "*Translated into the words of optical astronomy this is equivalent to observing stars from within a furnace (like the one in which the VLT blanks have been cast) rather than from a dark astronomical site.*" (Käufl, 1993). At least part of the technical remedy is cryogenics. The study of heat sources in the sky, thus, becomes cool science, literally speaking. The engineering task is to build complex opto-mechanical instruments that can operate flawlessly in ultra-low temperatures. But there is more to it. Even without the local effects the excess radiation from space is sparse. Observing an object in the infrared therefore requires a constant check on the general sky background, which is achieved by switching between the field of observation and a neighbouring field. This is best achieved with the help of a so-called "chopping secondary", i.e. by moving the secondary mirror frequently — and rapidly — between two positions. This is not in itself trivial, considering the mass of the mirror and the requirements in terms of constant optical precision, but it is feasible.

Infrared astronomy, however, was not just a struggle for or against photons. It was also a struggle for detectors. The technological development received a boost in the 1980s in the context of the US Strategic Defence Initiative, also known as the Star Wars Programme. This however meant that whilst new detectors were being developed in America, the US military also controlled the market and obtaining export licenses was difficult. Since US observatories had better access and furthermore benefited from the Hubble Space Telescope programme[15], ESO and European astronomers were put at a disadvantage. In Europe, the French and UK carried out military development programmes in this area as well, but obtaining access to these developments for civil applications was exceedingly difficult.

ESO's first infrared imager offered as a common user instrument was IRAC, the InfraRed Array Camera, which was mounted on the MPG/ESO 2.2-metre telescope in 1988. The camera had a 64 × 64 pixel detector. It was followed in 1992 by the Thermal Infrared Multimode Instrument (TIMMI). Developed at the Laboratoire d'électronique et de technologie de l'information (LETI/LIR) in Grenoble, the first TIMMI detector was based on the detector of ISOCAM[16], flown on the ESA Infrared

[15] In particular, the development of the NICMOS instrument for the Hubble Space Telescope (HST).

[16] Despite the difficulties regarding military developments, Pierre Léna had started discussions with Commissariat à l'Energie Atomique (CEA) about supplying the detectors for ISOCAM. Together with François Sibille, he carried out important preparatory work for this project. The principal investigator of ISOCAM was Catherine Cesarsky, who became ESO Director General in 1999.

Space Observatory (ISO). TIMMI was described as a kind of infrared EFOSC (Käufl *et al.*, 1992), and like its next of kin placed at the Cassegrain focus of the 3.6-metre telescope. It provided for imaging as well as some long-slit spectroscopy, and as such was probably the first instrument of its kind in the world (Käufl, 1994). The TIMMI instrument was built by the Service d'Astrophysique of the French Commissariat à l'Energie Atomique. In view of ESO's VLT instrumentation plans, the TIMMI contract therefore also served to test the waters regarding procurement of scientific instruments from national institutes, a topic to which we shall return later. As part of the La Silla upgrade effort, described in Chapter III-8, TIMMI was replaced in 2000 by a completely new camera — now with a 240 × 320 pixel detector — this time built by the Astrophysikalisches Institut und Sternwarte of the Friedrich-Schiller-Universität (FSU) in Jena, Germany[17]. The Universität Wien was also involved and contributed software. TIMMI 2, as the instrument was called, was discontinued in June 2006.

A further instrument for the near-infrared (wavelengths in the range 1–2.5 μm) was SOFI, an imager/spectrometer: i.e. yet another "infrared EFOSC". SOFI was installed at the NTT on La Silla in early December 1997 (Moorwood *et al.*, 1998). From a semantic point of view, the name constituted a bit of a paradox: SOFI stands for "Son OF ISAAC", the larger and more complex Infrared Spectrometer And Array Camera (ISAAC) infrared instrument, which was being built for the VLT. As we shall also see later, developing advanced instrumentation can take a long time, but SOFI was in fact realised by ESO in less than two years and in parallel with the ongoing ISAAC development. For that reason it was also finished before its father, hardly a normal father–son relationship! SOFI, however, was a son in spirit and to some extent even in parts. It was built in the context of the NTT Upgrade Plan to maintain the NTT, and La Silla, as a competitive facility, albeit at low cost and within a minimum use of time. This does not mean that it was a secondary instrument in terms of its scientific capabilities. By the late 1990s, infrared detector development had made great advances, opening new opportunities for observers. Thus SOFI provided much improved performance over IRSPEC, its predecessor at the NTT, both in terms of resolution, limiting magnitude and sensitivity.

It is difficult to overestimate the importance of the increased focus on instrumentation. In fact, improvements in instrumentation optics, allowing for higher

[17] This project was funded by means made available to institutes in the former East Germany in connection with the German reunification effort.

efficiencies, and dramatically increased sensitivity of detectors and the associated electronics had a significantly stronger impact than the growth in telescope size. The increase in sensitivity of existing telescopes was equivalent to a fivefold growth in their size (Gilmozzi, 2006). This development had, however, only just started. As we shall see when we deal with instrumentation for the VLT in Chapter III-2, enormous strides forward could still be made, a situation that prevails to this very day. Yet, as we build ever more efficient instruments, fitted with detectors that exploit almost 100% of the light, it also becomes clear that the next leap forward will only be possible with larger telescopes. Much larger telescopes. We will come back to this issue later in the book. For now, we shall take a look, not at a larger telescope, but at one in an unusual place.

Chapter II-4

Hubble at ESO

"ESO has provided much more than an operational infrastructure for the ECF. It has created a rich and active scientific environment for the staff and enabled the exchange of ideas on a wide range of technical issues."

Fosbury & Albrecht, 2002.

ESO is devoted to ground-based observational astronomy, but there are clear synergies with space facilities as well. As the project for the Space Telescope (later to be named after Edwin P. Hubble) began to take firm shape, both the synergies with observatories on the ground and the need for astronomers to be involved early on became evident, in the US as well as in Europe. The 2.4-metre Hubble Space Telescope is a joint NASA/ESA project, although with a 15% share, ESA is a relatively small partner. The main facility to handle the data is the Space Telescope Science Institute (STScI) located on the campus of the Johns Hopkins University in Baltimore, Maryland (with Riccardo Giacconi as its first director). To assist European astronomers, however, a service facility in Europe was considered necessary. In 1980, ESA invited bids to host and co-sponsor this facility, and on 26 June 1981, it selected ESO over rival proposals by the Royal Observatory Edinburgh, the Institute of Space Astrophysics at Frascati and the Institut d'Astrophysique de Paris. As reasons for choosing ESO, Bob Fosbury and Rudi Albrecht wrote: *"Strong points of the ESO proposal were the large, although not 100%, overlap of the ESA and ESO scientific communities and the fact that ESO was already operating a major multinational observatory in Chile using an operating concept similar in many respects to that foreseen for HST."* (Fosbury & Albrecht, *ibid.*). So, on 23 February 1983, the Directors General of ESA, Erik Quistgaard, and ESO, Lodewijk Woltjer, signed the agreement for the Space Telescope European Coordinating Facility (ST-ECF), and in April 1984, the facility was set up at ESO's Headquarters in Garching. The ST-ECF was headed by Piero Benvenuti, who had worked in the International Ultraviolet Explorer (IUE) project. It had an original staff complement of fourteen, seven from ESA and seven from ESO.

The ST-ECF was intended *"to enhance the capabilities within Europe for the scientific use of the Space Telescope and of its data archive."* (Benvenuti, 1984). The initial tasks of the ST-ECF were *"coordination of development of ST-related data analysis software in Europe and with the Space Telescope Science Institute in the US, [developing] original application software for the reduction and analysis of ST data, [creating] an efficient means of archiving, cataloguing, retrieving and disseminating non-proprietary ST data,"* as well as providing advice and support for European ST users.

≈

The telescope was originally supposed to be launched on the NASA Space Shuttle in 1983, but the project suffered serious delays. For a long time the launch was scheduled for October 1986, so that the ST-ECF was established well before the planned start of operations. But on 28 January of that year the Space Shuttle *Challenger* exploded during launch, with the tragic loss of seven astronauts, and the entire shuttle fleet was grounded for 32 months. The launch of the Space Telescope finally occurred on 24 April 1990 to much public attention. This was not simply because of the impressive sight of the launch itself, but because of the massive public relations campaign that NASA had conducted in support of the project. On 20 May 1990, the telescope saw official first light, initially with a one-second exposure, followed by another of 30 seconds of NGC 3532, an open cluster of stars in the southern constellation of Carina. First light images are always a special experience. They are not intended to be used scientifically. More than anything, they provide the first glimpse of how the telescope is performing, of great interest to scientists and engineers alike. They are of course also widely disseminated by the media and of interest to the public in general. At this point the scientists and engineers may subsequently spend quite a bit of time carefully analysing the images to understand better how well the telescope works and how it can be further fine-tuned. But the requirements of the media in terms of immediate information, in simple straightforward language, are difficult to reconcile. And so it happened that the first reports about Hubble's first light were quite upbeat. This was misleading, for things were far from well. A second observation carried out some days later, a 100-second exposure with Hubble's Wide-Field Planetary Camera, confirmed the growing suspicion that something was seriously wrong. On 17 June 1990, first light was then achieved for the second on-board instrument, the Faint Object Camera (FOC), an ESA contribution to Hubble.

According to Eric Chaisson, then Director of Educational Programs (and also handling media requests) at the Space Telescope Science Institute, the STScI scientists were acutely aware of ESO's initial success with its 3.5-metre New Technology Telescope (see Chapter II-9), which had delivered the then sharpest image of a celestial object ever obtained. *"There was a bit of a race between ESA and ESO for the sharpest images of the sky,"* he claims (Chaisson, 1994). This is a bit of stretch, though. Firstly, space-borne and ground-based facilities are clearly complementary; secondly if there had ever been competition in the field of optical imagery between the ESO NTT and the Hubble Space Telescope, it was certainly one of (a very modest) David against (a mighty) Goliath[1]. However, in this phase, it was indeed the NTT that took home the points — a fact that, as we shall see later, was not without irony. In any event, the first light images obtained with Hubble's FOC did not bring relief to the NASA/ESA team in Baltimore. As it turned out, the telescope's primary mirror suffered from an optical flaw, the primary mirror being too flat along the edges by a minute, but decisive 2 μm. This caused a spherical aberration, which meant that only about 15% of the starlight was concentrated within 0.2 arcseconds, the rest being spread over a much larger area (Wilson, 1999). A layman might have described the images simply as fuzzy or out of focus, but things were more complicated than that. And that proved to be a good thing.

As soon as the problem was recognised, a frantic search for solutions began. Between August and October 1990, an *ad hoc* panel of experts was formed to consider possible solutions. With Hubble in orbit above the Earth, there was obviously no possibility of a quick fix to the problem. The way forward therefore branched into two strands — what could be done immediately in terms of processing the degraded images and what could be done to cure the problem by *in situ* treatment, which would have to await the first service mission with NASA's Space Shuttle. Two ESO scientists played important roles in that connection, Ray Wilson and Leon Lucy, of whom the latter would soon join the ST-ECF team.

For the short-term solution, the particular pattern of the stellar images, caused by the spherical aberration, lent itself well to analysis and the application of the numerical image-processing technique known as deconvolution. Some of these deconvolution methods are closely related to those used in work on another problem that was recognised even before the launch of Hubble. Astronomers call this *"undersampling*

[1] In Chaisson's estimate, the NTT cost 1/100 of Hubble, though this cannot take the launch cost for the service missions into account.

of the point spread function (PSF) by the detector array", meaning that the detector is not able to capture details as fine as those that the telescope can actually see. The reason for this was partly that at the time the telescope was designed, CCD detectors were coarse, small and with relatively few pixels. To cover a reasonable field, detectors were mounted next to each other in an array[2]. Another reason had to do with the wish to minimise "detector noise". In any event, it was thought that this problem could be corrected partly through image reconstruction, and therefore scientists had been thinking about this for a while. Lucy had been working with another ESO colleague, Dietrich Baade, and in an article in *The ESO Messenger* in September 1990 they proposed their method as a powerful way of dealing with the gravely affected Hubble images (Baade & Lucy, 1990). During the previous year, they had already demonstrated the efficiency of their method in connection with the NTT first light image, which featured a treated version as an add-on to the already outstanding image, produced with the NTT's active optical system. The methods developed by Lucy, Richard Hook and Hans-Martin Adorf at the ST-ECF to reconstruct full-resolution images from dithered exposures from the Hubble cameras became a standard operating procedure for Hubble and strongly influenced the choice of pixel scale for the telescope's subsequent cameras (ACS and WFC3).

With the application of the Lucy method and others developed elsewhere, it became possible to reconstruct the Hubble images and thereby achieve the originally foreseen resolution, although crucially it could not make up for the loss in contrast and sensitivity. The work of those early years, thus, helped to deal with a serious threat to the Hubble programme. But it also had a long-term effect. Appropriately, Bob Fosbury and Rudi Albrecht, wrote: *"While, in retrospect, those three years of aberrated HST operation have been largely eclipsed by the superb optical performance subsequently achieved, the mathematical infrastructure triggered by those early needs has matured and grown into many areas of astronomical data analysis, including spectroscopy as well as imaging."*

As regards possible longer-term solutions to Hubble's ailment, Ray Wilson carried out a thorough analysis of the first images. Transforming what opticians call the longitudinal aberration into the wavefront aberration — which is the physical measure of the aberration — was necessary to understand the precise nature of the error and therefore to work out any possible corrective measure. In the end some 15 different

[2] To cover the gaps between the detectors the technique of "dithering" was developed, i.e. combining multiple, slightly shifted exposures.

options to solve the problem presented themselves. They ranged from a group of solutions to provide a full-field correction at full aperture to providing corrections for the individual on-board instruments. A group of options was designated radical solutions. The first group contained proposals such as placing a helium lens over the telescope or an aspheric mirror placed at an angle of 45 degrees above the telescope, possibly as a free-flying module. Based on his mathematical analysis, Wilson provided an elaborate overview with the pros and cons of each solution. One option, however, did not need scientific computation. It was listed under the radical solutions and simply said: *"Build new HST!"* (Wilson, 1990)[3].

In the end a relatively simple solution known as COSTAR was chosen. COSTAR, or the Corrective Optics Space Telescope Axial Replacement, was designed to correct the spherical aberration for three of Hubble's instruments, the Faint Object Camera, the High Resolution Spectrograph (HRS), and the Faint Object Spectrograph (FOS). It was popularly described as a "contact lens", but in reality consisted of several pairs of small mirrors that would correct the flaw before feeding the light into the science instruments, in what opticians call pre-focal plane correction[4]. New instruments, developed later each contained their own corrective optics, so that COSTAR could be removed during the 2009 servicing mission.

As Wilson later acknowledged: *"What determined what would be done, what could be done, was not people doing equations with optics on the Earth, but the astronauts up in space ... the COSTAR solution was chosen because it was the only one they could carry out."* COSTAR could be slipped in as replacement of the High Speed Photometer, and thus did not require high-precision work during the EVA — or spacewalk — beyond what was possible for the astronauts with their thick suits and gloves. The first servicing mission was carried out in December 1993, and COSTAR was installed, as well as a new, corrected, camera called WFPC2. Among the astronauts was the Swiss astronomer Claude Nicollier, who had in the past also observed at La Silla. It was therefore a particular pleasure to receive him and his colleagues at the ESO Headquarters on 16 February 1994, as the team made a high-profile visit to Europe.

[3] Much of Wilson's report is found again in the official NASA/STScI report, published in 1991. The radical solutions, however, are not!

[4] Wilson provides a brief overview of the options for correcting the spherical aberration in the March 1991 issue of the *ST-ECF Newsletter*.

On Wednesday, 16 February 1994, the STS-61 crew that successfully repaired the Hubble Space Telescope during a Space Shuttle mission in December 1993 paid a visit to the ESO Headquarters. Among the crew members was the Swiss scientist Claude Nicollier, who had previously carried out observations at La Silla. Seen here are Nicollier and Harry van der Laan. The ESO flag was flown on the Shuttle mission and is now on display at the ESO Headquarters. (Photo: Hans Hermann Heyer)

The ST-ECF was born before the internet-based worldwide web, originally developed at ESO's sister organisation CERN, changed so much about the way that we communicate and work. Not surprisingly, among the first professionals to understand and exploit the new opportunities offered by the web were scientists and so, by the mid 1990s, some of the original functions of the ST-ECF were no longer necessary. Of course, it was still important for ESA to possess in-house expertise as regards Hubble and its instruments as well as an independent data analysis and curation capability. Following a review by Len Culhane, Rolf-Peter Kudritzki and George Miley, the ST-ECF took on new tasks including work on the *Guide Star Catalog II* (GSCII) and post-operational archive calibration. The ST-ECF developed unique know-how, for example, in the area of dealing with slitless spectroscopy data, from Hubble's ACS[5] and NICMOS[6] instruments, delivering science-ready data to what has become known as the Hubble Legacy Archive. At the same time, not having

[5] Advanced Camera for Surveys

[6] Near Infrared Camera and Multi-Object Spectrometer

day-to-day operational responsibilities, it could act in some respects as a think tank. In a sense, the ST-ECF moved from being fundamentally a Hubble outlet, providing an interface to the European astronomical community, to contributing to the ongoing Hubble programme in an integrated manner. In this way the ST-ECF also helped to secure a strong European participation in the Hubble programme, perhaps even exceeding that which the financial contribution might seem to suggest. On the other hand, with the encouragement of NASA, the ST-ECF also began an outreach effort[7], aimed at demonstrating that Hubble included significant European participation. This effort would later be integrated into ESO's outreach activities.

With its iconic status, the HST generated much interest among visitors to the ESO Open House Days. In this picture, from October 2006, Michael Rosa and Lars Lindberg Christensen are surrounded by visitors, curious to learn about the telescope and the role of ST-ECF. (Photo: Hans Hermann Heyer)

≈

In 2010, the sun set on the ST-ECF. Amazingly, Hubble was still going strong[8]. It had arguably become the most iconic telescope in the history of astronomy, but it had seen

[7] Led by Lars Lindberg Christensen, who later took charge of ESO Outreach.

[8] Hubble's design lifetime was 15 years.

its last servicing mission and, in any case, had outlived the original need and reason for having a *pied-a-terre* in Europe. The next large space telescope, originally known as the Next Generation Space Telescope and later renamed the James Webb Space Telescope, will require a different operation model. Furthermore, by 2002, ESA had decided to move some of its activities from Garching to its own centre at Villafranca near Madrid, which in the past had been the operational centre for the IUE.

The location of the ST-ECF at ESO strengthened the ties between Europe's space-borne and ground-based observational astronomy. Like ESO, the ST-ECF had a Users Committee. In this photo, from 1998, the committee members, together with Tim de Zeeuw who was visiting ESO at the time, are taking time off to see for themselves the progress on VLT instrumentation, in the shape of the high resolution spectrograph UVES. From left to right: Reynier Peletier, Max Pettini, Tim de Zeeuw, Yannick Mellier, Lukas Labhardt, Sandro D'Odorico and Giovanni Zamorani. (Photo: Robert Fosbury)

What, then, did the ST-ECF mean to ESO? The ST-ECF was embedded in the ESO system, yet it always kept a high degree of independence. It operated in a fruitful symbiosis with ESO and, especially under the directorship of Giacconi, became an important cultural interface, between those familiar with space-borne operations in astronomy and those on the ground. The combination of the information technology (IT) revolution and the operational concepts and requirements of space operations strongly influenced ESO's thinking about the VLT, especially in the area of data management and curation of scientific data[9]. Indeed perhaps the most important contribution lay in the operational experience gained from constructing a science archive with the original data at STScI and a copy at the ST-ECF in Garching, the content of which would be made available to the worldwide community of scientists, released after one year of propriety usage by the original observing teams. It became the model upon which the ESO Science Archive was built and run, both as an archive in its own right and as an integral part of an overall operating concept for the VLT and some of the other telescopes at ESO. Through the ST-ECF, ESA/ESO relations were strengthened considerably, both institutionally and at the personal level. Thus the two organisations organised workshops and published reports about topics in astrophysics that could be addressed in a complementary way by each other — such as exoplanet research, fundamental cosmology, studies of star populations in the Milky Way, including their chemistry and dynamics, and on synergies between the Herschel mission and the Atacama Large Millimeter/submillimeter Array (ALMA) project.

[9] In fact, the long-term head of the ST-ECF, Benvenuti, also became the first — interim — head of ESO's Data Management Division, established during Riccardo Giacconi's directorship at ESO.

Comet Halley seen from La Silla on 21 March 1986. At that time the comet was 118 million kilometres from the Earth and 151 million kilometres from the Sun. (Photo: Reinhold Häffner)

Chapter II-5

History in Passing

"Comets drive people dotty."

Nigel Calder, 1980.

If members of the public were asked to list well-known astronomical objects apart from the planets in the Solar System, one object that would almost certainly come out at the top of the list would be a comet — Halley's Comet, named after the famous British astronomer Edmund Halley. It is a tradition to name comets after their discoverers, but Halley's great contribution lay not in discovering the comet of 1682, but in showing that it was a periodic comet that had been seen in the past. In fact, research by historians of astronomy has since revealed observational records of Halley's comet as far back as 164 BC (Tammann & Véron, 1985), and perhaps even earlier. But more importantly Edmund Halley's insight enabled him to successfully predict the return of the comet in 1758, 76 years later — an event that he unfortunately was not to enjoy himself. Since then, Halley's Comet has been with us as a recurrent feature and through its returns in 1835 and 1910 attained an almost iconic status. The return in 1985/86, thus, was not a surprise, yet it was eagerly awaited by those astronomers who take a special interest in the Solar System and its objects. And, of course, members of the public had also heard about the famous visitor, but few living people had actually seen it. This time Halley would be welcomed by an armada of astronomical telescopes placed all over the world and also spacecraft from Japan, Russia (USSR) and Europe. The most ambitious space project was the Giotto mission from ESO's sister organisation, the European Space Agency. To coordinate the global observation effort, the International Halley Watch was set up in 1982 as a network of observers across the planet and with NASA's Jet Propulsion Laboratory (JPL) as the central node.

The conditions for observing the comet were somewhat complicated by the fact that around the time of perihelion, the closest point to the Sun, the comet and the Earth would be on opposite sides of the Sun. On the other hand, the comet was predicted to reach its peak brightness while moving through the southern skies, a fact that naturally pleased ESO observers. It was therefore evident that ESO's telescopes at La Silla could make a major contribution to the worldwide effort. Perhaps of greatest importance were precise measurements to determine the orbit and the rotation of the comet's nucleus, crucial information needed to steer the spacecraft scheduled to perform fly-bys. For the Giotto probe, it meant penetrating the coma and sweeping past the nucleus at a distance of a few hundred kilometres. Without a correct orbit determination, the probe might not achieve this goal.

During the 1910 passage — or apparition as astronomers call it — the first detection was made by the Helwan Observatory in Egypt on 24 August 1909, eight months before the comet passed perihelion. By the time of the second apparition in the 20th century, however, progress in observation techniques allowed astronomers to follow the comet for much longer (and earlier). This allowed them to study the comet before heat from the Sun had caused the development of the coma, the characteristic giant cloud of gas and dust that surrounds the comet's nucleus, and the source of the brightness that allowed naked-eye observations during the passage. The first detection was made with the 5-metre Hale Telescope at Mount Palomar as early as 16 October 1982, no less than 40 months before perihelion. At that time, the comet was at a distance of 10.5 AU[1], i.e. beyond the orbit of Saturn, and racing towards the Sun. And on 10 December of that year, ESO reported its first detection[2] by Holger Pedersen, using the significantly smaller Danish 1.54-metre telescope and blind tracking. At the time, Halley had a magnitude of a mere 24.5, more than 8 magnitudes fainter than when it had first been spotted in 1909.

Meanwhile, the need for fresh positional information to refine the orbital calculations remained. In an article in *The ESO Messenger*, one of the observers, Richard West, provided a vivid description of this difficult work, and we shall simply let him speak here, describing this quite unusual observing run: *"During most of the months of June and July 1985, Halley was behind the Sun and could not be observed. From about 15 July, attempts to obtain images of Halley were made in various places, and it now appears that the first confirmed sighting was made at the European Southern Observatory on 19 July. On this date Halley was at declination +18 degrees and only 30 degrees west of the Sun. From the time the comet rose over the eastern horizon*

[1] An astronomical unit (AU) equals the mean Sun–Earth distance or roughly 149.6 million kilometres.

[2] First attempts with the 3.6-metre telescope and the 1.54-metre Danish telescope had been undertaken as early as 1980, but they had failed, the comet's nucleus being too faint for the detector, photographic plates at the 3.6-metre telescope, and the electronographic McMullan camera on the 1.54-metre (West, 1983).

at La Silla to the moment when the sky brightness became excessive, there was at most 20 minutes. The only telescope at La Silla which is able to point in this direction is the 40-centimetre double astrograph (GPO) — all others are prohibited [from doing so] by sophisticated computer control or limit switches without pity. The 40 centimetre is also the smallest telescope on La Silla, so the outcome of the attempt to observe Halley was very doubtful. At this very low altitude, accurate guiding is difficult because the refraction in the Earth's atmosphere changes very rapidly as the object moves away from the horizon. Moreover, the expected brightness of Halley was only 14.5–15.0 and it could under no circumstances be seen in the guiding telescope. Offset guiding at the rate of the comet's motion was therefore necessary.

Together with Drs Pereyra and Tucholke, I was not optimistic about the prospects on that morning, when we made the first attempt. A heavy bar, which supports the dome, obstructed the view and it was only possible to get one good 10-minute IIa-O plate. After processing, it was inspected carefully — it showed stars down to 16th magnitude or fainter, but there was no obvious comet image. Since the field is in the Milky Way, there were several very faint images; however, they all turned out to be faint stars when a comparison was made with the Palomar Atlas.

Another plate was obtained on 20 July, but this morning the wind was strong and gusty and the telescope could not be held steady. The limiting magnitude was therefore about 0.5 magnitudes brighter than on the preceding morning. Then came a snowstorm and with that the end of our attempts." (West, 1985) But they were not in vain. Using a special enhancement technique, the comet was in fact later found on both plates, allowing new positional determinations that were subsequently confirmed from other observatories. As a side note, it was gratifying that the venerable GPO, which had been in service during the early days in South Africa as well as at La Silla, was still found to be a very useful instrument, albeit for these rather special observations[3]. It is worth remembering at a time when observers sit comfortably behind flat-screen monitors in air-conditioned control rooms that observing with a telescope like the GPO meant working in an open dome, hardly a pleasure in a situation with strong winds and icy temperatures.

≈

[3] The GPO was finally retired in the middle of 1995. Some 15 000 photographic plates had been obtained with it. Belgian astronomer Olivier Hainaut and his French colleague Christian Coutures switched off the light (Ihle, 1995).

When Halley passed perihelion on 9 February 1986, it was again behind the Sun as seen from the Earth. This was just one month before the planned encounter with the Giotto probe and once more updated positional measurements were in high demand. Fortunately on 15 February, when the comet was seen no more than 15 arcminutes above the horizon, the first ground-based observations could be made from La Silla — once again with the venerable GPO (Laustsen *et al.*, 1987). The early sighting confirmed that the comet had survived its perihelion passage and allowed for higher precision determination of the orbit as Giotto and the other space probes were racing towards their encounter. Astrometric observations from La Silla continued right up to shortly before the encounter on 13 March.

The first observation of Comet Halley after perihelion. The comet is the tiny white dot just above the horizon.

As Halley moved closer to the Sun, it slowly began to brighten and eventually to unfold a spectacular dust and ion tail. However, some of the most spectacular observations were only obtained after perihelion. At the beginning of March 1986, the comet reached its peak brightness, with a magnitude of 2.6, while on 11 April it reached its closest point to the Earth.

While many observers focussed on improving our knowledge of the nucleus — its composition and structure, its jet-releasing vents and its rotation, the ESO Schmidt telescope was essential to study the large-scale phenomena, such as the evolution of the tails. But specially developed equipment was also used at La Silla. With a wide-field camera equipped with a CCD detector, Holger Pedersen obtained a truly spectacular time-lapse sequence of the rapid changes in Halley's tails between 23 February and 8 March.

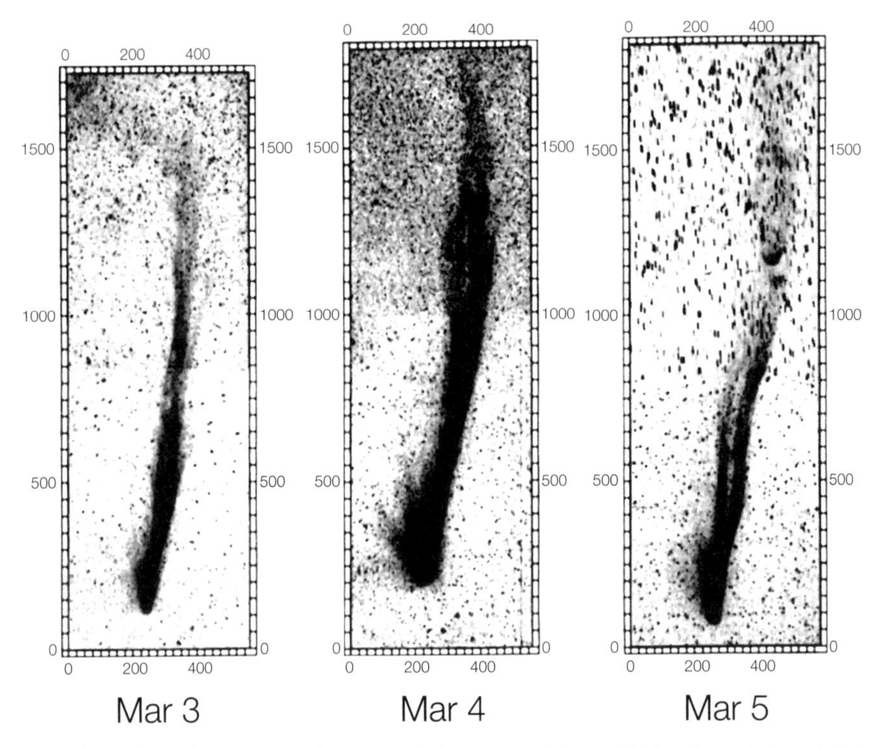

From Holger Pedersen's time-lapse observations: Variations in Comet Halley's ion tail in early March 1986.

On 13 March, ESA's Giotto space probe delivered a sterling performance in a dramatic close encounter with Halley. Not surprisingly, the probe was seriously damaged[4] by cometary dust particles, and its camera was destroyed. But before this happened, it

[4] Orginally thought to have been lost, Giotto survived and was brought back into action for a new cometary encounter. Although the camera remained dysfunctional, the spacecraft collected very valuable data about Comet Grigg-Skjellerup six years later, a feat of which ESA can justifiably be proud.

had captured stunning close-up images of the nucleus, leading to a revision of our ideas about this kind of body. From the point of view of this book, the Giotto experience also emphasised the value of close cooperation between ground-based and space-borne facilities with their complementary capabilities. With scientists from ESA and ESO working hand in hand during these intense months there is no doubt that relations between the two organisations was strengthened.

As Halley again moved away from us towards its icy aphelion, which it will reach in December 2023, observations gradually ceased. However, in 2003, using three Unit Telescopes of the Very Large Telescope Array, ESO astronomers caught a final glimpse of the famous celestial body. With a magnitude of a feeble 28.2, no other observations have ever been made of a comet that faint and far away — 4200 million kilometres or 28.06 AU from the Sun. However, with today's technology it would even be technically possible to follow the comet throughout its entire orbit.

≈

Of course, Halley's Comet generated strong public interest. Weather conditions in Europe were not favourable, but many people saw a comet for the first time. Some even went for comet flights on chartered airplanes, evading the winter clouds. ESO had just established an information service and a small embryonic group was beginning to form. The Halley apparition clearly offered a unique chance to launch this new service, and within the shortest space of time the group created an itinerant exhibition with spectacular images of the comet, obtained from La Silla. The first stop for the show was the head office seat of the Reuschel Bank in downtown Munich. The exhibition opened here on 20 March amidst great public attention. Much of the image material, including the time-lapse sequence, would later find its way into planetarium programmes in Europe. Perhaps coincidentally, the 1986 apparition of Comet Halley became the moment of birth for ESO's public outreach activities to which we shall return in Chapter IV-10.

Halley's Comet has occasionally been described as the comet of the century. If brightness is the determinant, Halley does not deserve this title. The brightest comet of the 20th century was Comet Ikeya-Seki (C/1965 S1), but other comets have stood out, including Comet West (C/1975 V1), Comet Kohoutek (C/1973 E1), Comet Hyakutake (C/1996 B2) and Comet Hale–Bopp (C/1995 O1). Even so, Halley visited us *twice* in the 20th century and it is bound to remain the archetypal cometary event in the public perception. Yet if 1986 had been a remarkable year in terms of the public awareness of astronomy, the following year would be no less interesting, although this time the events were completely unexpected.

Supernova 1987A, the bright star at the centre of the picture, observed with the ESO Schmidt telescope. The supernova is embedded in a complex system of ionised hydrogen nebulae, called HII regions by astronomers, with the Tarantula Nebula (upper left), as the most prominent object.

Chapter II-6

An *Annus Mirabilis*

"I don't think that any astronomer slept for more than 2–3 hours a day during the first two weeks."

Patrice Bouchet[1]

For Europeans, the month of February is associated with the depths of winter, with rain or snow depending on the location. But in the southern hemisphere, it is summer with warm days and clear skies. February is a good month to observe in. Naturally, the nights are not as long as the winter nights, but the observing conditions are usually excellent. In the language of the ESO Annual Report, *"the October-to-April period includes the meteorologically most favourable months and also coincides with the Magellanic Clouds season."* The month of February 1987 was no exception and La Silla was as vibrant as ever, with 14 telescopes in operation.

To accommodate the visiting astronomers, La Silla has a hotel and a large canteen. Since the observatory is open 24 hours a day, the canteen is not only open most of the time, but also geared to serve different kinds of meals at odd times. In the afternoon, at tea time, astronomers normally show up, some to have breakfast and all to have a chat before preparations for the work during the coming night begin. From the canteen there is a wonderful view of the surrounding mountains and, 27 kilometres away, another string of telescope domes can be seen: the Las Campanas Observatory, operated by the Carnegie Foundation. While sipping their tea in the afternoon of 24 February, how many astronomers took a casual glance at Campanas, with its domes reflecting the afternoon sunlight? We shall never know, but if some of them were still sleepy from the previous night's observations, they would soon wake up and look across the valley.

[1] Interview with the author on 16 October, 1992.

The canteen at La Silla, seen sometime in the 1980s. (Photo: Albert Bosker)

Before tea, Bo Reipurth, who was the Astronomer on Duty during that week, sat in his office busily preparing for the night. Stefano Cristiani came into the office and handed him a telex[2]. Reipurth took a quick glance at it. *"Well, another supernova,"* he said and gave the telex back. Occupied with operational issues, he had clearly not yet grasped what this meant. *"Take a look at the magnitude...,"* Cristiani suggested. Then the implications dawned on Reipurth. He went over to the restaurant with the telex.

The telex contained a note from the IAU Central Bureau for Astronomical Telegrams with the number Circular 4316. It said: *"SUPERNOVA 1987A IN THE LARGE MAGELLANIC CLOUD — W. Kunkel and B. Madore, Las Campanas Observatory, report the discovery by Ian Shelton, University of Toronto Las Campanas Station, of a mag 5 object, ostensibly a supernova, in the Large Magellanic Cloud... ."*

Supernovae of the core-collapse type are massive stars — at least eight times heavier than the Sun — that have come to the end of their life-cycle. It is a little similar for

[2] With three large observatories in the IV region of Chile, direct contact between them could be expected on such an occasion. An alert call had also come in from Mark Phillips at Cerro Tololo to Patrice Bouchet, who had just arrived in Santiago after an observing run at La Silla. Needless to say, Bouchet made a dash back for La Silla.

human beings: being overweight can shorten your life and the end can be unpleasant. In the case of supernovae, the end is apocalyptic, as the star succumbs to a giant explosion. During the last phase, the temperature inside the star rises dramatically, which leads to a process known as nucleosynthesis, the building up of heavy elements originally from hydrogen and helium, but already transformed in the nuclear furnace to carbon, oxygen, neon and silicon. Supernovae are thought to be cosmic factories in which a wide range of elements are forged. Most of the heavy elements that we find on Earth were formed inside a supernova. With the explosion that rips the star apart, the outer layers are ejected into the surrounding interstellar space, carrying some of these elements and thereby enriching the local environment. Some of these elements may ultimately become part of the building materials for a new star. During the explosion the star increases its brightness so much that it can shine as brilliantly as a whole galaxy. Detecting supernovae is therefore not in itself difficult. And, given the number of stars in the Universe, supernovae are frequent occurrences. Yet they also happen within a gigantic volume of space. Detecting them at the time of their explosion is thus to some extent a matter of luck. Supernovae are also common in our own galaxy, the Milky Way. Their frequency is thought to be around one every few decades. However, they may well occur in places that are out of sight for us, behind the thick dust clouds that are concentrated in the plane of the Milky Way. Seeing a supernova with the unaided eye is therefore extremely rare. But it has happened. The previous occasion was in 1604, when a supernova was observed by Johannes Kepler. A few years earlier, in 1572, Tycho Brahe — Kepler's mentor — had seen one, the famous supernova in Cassiopeia. Tycho noticed the phenomenon and described it as a "new star" (*stella nova*), because the brightness increase enabled him to see it for the first time. Derived from Tycho's description, the term supernova was introduced in the 1930s by Walter Baade and Fritz Zwicky, a Swiss astronomer, who like so many other scientists from Europe had emigrated to the US.

The situation that day in February 1987 was clearly exceptional: at a distance of only 160 000 light-years, this was the closest recorded supernova since the telescope was invented. Furthermore, no supernova had been discovered at such an early stage before. And this time a battery of advanced telescopes with sophisticated instruments was ready.

However, a rather curious problem showed up. Most of the telescopes and indeed the detectors and instruments had been developed with much fainter objects in mind and not for observing objects as bright as this supernova. Before night fell, astronomers and technicians had to come up with some fairly creative solutions at the telescopes to cope with this excessive amount of light. On the other hand, it also meant that smaller telescopes could also be used with some advantage[3].

"After sunset, we gathered in a small group waiting for full darkness to set in. As the evening glow faded, the Large Magellanic Cloud emerged, and in it appeared a faint reddish star. At that moment I realised that it was the most distant star I'd ever seen without the help of a telescope, its light having travelled for 160 000 years before reaching us," Reipurth later recalled (Reipurth, 2010). But then a hectic observing night began. Telescopes were trained on the target, shutters opened, photons were collected, probing eyes were cast on the first images, spectra and photometer readings as they appeared on the screens in the control rooms.

Wide-field images of the Large Magellanic Cloud taken on 23 and 25 February 1987. These were taken with a Hasselblad camera riding piggyback on the GPO and the Danish 1.54-metre telescope respectively. The supernova is seen as a bright star to the right and below the Tarantula Nebula in the right-hand photo. (Photos by the author)

One of the first questions was, which star (and which kind of star) had exploded. If that star had been studied earlier, it would provide important information about how stars evolve into supernovae. Here, the ESO/SRC survey came in handy. At Garching, Richard West made the first identification on ESO Schmidt plates obtained before the event, as soon as the first astrometric information became available. It turned

[3] Since smaller telescopes are less likely to be heavily oversubscribed, it is easier to carry out long-term observations with them. In the case of SN1987, the 0.6-metre Bochum telescope in particular played an important role, following the supernova for a year.

out to be a blue supergiant, known to astronomers as Sanduleak -69 202. This was a surprise, since supernovae had been thought to originate from red supergiants. Another burning question was — what kind of supernova was this? Supernovae come in two different main types, and until this moment, no Type II had ever been seen in an irregular galaxy, such as the Large Magellanic Cloud. But during the very first night at La Silla, the astronomers concluded that here was another surprise — this was indeed a Type II. There was more to come. Later, in *The ESO Messenger*, John Danziger and Patrice Bouchet captured the early moments of this remarkable episode: *"During that initial period was the news, informally propagating at that stage, but soon officially revealed in IAU Circulars ... that neutrino bursts had been detected*[4]*, providing observational evidence that the theory was on the right track, but also giving a precise time for the collapse of the core after the physicists had clarified which reported events were real*[5]*. These neutrinos certainly played a part in giving an extra incentive push that we were on to something special... ."* (Danziger & Bouchet, 2007) Indeed. It was in fact the first time ever that neutrinos from a celestial body other than the Sun had been detected.

≈

Given the unpredictability of supernovae, it was a lucky coincidence that a number of visiting astronomers specialising in supernova research were present at La Silla, including Henning Jørgensen, Patrice Bouchet and John Danziger[6]. Also ESO's Director General, himself also an expert on the Crab Nebula[7], was at La Silla at the time (Woltjer habitually spent an extended period at the observatory at the beginning of each year). As we have discussed in Chapter I-5, observing time was based on applications that had been vetted by the Observing Programmes Committee. But allocation of telescope time is the prerogative of the Director General. He or she will normally follow the recommendations by the OPC, but this was no normal situation and observers were therefore requested[8] to abandon whatever programme they had

[4] To detect neutrinos, large underground facilities are used, often in former mines or other large subterranean cavities. In the case of SN1987a neutrinos were detected at the Kamiokande observatory in Japan, the Irvine-Michigan-Brookhaven detector in the USA and at the Baksan Neutrino Observatory in the former USSR.

[5] For the detection of the neutrinos, Masatoshi Koshiba was awarded the Nobel Prize in Physics in 2002.

[6] The main results of the observations were presented in a paper by Danziger *et al.*, which later also became the contribution of the ESO astronomers to the VLT Time Capsule (See Chapter III-11).

[7] A remnant of the supernova observed in the year 1054.

[8] Danziger and Bouchet use the term encouraged rather than requested, but it really meant the same!

planned in favour of studying the supernova. Most observers embraced the order with enthusiasm, as did astronomers elsewhere at southern hemisphere observing sites.

The excitement is clear also from *The ESO Messenger* of March 1987, published only 20 days after the event, in which two photographs of the LMC region were shown, before and after the explosion. Under the heading "Chronology of a Once-in-a-Lifetime Event", Richard West, the editor, wrote *"Only 48 hours separate the two photos ... but during this brief interval an event happened that excited an entire generation of astronomers. Reflecting the importance of this supernova, the Central Bureau for Astronomical Telegrams issued no less than 15 IAU Circulars in the course of nine days only, breaking all records in the history of astronomy. At ESO, these circulars were read through a computer link to the Headquarters in Garching as soon as they were issued, and immediately sent on to the observers at La Silla by telefax. In this way, and also by numerous telex messages and phone calls, the scientists were kept informed about what was going on in other places."*

After the third night, Woltjer — in John Danziger's *Messenger* article described as "the admiral" — decided to have dedicated daily meetings of all the observers. To continue the metaphor, the admiral steered his fleet effectively and efficiently — with an authority that few others could match. The meetings lasted for hours and one senses the excitement in a (later) statement by Patrice Bouchet: *"The follow-up on the observations.... I hope that every future generation of astronomers will experience something like that ... it was fantastic!"* If the internal exchange of information functioned flawlessly, there was of course an equally intense exchange of news between the observatories. Richard West has already mentioned the IAU Circulars, but naturally information was exchanged through all possible means — also reaching the US media. Here, the admiral was more cautious. He sensed that, with the largest telescope park in the southern hemisphere, this was not simply ESO's chance to demonstrate its scientific clout — a chance he had no doubt longed for — but also that it was important to ascertain that the information to be released was correct and its origin properly acknowledged. Amidst the understandable excitement during those hectic days, not all observers may have agreed with this, but Woltjer had steeled himself — and he was not alone. *"The decision was correct. Others would have done exactly the same,"* one of the senior astronomers who were present later confided to the author. For the same reason, Woltjer also pushed for fast publication of the results in the way scientists normally do — in a peer-reviewed journal. And so, as early as May 1987, just three months after the supernova had been discovered, six papers with 38 authors appeared in *Astronomy & Astrophysics,* covering astrometry, optical and infrared photometry, polarimetry, optical and infrared spectroscopy and high-resolution spectroscopy of this unique object. Aside from information about the supernova itself — the precursor star, the explosion and the material ejected into the interstellar medium — the supernova had also acted as a powerful light source in the LMC, much as a searchlight does. In this way it was possible to obtain valuable information about intervening matter in the halo of our own galaxy. This information came through high-resolution spectroscopy with the Coudé Auxiliary Telescope[9], the 1.4-metre telescope that was connected with the 3.6-metre telescope, feeding the CES. ESO also organised the first full-scale international scientific conference, with 200 participants, at the Garching Headquarters, as early as July that year.

Photo left: The participants listening to a talk by Paul Murdin at the VLT Conference in Garching. Photo right: Wilhelm Kegel, Bruno Leibundgut, Gustav Tammann and Bob Kirshner in discussion. (Photos: Herbert Zodet)

≈

What happened to the supernova? In the first instance, the brightness increased to magnitude 4.5 on 28 February, then dropped briefly. It rebounded to reach 4.0 in late March and continued until it peaked on 10 May at a magnitude of 2.8. This is thought to have been driven by the radioactive decay of nickel and cobalt, and the subsequent fading matched the half-life of cobalt-56 (Danziger & Bouchet, *ibid.*; Wheeler, 2007).

In a review article, 20 years later, Claes Fransson and his colleagues summarise the outcome of SN1987a so far. *"It provided several observational firsts, like the detection of neutrinos from the core collapse, the observation of the progenitor star on archival*

[9] Especially in the study of interstellar lithium.

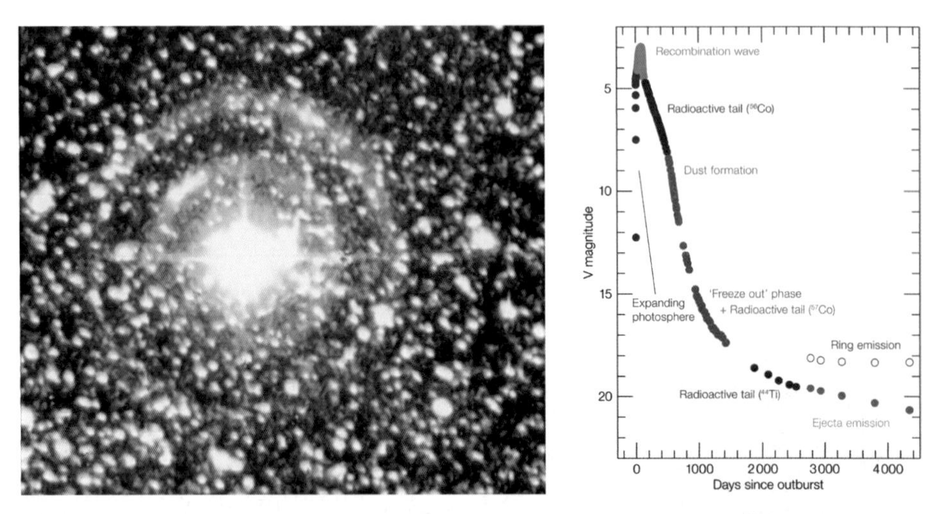

Picture left: Light echoes from SN1987A, observed with the NTT. Emanating in all directions, some of the light from the supernova was reflected by interstellar matter. By observing the resulting light echo it was possible to construct a three-dimensional view of the supernova surroundings. Picture right: The light curve of SN 1987A over more than ten years. (Source: Nick Suntzeff)

photographic plates, the signatures of a non-spherical explosion and mixing in the ejecta, the direct observation of supernova nucleosynthesis ... observation of the formation of dust in the supernova, as well as the detection of circumstellar and interstellar material." (Fransson *et al.*, 2007). Several of these firsts were due to ESO observations, including the observations of the cobalt content (observed with the 1-metre telescope) and the detection of dust in the envelope of the supernova on day 530 after the explosion was detected. Thus the supernova has remained an interesting object to observe, among others, with several VLT instruments, such as UVES, ISAAC and SINFONI[10]. And it still retains a secret or two: according to our current understanding of the ultimate fate of a supernova, it may end as a black hole or as a neutron star, perhaps in the form of a pulsar — an extremely compact, rapidly spinning object. Inspite of intense observations, however, nothing has been found so far.

While the supernova electrified the world's community of astronomers, the public impact in Europe was more modest, although far from negligible. By pure coincidence, a seasoned European science journalist, who shall remain unnamed here, was present at La Silla at the time. His objective was to write about a visiting astronomer specialising in supernova research. In the morning after observations of SN 1987A had begun at La Silla, the author of this book met him happily walking towards the

[10] See Chapter III-2

restaurant and a hearty breakfast. *"So, have you filed your story?"* I asked. *"No,"* he said. *"The supernova is not interesting. It can't be seen from Europe!"* The author's reaction shall not be reported here, but it must have left a certain impression on him — two hours later he returned. *"I just sent the article,"* he said.

≈

The famous gravitational arc in the galaxy cluster Abell 370, observed with the FOcal Reducer and low dispersion Spectrograph (FORS1) on the VLT.

SN 1987A was handed to astronomers on a silver platter, not a frequent occurrence. Normally, astronomy is associated with carefully planned, often difficult, painstaking observations followed by months of meticulous analysis of the data. This is the way most great results come about, and the following is no exception. Thus the year 1987 saw the publication of another important discovery, explaining the origin of the mysterious luminous arcs that had been observed in clusters of galaxies. French astronomers had carried out spectroscopic measurements of such an arc, in a cluster known as Abell 370, with the ESO 3.6-metre telescope. The result showed a redshift similar to that of a much more distant object. It was concluded that the arc phenomenon was the result of bending of the light from a distant galaxy in an intermediate gravitational field, leading to a displaced and distorted image of the galaxy in question (Soucail *et al.*, 1988). Already predicted by Einstein, such an effect would be observable if a heavy object lay almost exactly in the line of sight between the object from where the light originated and the observer. Accordingly the phenomenon became known as an Einstein ring caused by gravitational lensing. Aside from providing further evidence that Einstein's general theory of relativity is correct, it had the welcome effect that, thanks to the optical enhancement by the lens, such natural lenses enabled astronomers to study the furthest reaches of the Universe, so to speak, for free.

Gravitational lenses were undoubtedly a hot topic at the time. Later in the same year, another group, led by the Belgian astronomer Jean Surdej, and using the MPG/ESO 2.2-metre and 3.6-metre telescopes at La Silla, had found a gravitationally lensed quasar in the southern constellation of Cetus. The first of this kind had been discovered back in 1979, but the new result was the first one found purely with optical observations and based on a systematic search programme carried out at La Silla (Surdej *et al.*, 1987)[11].

≈

The year 1987, however, had more to offer. At La Silla yet another telescope was added to the already impressive park. This new telescope was of a very different kind from those already gracing the mountain ridge. And although it constituted very good news indeed, it almost started as a disaster.

Peter de Jonge, who had worked at ESO in Chile in the 1970s, had become director of IRAM, the Institut de Radioastronomie Millimétrique in Grenoble, founded in 1979 by the Centre National de la Recherche Scientifique (CNRS) of France, Germany's Max-Planck-Gesellschaft and the Spanish Instituto Geográfico Nacional. He was keen to establish close connections between his new institute and ESO. IRAM was now building an array of 15-metre telescopes for observations in the

[11] Soon after, Surdej and his team discovered the Cloverleaf Quasi-stellar Object (QSO) at ESO. This object, formally designated as GL 2 = H 1413 + 117, is a quadruply lensed quasar that was first observed with the MPG/ESO 2.2-metre telescope. The scientific paper, published in *Nature*, was also selected for the VLT Time Capsule (see Chapter III-11).

submillimetre/millimetre domain to be placed at the 2550-metre-high Plateau de Bure near Grenoble[12]. It appeared possible to add one more antenna to the production at a very reasonable cost and thus the idea arose to place it in the southern hemisphere. At the same time, Swedish astronomers were thinking about participation in a major radio astronomy project. These ideas were given a boost by the arrival from the UK of Roy Booth as director of the Swedish Onsala Space Observatory (Onsala Rymdobservatorium, OSO). In August 1982, at the IAU General Assembly in Patras, Greece, Booth and de Jonge discussed the idea and the prospect of a collaboration between OSO and IRAM. In subsequent conversations with Woltjer, they gained his support, leading to a three-way partnership, now with ESO, and the decision to place the new telescope at La Silla. In June 1984 Council approved the necessary agreements with Sweden and IRAM. Accordingly, IRAM oversaw the installation of the telescope, OSO was responsible for operations and the facility was incorporated into La Silla like a national telescope. With its 15-metre diameter dish, the Swedish-ESO Submillimetre Telescope (SEST), as it was called, became the first large millimetre/submillimetre reflector in the southern hemisphere. It also marked ESO's first step away from being a purely optical astronomy observatory.

In early 1987, SEST assembly at La Silla was completed. The assembly had begun in the previous year and progressed well. The 176 panels had been mounted to form the collecting dish, followed by the subreflector (the equivalent of the secondary mirror of an optical telescope) on its four supporting legs, a quadrupod. For the purpose of mounting, a scaffold had been erected at the north side of the telescope. The mounting then followed easily by rotating the telescope in azimuth (around the vertical axis). Then, on Saturday 22 November, when most of the assembly team had gone down to La Serena for the weekend, an accident happened. La Silla is located at 29 degrees southern latitude. At noon, the Sun is close to zenith during the southern summer. At an elevation of 2400 metres, the solar intensity is immense, and a 15-metre dish with highly reflective panels will collect a lot of it. For that reason, the mounting was not carried out close to midday. On the previous day, the quadrupod with the subreflector had been mounted, although only fixed in a preliminary way. The work continued in the morning, when a handset used to rotate the telescope from the top of the scaffold broke. Unable to rotate the dish, heat started to

[12] Europe's involvement in millimetre astronomy involved a range of facilities. These included the array on the Plateau de Bure, the 30-metre IRAM antenna at Pico Valeta in Spain, the 20-metre at Onsala in Sweden, the 15-metre James Clerk Maxwell Telescope (JCMT) at Hawaii, and others (Shaver & Booth, 1998). But, as already mentioned the 15-metre SEST was the only one in the southern hemisphere, in stark contrast to Europe's engagement in optical astronomy.

build up. While trying to repair the handset, the engineer noticed smoke from edge of the subreflector, prompting him to alert the fire brigade. The firefighters quickly arrived on the scene[13], but due to strong winds and problems with the water pressure, they could not reach the 15-metre dish and the subreflector in its zenith position. It was necessary to lower the dish towards the horizon. This was not easy, however. Firstly, the quadrupod was not properly fixed. Secondly, the scaffold was in the way. Thirdly, to clear the scaffold the engineer had to operate the telescope manually from inside the pedestal (out of sight of the telescope). Working under stress, when tilting the telescope dish, the engineer drove the dish into the scaffold. Meanwhile, the fire caused molten aluminum from the upper structure to drop onto the surface of the dish, burning holes in the panels. Luckily, the fire could be contained. After replacing the parts that had been destroyed and repairing those that had only been lightly damaged, the telescope was handed over to ESO on 13 March 1987 with first light achieved just 11 days later (Booth *et al.*, 1989). A unique facility, it would constitute an important element of the observatory until it ceased operations, with the last scheduled observation on 26 August 2003. We shall briefly return to the SEST again in the context of the ALMA project.

After the fire. Photo left: Preparing the new subreflector for the SEST. Photo right: Albert Grewe measuring the accuracy of the reflecting panels with a theodolite. (Photos by the author)

≈

In December 1986, Lodewijk Woltjer had informed Council of his wish to pass on the baton as ESO Director General by 1 January 1988. He did so with the intention

[13] Fire has been an extremely rare occurrence at La Silla. However, on 25 October 25 1992, the dome of the 1-metre telescope caught fire. Luckily neither the telescope itself nor the delicate instruments suffered any damage.

of achieving a decision to construct the Very Large Telescope before his departure. In March 1987, ESO submitted the formal proposal, in the shape of a 342-page document, entitled "Proposal for the Construction of the 16-metre Very Large Telescope". In the scientific community, it became known simply as the Blue Book. According to the proposal, the VLT would comprise a linear array of four independently mounted 8-metre telescopes, each with two Nasmyth foci, and a common incoherent focus, hence the term "a 16-metre telescope"[14]. There would also be a coherent focus and this opened up the possibility of using the four telescopes as an optical interferometer with a baseline of 104 metres. The primary mirrors would be thin meniscus mirrors and both the primary and secondary mirrors were to be actively controlled — an extension of the system that would be used at the NTT, at the time under construction. While Cassegrain foci were possible, they were not foreseen to be implemented "*until the need for it is proven*", and therefore also not budgeted for. Importantly, the telescopes were designed to work without adaptive optics (AO) — understandable, since the first AO systems were only under development at the time[15]. As an interesting feature, the telescopes were to be operated in the open air with a fixed, giant wind screen providing the necessary protection during observations and inflatable domes to be used in case of bad weather[16]. The proposal did not contain a dedicated instrumentation plan but referred to *"a preliminary model instrumentation package [that] was outlined at the Venice Workshop in October 1986"*. The ideas included a focal reducer with spectroscopic capability — a VLT EFOSC, a multimode instrument (reference was made to EMMI, the instrument which was planned for the NTT), a high-resolution instrument for the incoherent combined focus[17], infrared instrumentation including cameras optimised for two pass-bands (1–5 μm and 8–20 μm), a photometer and spectrographs for medium- and high-resolution work. Despite the strong interest in the interferometric option, expressed especially by French astronomers, no instrumentation for interferometry was mentioned — a testimony to the experimental character of optical interferometry at the time.

The proposal set mid-1994 as the target for first light, with completion of the entire facility in 1998 and finally provided a budget estimate of 382 million deutschmarks

[14] Together, the four telescopes would have the same light-collecting surface as a 16-metre telescope.

[15] Even so, it was clear that adaptive optics was likely to play a big role and a whole chapter in the proposal was devoted to this technology.

[16] Open-air operation remains a fascinating possibility for telescope designers. Later, the OverWhelmingly Large (OWL) Telescope project for a 100-metre telescope also pursued this idea.

[17] In the end, this is only planned to be realised with the second generation VLT instrument ESPRESSO.

at 1986 values. Optimistically, it stated that it would not be necessary to add more than a dozen people to the Telescope Project Division staff (55 people at the time).

Woltjer's ambition, however, went beyond the VLT approval. He also pushed to choose Cerro Paranal as the site for the new telescope, as the preliminary results from the long site-testing campaign had convinced him of the superior conditions that prevailed there. The Blue Book, however, did not press the matter, but rather it stated that *"it would seem ... advantageous to postpone a definite choice of site until some time in 1990, so that the available data on which a decision is to be based will be as complete as possible."* This notwithstanding, under the title "A Time for Change", published in *The ESO Messenger* in June 1987, Woljer presented his thoughts about ESO's future, including the two-site scenario. Appearing on the front page of the journal, his article featured a picture of the Paranal mountain with a truck climbing the steep slope of the eastern side, testifying to the accessibility of the site. Richard West, who as editor of *The ESO Messenger* and author of the ESO press releases excelled in masterly chosen, and often quite subtle, plays on words, selected a seemingly innocent caption: The road to Paranal.

≈

On 5 October that year, the Council came together in Paris to celebrate ESO's 25th anniversary. The event took place in the distinguished surroundings of the Kléber Centre des Conférences Internationales, where the ESO Convention had been signed two and a half decades earlier. On this occasion the French hosts presented an extraordinary gift to ESO's Director General, an original print of the book by P. S. de Laplace, *Système du Monde*. ESO's birthday present to itself was a coffee table book entitled *Exploring the Southern Sky*, with Svend Laustsen as the main author and published by Springer Verlag, appearing in English, German, French, Spanish and Danish language versions.

In parallel with the formal celebration, ESO organised a large exhibition at the prestigious Palais de la Decouverte on Avenue Franklin and close to the main artery of Paris, the Champs Elysée. The exhibition, as well as the anniversary event and the book, was not only a celebratory gesture, but also served to present the Very Large Telescope project to a wider audience, including key decision makers.

In terms of European research infrastructures this time stands out as quite unique, possibly because the governments of several of the major countries on the continent had prominent science ministers (Curien, Ruberti, Riesenhuber). Within a few years, major steps and decisions were taken about the Large Hadron Collider at CERN, the establishment of the European Synchrotron Radiation Facility (ESRF), the European Transonic Windtunnel, European participation in the International Space Station, the Ariane V programme, ENVISAT and other large-scale initiatives. And of course the VLT.

As regards the latter, needless to say the scientists on the ESO Council were all highly supportive of the proposal. But ESO also benefited from the strong commitment and close cooperation between three influential science administrators: Umberto Vattani of Italy, Jean-François Stuyck-Taillandier of France and Christian Patermann of Germany. Even so, in the run up to the decision about the VLT it had become clear that, whilst there was strong support in most Member States, a positive decision could not be regarded as a foregone conclusion. Unexpectedly, in France the situation was perhaps more critical than elsewhere.

Hubert Curien, one of the most prolific science ministers and an ardent supporter of European cooperation in science, had prepared the way for the forthcoming decision, with French support for the VLT. He was backed by the President of the Republic, François Mitterand, whose science advisor, Jean-Daniel Levi, was very much in favour of the VLT project. But by 1986, the socialist government had lost in parliamentary elections and the new French Prime Minister was Jacques Chirac. This was the difficult epoch of "co-habitation", where the President and his Prime Minister represented opposing political camps and the policies and decisions of the previous government were questioned, including the French position regarding the VLT project. The Science Minister was also new in office and he did not express himself very clearly on the subject. Shortly before the decisive Council meeting in December 1987, Pierre Léna, one of the two French Council delegates, was summoned to the Prime Minister's Office, where he was received by an advisor. Léna knew that there was some opposition there to the VLT project, but during the two-hour long conversation he reiterated the arguments in favour of the VLT. The meeting ended at 9 pm. The next day, Léna received a call asserting that the Prime Minister had agreed to the VLT. The vote of the delegate from the Ministry of Foreign Affairs to the ESO Council would be positive. Encouraging signals had also been sent from other European capitals.

The stage, it seemed, was set for a positive decision on the VLT, but it would turn out that there were more hurdles to overcome.

The ESO Council on 8 December 1987. From left to right: Kurt Hunger, Gerhard Bachmann, Lodewijk Woltjer, Harry van der Laan, Christian de Loore, Noel Vercruysse, Jean-François Stuyck-Tallandier, Pierre Léna, Marcello Griccioli, Jan Bezemer, Willem Brouw, Giancarlo Setti, Massimo Tarenghi, Daniel Enard, Richard West, Henning Jørgensen, Henrik Grage, Christian Patermann, Bengt Westerlund, Mats Ola Ottosson, Peter Creola, Marcel Golay, Martin Huber and Per-Olof Lindblad. (Photo by the author)

Chapter II-7

Decision Day

"Europe Decides To Build The World's Largest Optical Telescope"

ESO Press Release, 8 December 1987.

On 8 December 1987, Council convened at the ESO Headquarters. The main point on the agenda was the proposal to construct the VLT as it had been presented in the Blue Book. The discussion began with a *tour de table* to assess the general mood. First to speak was Jan Bezemer of the Netherlands, who declared that his government was ready to approve the proposal. Next to speak was Jean-François Stuyck-Tallandier on behalf of the French delegation. France would be in favour, but hoped that a decision could be obtained with unanimity. Italy followed suit, but stressed that they saw a *need* for unanimity. The Belgian delegation declared its support as well, but could only vote *ad referendum* (i.e. subject to later confirmation), since for internal political reasons (a forthcoming election), the final decision on the part of the Belgian government had been deferred. Switzerland and Sweden were in favour. Germany, represented by Christian Patermann, declared that Germany, too, would support the project. The German government did not insist on unanimity, but could, on the other hand, not cover the contributions of countries that might not be able to participate — basically a position that Switzerland had also assumed. It looked good. Last to speak was Henrik Grage from Denmark. He stressed that whilst the Danish delegation strongly supported the VLT as such, he was not authorised to commit his country financially[1]. With a share of 2.9% Denmark was the smallest contributor in real terms, but still, with this statement unanimity seemed out of reach. Would the proposal fall?

---

1 The Danish astronomical community was fully behind the VLT project, but in the wake of general prioritisation discussions in the country's research council, the VLT had not fared well, partly because of Denmark's engagement in the Nordic Optical Telescope (NOT), which occurred at the same time and for which more resources had become necessary than originally foreseen.

The wish for unanimity was not just a question of cost-sharing, but also linked to the ESO Convention. The Convention had described a well-defined initial programme. But wisely, Article II, paragraph 3 allowed for a supplementary programme, which could be approved with a two thirds majority. Countries that did not vote in favour of the supplementary programme would neither be required to contribute to it financially, nor allowed to use the extra facility. Whilst the option was technically there, using it would change ESO fundamentally by breaking the coherence of the organisation, increase the administrative overhead and complicate matters tremendously. In other words, proceeding without unanimity was not an attractive option. But, on the other hand, time was short — a solution had to be worked out without delay.

One reason was that shortly before, in the middle of October, the stock market had crashed. Within a five-day period, the Dow Jones Industrial Average had suffered a decline of a staggering 31%. Though the markets recovered quickly, the event sent tremors through the financial system and created considerable nervousness everywhere — certainly also in the finance ministries of ESO's Member States. Postponing the decision about the VLT, thus, would be a risky bet.

A solution was formally proposed by the Swiss delegate, Peter Creola[2]. The decision could be split into two resolutions, voted on independently. The first would be to build the VLT, as proposed and within the foreseen budget. The second vote would address the financing. Unanimity could be obtained for the first resolution and the second resolution would basically declare that 2.9% of the financing was — at that moment — not covered. Denmark would be invited to join the project in the course of 1988. It was a simple solution that received immediate support from the French and the German delegations. Nonetheless, the drafting of the resolution texts took time. Council was anxious not to send wrong signals either to Denmark, whose participation was still hoped for, or to any of the other countries who might renege on their promise, using the argument that other countries had not committed firmly. And, as Pierre Léna remarked, it was important to announce in a clear manner to the outside world *"that the VLT was started"*. As expected, the first resolution received unanimous support, and after much fine-tuning, the second resolution passed as well, with seven votes in favour and — understandably — Denmark abstaining[3]. The

[2] This solution had been informally discussed already at the preceding Committee of Council meeting on 1 December with the full support of Mr Grage.

[3] As hoped, Denmark joined the VLT project fully in 1989, restoring the "musketeer principle" that has been followed by ESO over its lifetime.

factual minutes of the Council meeting fail to capture the intensity of the discussions as the delegates, guided by the President, Kurt Hunger, and assisted by the Director General, struggled to find a way forward. But later, in an interview Harry van der Laan, who attended the meeting as incoming Director General, recalled: *"It was a very long Council meeting, interrupted several times by coffee breaks, by the necessity of consulting ministries back home, with an excellent lunch to improve the atmosphere, and when we all thought — and Prof. Hunger, the President of Council thought — we are ready for a vote — this was after many* tours de table *we had to deal with particularly thorny issues, to try to mitigate conditions and provisional acceptance that some delegations wanted to make ... when we were ready for the vote, at the very last second, the Italian delegate, Griccioli, raised his hand ... Christian Patermann [the German government delegate], I saw him tense, looking extremely disturbed, when the Italian delegate said: 'Mr President, I still have a very important issue!' There was dead silence around the table. And then he said: 'Mr President, what about the press release?' This was such an incredible statement ... that it totally broke the tension, everybody roared with laughter. Patermann couldn't contain himself and he laughed the loudest of all."* Keen to bring the meeting to a successful conclusion, Peter Creola refocused the attention of all: *"Look, Gentlemen, suppose an Earth tremor occurs in any minute, with us not having taken the decision!"* (Léna, 2012). And then the vote was cast.

Griccioli's successful intervention to relieve the tension of the meeting reveals a deeper change within ESO and within science in general. Attempts to involve the public at large were practically non-existent in the early years of ESO. Europe's great adventure in astronomy happened within the confines of the professional community of scientists and, perhaps, to some extent, the rather limited world of advanced amateur astronomers. But these were times of change and the need for a much more extrovert communication culture had been recognised. By 1987, press releases and thorough public information had begun to be seen as an integral part of ESO's activities.

≈

The decision to build the VLT did not go unnoticed across the Atlantic. Under the title *"European Telescope Will Be Biggest"*, Malcolm Browne wrote in the New *York Times*: *"American astronomers who had hoped to build a telescope of equivalent power will be forced to watch from the sidelines."* He continued: *"Dr Peter A. Strittmatter, director of the University of Arizona's Steward Observatory, said that American scientists now face a crisis.... We congratulate European astronomers for ESO's decision to build a 16-meter telescope, but we can't help envying them."* But competition stimulates business. American plans to build a 15-metre New National Technology Telescope had just been scrapped, although in September 1987 they had been replaced by a recommendation to build two 8-metre telescopes (one to be placed in each hemisphere) (McCray, 2004), a much more moderate proposition. But ESO's decision may have boosted another large telescope project that had been underway at the University of California since the late 1970s — the 10-metre W. M. Keck telescope, which would enter into operation in 1993, with an interconnected twin telescope achieving first light in 1996. Others in the US made plans, too, and Japan would also soon be joining the company with its 8.2-metre Subaru Telescope. What was on the drawing boards at the various design bureaus, however, was more than a number of large, new telescopes. It was blueprint for a new golden age of astronomy. Especially gratifying to European astronomers was that for the first time in almost 100 years, the old continent was playing a key role.

The Paranal area, seen in early 1991, before construction began. (Photo by the author)

Chapter II-8

A Mountain in the Middle of Nowhere

*"If the stars were visible from just one place on Earth,
people would never stop
travelling to that place in order to see them."*

Lucio Anneo Seneca, *Naturales Questiones.*

To almost every living soul on Planet Earth, Cerro Paranal is a place in the middle of nowhere, which is of course one of the prime reasons why it is of interest to astronomers. Paranal is located some 130 kilometres south of the Chilean port town of Antofagasta and about 12 kilometres inland from the Pacific Ocean. Standing more than 2600 metres tall, it forms an almost perfect cone. Cerro Paranal is part of the range of mountains that runs along Chile's Pacific coastline — relatively low mountains as opposed to the majestic Andes range further to the east that also demarcate Chile's eastern borders. Between the twin mountain ranges stretches the Atacama Desert, thought to be the driest place on Earth: the average rainfall in the region is just 1 millimetre per year. Some weather stations in the Atacama Desert have, in fact, never recorded rain. Paranal is indeed an inhospitable place, without water, without vegetation, without animal life. The indigenous people, generically known as the Atacameños, live a simple life along the eastern fringes of the desert. And, along the coast, there are a few towns with seaports. But the main desert is empty. Or almost so. In the 16th century, Spanish *conquistadores* came through this region in their attempt to expand the Spanish empire southwards. Later, others have traversed the arid land under the blazing Sun. The trails of many of those travellers can still be found, for little changes in this landscape which seems locked in a time-warp. Yet it is anything but static. During the early hours of the day or towards sunset, this seemingly monotonous landscape is modulated by long shadows and intensely glowing colours as the low rays of sunlight are reflected by minerals embedded in the topsoil of the desert. Minerals are the treasures of the Atacama. And as the Sun sinks below the horizon, another treasure unveils itself — the breathtaking night sky with the centre of the Milky Way overhead and the twin Magellanic Clouds vying for attention from their positions to one side of the Milky Way's band of stars. Standing atop Paranal is like having a front seat to the intensely beautiful, yet enigmatic view that the Universe offers us. Some describe Paranal as out of this world, but the truth is that in a very basic sense, it may be more in tune with the world than we think[1]. Yet it takes more than beauty to seduce levelheaded astronomers. In their quest for finding the best observational sites, terms like sky transparency, water vapour column density, atmospheric turbulence, temperature gradients, number of photometric and spectroscopic nights make astronomers' hearts beat faster. Of course, seasonal variations and long-term climatic effects are important, too. Establishing what these might be required meteorological studies over long time periods, clearly impossible in the case of northern Chile, as meteorological records have not been obtained for much longer than half a century in this sparsely populated part of the world. Nonetheless, some information existed and piecing the various bits of information together enabled astronomers to obtain a general idea of the prevailing conditions. It looked promising. One such piece came from the records of the Montezuma Station, near San Pedro de Atacama, which was operated from 1923 and into the 1950s by the Smithsonian Institution, to undetake solar observations. Other information constraining the conditions in the area of interest came from work in the 1970s by John Warner, observing from a site near La Paz. Also general findings by Jürgen Stock, whose pioneering work in the early 1960s on behalf of AURA cannot be overestimated, played a role. In 1968, he reported in the *ESO Bulletin*: *"There are a number of mountains of sufficient elevation south of the town of Antofagasta and very close to the coast. The abrupt rise from the Pacific Ocean on one side, and a large flat plain, more than 1000 metres lower, on the other side give these mountains rather special conditions."* (Stock, 1968). As space satellite imagery became widely available, Arne Ardeberg, then Director of La Silla, used this new tool and, finally, an *in situ* inspection of the area in April 1983 by Lodewijk Woltjer and a small group of senior ESO staff enabled ESO to home in on a few specific, highly promising sites. A description of the expedition, in which Paranal was "discovered" is given in *The ESO Messenger* (Woltjer, 1991). Shortly afterwards, a small site-testing team set up shop on Paranal. The core members of the team were Francisco Gomez and his sons Francisco and Italo together with Julio Navarrete. The team had two containers at their disposal, one with beds and one called "the office" with the instruments. For seven years, they carried out continuous meteorological and seeing measurements, first under the leadership of Arne Ardeberg and subsequently of Marc Sarazin, with

[1] The Atacama Desert has also been struck by a meteorite, known as the Vaca Muerte metorite. The impact is believed to have occurred 3500 years ago. In the late 1980s and early 1990s, two ESO astronomers, Harri Lindgren and Holger Pedersen, assisted by Canut de Bon of the Universidad de La Serena, collected 77 meteorite fragments in the desert, amounting to a total of 3400 kg (H. Pedersen *et al.*, 1991).

Hans-Emil Schuster in charge of the logistics. The team received supplies twice a month, but other than that was left on its own. Until 1987, when the road was built and it became possible to bring water to the summit, water was kept at the foot of the mountain, so that people had to walk down to wash.

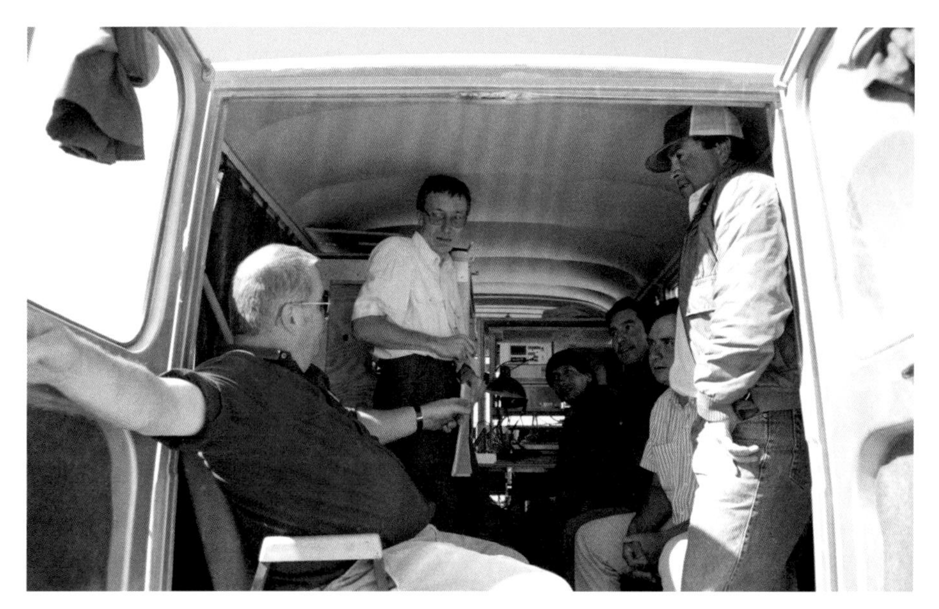

Site testing meeting at Paranal. From right to left: Hans-Emil Schuster, Marc Sarazin, Roberto Alcayaga, Francisco Gomez Campiano, Alfonso Vargas and Julio Navarrete.

≈

Life in the sun-scorched Atacama could be hard, but not necessarily one that would be considered particularly dangerous. Yet the desert can be treacherous. And if rain is indeed a rare occurrence at Paranal, it is different along the coast. Few people have had more experience in moving around in this arid landscape than Arne Ardeberg. On Monday 27 July 1987, Ardeberg, with a driver, had left Paranal in a southwesterly direction to carry out further site-testing measurements in the area. Earlier in the day, there had been sudden and very heavy rainfall, turning the part of desert close to the coast into a field of mud. Eventually, the heavy-duty four-wheel drive vehicle got stuck. They decided to head back towards Paranal on foot, but after a while fatigue set in and Ardeberg felt the need to rest. The driver continued, however, with Ardeberg set to follow somewhat later. By midnight, the driver reached the summit. But by the next morning, Ardeberg had still not shown up and the small

Paranal team became seriously worried. They contacted Daniel Hofstadt at La Silla, who initiated a search, including soliticiting help from the armed forces. However, the heavy rainfall had also hit Antofagasta causing a major mudslide and the emergency services were all engaged in the rescue operation in the town. On Wednesday, the weather conditions improved and ESO hired a helicopter in Santiago and sent it up to the Paranal area. The search focussed on an area to the south and southwest of Paranal, the desert area in which Ardeberg was likely to be, but it did not yield any results. Eventually, the Chilean Air Force freed up capacity in the shape of an additional helicopter. The *carabineros* also put together a search team of some 60 officers, in addition to two ESO teams, combing the desert in a systematic way. Still without success, it was decided to expand the search area towards the north. This time they were lucky: after four days, Ardeberg was found in a location far north of Paranal — alive. *"I'm fine, I've no problems!"* was his comments according to the Chilean daily *El Mercurio*, as he got out of the helicopter at Antofagasta's air force base. *"His 'cold blood' and extraordinary capacity to survive allowed him to go on for four days,"*[2] the newspaper concluded (Guicharrousse, 1987). Newspapers are never shy of drama, of course, but this had indeed been a close call. Luckily, this remained an isolated incident during the years of site testing.

≈

Apart from recording basic meteorological parameters, the initial site testing involved a visual assessment of the photometric conditions[3]. The work would soon receive a boost with the arrival of a young Frenchman, Marc Sarazin who joined ESO in 1984. He came from the Engineering School of Physics in Marseille, and had worked on problems of the propagation of light in atmospheric turbulence. The recruitment of Sarazin and the conscious investment in a long-term research activity in this field would pay off handsomely. Not simply in terms of assuring that the best site would be chosen, but also in developing a comprehensive understanding of the complex phenomenon of "seeing", whether high up in the atmosphere or at ground level, including the issue of dome seeing. To improve this knowledge, both about the quality of the candidate sites and to try to understand some of the local phenomena, a dedicated

[2] *"Su 'sangre fría' y extraordinaria capacidad de supervivienca le permitieron mantenerse cuatros días."*

[3] Woltjer quickly became convinced of the merits of the site and he even suggested moving the NTT, at the time under construction. The place that he had selected for the NTT, however, was not on Paranal itself, but on a close-by, but lower mountain north of Paranal, which he dubbed the NTT Peak. Within ESO it kept that name for several years, even if the NTT never moved there. Today, the peak is home to a telescope of a similar size, albeit different in every other way: the 4.1-metre VISTA telescope.

effort under the project name of the La Silla Seeing Campaign, or LASSCA for short, began in January 1986. The campaign consisted of a wide range of measurements, including classical data obtained with the meteorological station and an acoustic sounder (SOund Detection And Ranging — SODAR) to reveal perturbations caused by small temperature variations in the air between 30 metres and 800 metres above the site (Sarazin, 1986). Furthermore, a radar sounding receiving station provided by the French Centre National de Recherches Météorologiques was used to track balloons flying high into the troposphere, the realm of jetstreams. Importantly, the tests also included three telescopes at La Silla, the MPG/ESO 2.2-metre telescope, the ESO 1.52-metre telescope and, finally, the ESO 50-centimetre telescope. At the 2.2-metre, Gerd Weigelt and his team observed a number of single and double stars, using a method with ultra-short exposures, known to professional astronomers as speckle interferograms, a technique that had been proposed by Antoine Labeyrie in 1970 and developed further by Weigelt. For astronomical observations, speckle interferometry would *de facto* remove the effects of the atmosphere, but the technique could only be applied in particular cases and only if the sources were reasonably bright[4]. Used in the context of LASSCA, however, the interferograms would serve a different purpose, showing the diffraction patterns of the refractive index changes in the atmosphere. Simultaneous observations with the 1.5-metre telescope, carried out by Jean Vernin and Max Azouit looked at the scintillation, i.e. the twinkling of the stars, whilst François and Claude Roddier mounted an interferometer at the 50-centimetre telescope. This first concerted effort at ESO was conducted as part of the work by the VLT working group for site evaluation, at the time under the chairmanship of Harry van der Laan.

Sarazin soon replaced the system of visual and photographic assessment of the seeing by developing an automated 35-centimetre telescope known as a Differential Image Motion Monitor, or DIMM[5,6]. The entrance of the DIMM was fitted with

[4] Speckle interferometry built on work by the American physicist David Fried in the 1960s. Speckle imaging basically means making exposures of a few milliseconds of stars, thereby obtaining images of the star in a "frozen" atmosphere. For the purpose of studying astronomical object, these images can be added and/or analysed to reconstruct high-resolution pictures of the source. Speckle interferometry was already used for observations of binary stars with telescopes at La Silla in the late 1970s (Ebersberger &Weigelt, 1978).

[5] The principle of the DIMM goes back to the 1950s (Hosfeld, 1954). François Roddier developed the mathematics to needed get absolute measurements (Pedersen *et al.*,1988), and with the advent of CCD technology, such measurements became feasible.

[6] The DIMMs were built by the Belgian company AMOS in Liège. It was their first contract for an optical telescope, but since then AMOS has become a major supplier of precision optical instruments in astronomy, including the movable Auxiliary Telescopes for the VLT.

two circular holes, creating two images of a star on the CCD detector. Under perfect conditions, the images would be identical as regards their position. However, the distortion of the wavefront, caused by turbulence in the atmosphere, would mean that the two spots would shift relative to each other, allowing for a precise determination of the shift and thus of the amount and frequency of changes in the atmosphere. Since the measurements would be of the relative image motion, the result would not be influenced by tracking problems, e.g., induced by the wind (Roddier & Sarazin, 1990). In Chapter III-3, we will discuss how this technique can be used not only to study the atmosphere, but also to correct for the effects of the turbulence on the images.

Site tests at Cerro Vizcachas. The 3.6-metre telescope at La Silla can be seen in the distance. (Photo by the author)

A prototype of the instrument was tested at La Silla in 1986, and the first DIMM was installed on Paranal in April 1987. It was followed one year later by one at Cerro Vizcachas, the peak next to La Silla and at the time a serious candidate site for the VLT. Before that, as is the custom at ESO, the instrument was tested in Garching. It passed the tests, but Garching itself certainly did not, according to Sarazin, who wrote tongue-in-cheek: *"The ESO parking lot in Garching must be eliminated from the list of candidate sites for the VLT. During the night of 19 April, the second seeing monitor underwent final full-scale tests before being shipped to La Silla. It was an opportunity to verify that the seeing in Garching is not of the quality that the VLT deserves. During the few minutes we had before the instruments were covered with condensation, an average seeing of 3 arcseconds was measured. (Not all that bad for Europe, though!)"* (Sarazin, 1988).

Later site tests[7] also included La Montura (four kilometres northeast of Paranal) and Cerro Armazones, 20 kilometres east of Paranal — the mountain that would later be chosen for the ESO European Extremely Large Telescope (E-ELT) project. ESO focused strongly on sites in Chile, but that did not mean that others outside Chile were excluded. In fact, at the suggestion of Pierre Léna, site tests were carried out on the French island of La Réunion in the Indian Ocean. Later, tests were also done in Namibia. However, the results could in no way compete with those obtained in the Atacama Desert.

The data collected with the DIMMs covered a period of one and a half years, but other measurements were made as well. Firstly we remember the year-long visual assessments made by the Chilean site-testing team, which had actually yielded data of remarkable reliability and consistency[8]. Secondly, the altitude distribution of the turbulent layers was studied with three instruments: a scintillometer (for high-altitude effects), the acoustic sounder (SODAR) for the 30–800 metre elevation range, and finally microthermal sensors for monitoring local effects close to the ground. In May 1990, the results were presented in a report by the Site Selection Working Group, now chaired by Jean-Pierre Swings[9]. The data spoke for themselves: apart from its principal asset, the low amount of water vapour in the air, Paranal offered up to 70 clear nights per year more than La Silla. Importantly, these were the longer winter nights which therefore offered much more observing time. In terms of seeing, Paranal also turned out to be superior, though interestingly, it also became clear that La Silla was in fact a better site than had been thought so far. The median seeing for La Silla was 0.76 arcseconds and Paranal bettered it by reaching 0.66 arcseconds. But when considering nights with exceptional seeing (better than 0.5 arcseconds), with

[7] A dedicated campaign, similar in scope to the LASSCA campaign (though without major telescopes), was conducted in 1992 in the Paranal area under the name PARSCA 92. Since, Paranal itself was subject to heavy construction work at that time, many tests were carried out from the nearby NTT Peak (Sarazin, 1992).

[8] One reason for the consistency was certainly that the measurements were carried out over the whole period by the same small group of people.

[9] Succeeding Harry van der Laan in this function.

16% of all nights, Paranal came out as a clear winner with 2.4 times as many such nights as La Silla. From a science point of view, this was it! But Paranal was bound to pose a number of operational and logistical problems as well as financial issues. A second site selection working group[10] was therefore established to consider the full range of questions.

≈

As we shall discuss later, Chile is prone to earthquakes. For that reason, important seismic studies were carried out by the Institut de Physique du Globe in Paris by Raul Madariaga, a geophysicist who was well acquainted with the geology of Chile. The result was promising, as it showed that although also Paranal is also in a seismically active area, the rock there — solid granite — was considered to be more favourable in case of earthquakes than the geology in other areas in Chile.

Finally, a multi-disciplinary study by Michel Grenon of the Observatoire de Genève provided further information, in particular about the long-term evolution of the local climate. Grenon had conducted botanical studies in the area since 1971, and he pursued an independent and highly interesting avenue of enquiry, using *"biogeography as a tool to define with a high spatial resolution the integrated climatic properties over various time scales depending on the life duration and propagation times of living beings considered."* (Grenon, 1990). His conclusion further underpinned the notion that ESO was on the right track in its efforts to find a home for its new telescope. *"All the climatic indicators considered here, biogeographic and meteorological, lead to the conclusion that Paranal mountain is located in the best possible area of South America for the settlement of a modern astronomical observatory,"* were his closing words.

On 4 December 1990, Council unanimously voted in favour of Paranal as the site for the VLT. During the meeting, snow was falling gently outside the Garching Council Room, perhaps indicating the first signs of Christmas. If so, ESO had just given itself the greatest Christmas present it could think of.

The primary purpose of the site tests was to ensure that the VLT would be put in the best possible location. But, as the understanding of the many aspects of "seeing" improved, in particular as regards the construction of telescope enclosures, this

[10] Comprised of Jean-Pierre Swings (Chairman), Immo Appenzeller, Arne Ardeberg, Gerard Lelievre and Sergio Ortolani.

also led to major improvements in the performance of existing telescopes. In this sense, the tests not only led to the choice of a new, and better, site, but also to major improvements at La Silla, to which we shall return later. The atmospheric models that were developed also took ESO on a path towards predicting observational conditions, much like the weather forecast, and formed a fundamental element of what later became known as flexible scheduling. The unique expertise built up by ESO was made available to others and Sarazin subsequently took part in other site characterisation efforts, e.g., in Uzbekistan, Argentina and Morocco. He also chaired the site selection committee for the US Large Synoptic Survey Telescope (LSST) project.

The ESO 3.6-metre telescope and the NTT at La Silla in 1989. (Photo by the author)

Chapter II-9

NTT First Light

"A Revolution in Ground-Based Direct-Imaging Resolution"

ESO Press Release, 11 May 1989.

In 1988, the La Silla skyline changed. A conspicuous new building appeared on the hill that had been occupied by the Swiss 0.4-metre telescope in the past. It looked very different from any other dome, with an appearance that reminded some of a ship, others maybe of a somewhat peculiar shoebox: the New Technology Telescope. The NTT was a 3.5-metre telescope[1] on an alt-az mount. The primary mirror was a meniscus, 24 centimetre thick and weighing 6 tonnes, resting on a system of actuators[2]. The thickness was chosen as a compromise between a fully rigid classical mirror and a very thin mirror that would need active support at all times. Both the shape of the NTT primary and the position of the secondary mirrors were actively controlled. In a break with previous practice, the telescope had neither a Cassegrain, nor a prime focus. Instead it featured multimode instruments placed at the twin Nasmyth foci. The telescope was housed in a very compact and well-ventilated enclosure[3] to avoid dome seeing, an ailment that had been the curse of the classical telescopes. For the same reason, and as is now customary, all heat-generating sources had been removed from the observing floor. Since the compactness meant that the telescope could only move in altitude within the building, the enclosure was designed to rotate with the telescope. Technically, but also in a very visual way, the NTT signalled the dawn of a new age of telescope design.

From the outset, it was not at all obvious that this new telescope would become the pioneer of a new generation of telescopes. In fact, when idea of the telescope first

[1] ESO astronomers colloquially talk about the 3.6-metre telescope — meaning the classical telescope at the very top of the La Silla mountain — and the 3.5-metre NTT. But this is a bit curious since the 3.6-metre features a primary mirror of 3.57 metres diameter, whereas the mirror of the NTT is actually 3.58 metres in diameter.

[2] 75 axial actuators, three fixed points and 24 lateral actuators.

[3] Inspired by the Multiple-Mirror Telescope building on Mt. Hopkins.

arose — with Wolfgang Richter as the key engineer — the most novel feature had been the alt-azimuth mount, an idea that, as we have seen, had also begun to be used elsewhere. Soon, however, Ray Wilson's ideas about active optics support systems were adopted as an integral part of the project. The early phase also brought a reminder of the troubles of the 3.6-metre telescope project: originally a German consortium had been tasked with designing the telescope, but in the end their proposal did not meet with satisfaction at ESO. Massimo Tarenghi, the NTT project leader, consulted Wolfgang Richter, who had stayed in Geneva when ESO moved to Garching. In the end, ESO decided to revise the design completely with the help of an internal group, with Rainer Grothe and Sergio Lopriore as the main people involved. The contract for the main mechanical structure was awarded to Innocenti-Santeustacchio (INNSE) in Brescia, Italy, while Carl Zeiss won the contract for the delivery of the mirrors. The primary mirror was cast on 25 July 1984 at the Schott Glassworks in Mainz and transferred to Carl Zeiss Oberkochen in June 1986 for the optical figuring.

The NTT primary mirror at Zeiss, mounted on the support structure, seen shortly before delivery to ESO. (Photo by the author)

Around this time the building was completed and the main mechanical structure arrived in Chile from Genoa. On 13 July 1988 the NTT primary mirror was presented to a wider audience at a press conference at Carl Zeiss in Oberkochen. Before this public event, the optical elements — the primary, secondary and Nasmyth mirrors had been subject to thorough testing by the manufacturer. The results were truly impressive, as reported in *The ESO Messenger*: *"... the results of the Carl Zeiss interferometric tests, established in a most rigorous way, are as follows for the whole optical train M1, M2, M3: (a) 80% geometrical energy within ca. 0.30 arcseconds; (b) 80% geometrical energy within ca. 0.125 arcseconds for the Intrinsic Quality*[4]*."* In a qualitative assessment, the project team wrote that *"the results were so good that they exceeded even the expectations of both Carl Zeiss and ESO in all respects."* (Wilson *et al.*, 1988). However, despite the statement about the *"the whole optical train"*, the telescope optics were not tested together before being shipped to Chile.

≈

The mirror arrived safely at La Silla in September 1988 and was installed at the NTT in January 1989, and everyone was getting ready for the moment of first light. In Chapter III-14 (in connection with the VLT), we shall discuss in more detail why this is so significant, but at the moment, it suffices to say that traditionally, first light is the milestone that marks the passage of a telescope into the ranks of active scientific instruments. First light is normally preceded by extensive tests of all the systems, and this was no different in the case of the NTT. So, as soon as possible, Ray Wilson took a first glance through the telescope at a real star. He had longed for this moment and his expectations were high — after all, the optical test results at Zeiss had been outstanding. Wilson is a truly kind and gentle person with a wide breadth of interests. But below his amenable appearance is a personality with marked opinions and attitudes, not infrequently deviating from mainstream thinking. *"Ray, the heretic,"* he sometimes calls himself, with a smile — and some pride. Perhaps these are the hallmarks of a perfectionist. So it is no surprise that his dream was to build the perfect telescope. But what he saw that night at the telescope was far from perfect. An expert optician, it did not take him long to recognise serious spherical aberration in the images. His first thought was that it was possibly caused by a large temperature gradient, in other words a transient effect. But he was worried as he went to bed. The next night brought certainty: the reason why the images suffered badly

[4] Meaning *"if five terms, to be controlled actively, are mathematically removed from the combined image forming wavefront"*.

from spherical aberration was a matching error between the primary and the secondary mirror: the figure of the primary mirror was flawed! We have already discussed the Hubble Space Telescope and the optical problem that initially plagued this telescope. At the time of the NTT commissioning, this was still not known, but it is sobering to realise that, within slightly more than a year[5], two astronomical telescopes — both highly praised, eagerly awaited by the scientific community, and indeed expected to become the standard bearers of astronomical science — would turn out to feature flawed optics. The error had essentially the same cause in both cases too: faulty test procedures leading to an incorrect figure for the primary mirror and hence a problem known as matching error. It is easy to criticise the tests, but the truth is that testing a Ritchey-Chretien Cassegrain telescope, with its combination of a hyperbolic primary and a convex secondary, is intrinsically difficult and even minute errors may have catastrophic effects. A key tool to probe the optical performance is known as the compensation system or reflective null corrector. This piece of equipment is itself difficult to manufacture because of the extreme precision required. Of course, the experts at Carl Zeiss knew how to make a null corrector, but sometimes human fallibility also puts in an appearance in the cathedrals of science and technology. It turned out that during the manufacture, a spacer error in the optical system of the null corrector had been discovered. However, the technician who was supposed to correct the error had mistaken a sign in front of a number and thereby doubled the mistake rather than eliminated it! In the end, the displacement was 1.8 millimetres. This may seem like a small amount, but in optics the result is dramatic[6]. In his book, *Reflecting Telescope Optics*, Wilson quotes an optical expert at Zeiss, Christoph Kühne, for the following remark: *"God invented the number, the devil invented the sign!"* (Wilson, 1996). However, in the case of the NTT, it was not divine intervention that made up for the error. Fittingly for a telescope that is "active", the solution lay at hand — in the active support system. We remember that the actuators underneath the primary mirror serve precisely to influence the shape of the mirror, so in principle, it should be possible to cure the fault by means of forces applied by the actuators.

[5] The Hubble Space Telescope was launched on 24 April 1990, one year, one month, and one day after the NTT first light.

[6] Zeiss had decided not to perform an additional test, known as the pentaprism test, which would have revealed the problem. In the parallel case of the Hubble Space Telescope, Perkin Elmer, the company responsible for the optics, had also not made this final check. In the case of the NTT, the error was corrected by the active optics system. This was not possible for the Hubble Space Telescope, its optical system being a passive one.

Lothar Noethe, Gaetano Andreoni and Ray Wilson in the NTT control room during first light.

"The Best Telescope Yet" was the headline about the NTT in Sky and Telescope, *the American popular astronomy magazine in September 1989.*

Since the start of the NTT project Woltjer had firmly supported Ray Wilson's ideas about telescope design, but he had imposed one condition on the project: the telescope also had to work in a passive mode. For this reason, with a thickness of 24 centimetres, and thus an aspect ratio of 1:15, the primary mirror was not as thin — and therefore also not as flexible — as Wilson had wished. But, fortunately, Wilson had allowed for sufficient range in the forces that could be applied by the support system. Using slightly less than 90% of the full range, it was possible to fully correct the error.

On the night that began on 22 March 1989, the corrected NTT saw first light with spectacular success, delivering *"the best ever recorded images in ground-based astronomy,"* as Wilson subsequently wrote in *The ESO Messenger*. The NTT team had chosen a perfect night, and in the telescope control room, Ray Wilson, Lothar Noethe and Jorge Melnick, among others, watched with growing excitement as the first images appeared on the computer screen. The Director General, Harry van der Laan, recalled: *"Lots of people were sceptical [in view of the many new design ideas]. First light was something we looked forward to with anticipation and also with some trepidation. But when the first results came in we could hardly contain ourselves!"*

Ray Wilson later described the moment as *"the culmination of his career"* (Wilson, 2002), which by a stroke of luck coincided with his birthday. The best image was obtained two hours after midnight. From La Silla, the image files were transferred to a control room in Garching, where Richard West and his team waited in suspense. Upon arrival of the first picture, a crowded field of 47 × 47 arcseconds in the globular cluster Omega Centauri, West withdrew to his office, packed with charts and photographic plates. Soon thereafter, he emerged with a happy smile: he had identified the field in the cluster and had found a photographic plate of the same field obtained with the 3.6-metre telescope during commissioning. The comparison was breathtaking: the NTT image displayed a resolution of 0.33 arcseconds against 1.0 arcseconds for the 3.6-metre telescope[7]. *"Some people said: Did they fake this? This can't really be true that the very first experiment more than meets the specifications,"* van der Laan continued. But true it was.

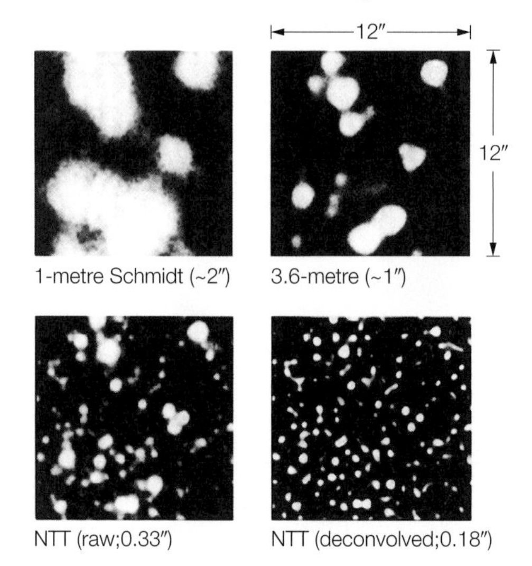

A result that raised eyebrows — the NTT first light image of a small part of Omega Centauri.

[7] It is a tribute to the new design, underpinned by technological advances, that while the NTT yielded a threefold increase in resolution, the construction cost actually amounted to roughly a third of the cost for the 3.6-metre telescope.

Impressive as it was, it could be bettered: in 1991, an observing team led by Reinhard Genzel obtained NTT images[8] with a resolution of 0.25 arcseconds[9] of a field in the Milky Way centre. The observations were part of a long-term study by the group of the motion of stars at the centre of the galaxy in order to determine the mass of the black hole then believed to reside there and now confirmed.

≈

The success of the NTT, in the first hours of its use, provided a strong boost to ESO at a time when the organisation was preparing for the VLT. Despite severe technical limitations at the commissioning stage in 1989, early observations included, among others, studies of planetary nebulae, the optical identification of X-ray sources and naturally observations of SN1987A. With the advantage of the superior optical performance, the small nebula of interstellar matter, excited by the ultraviolet flash at the explosion, could be imaged *"better than any reported to date"* (Gouiffes *et al.*, 1989).

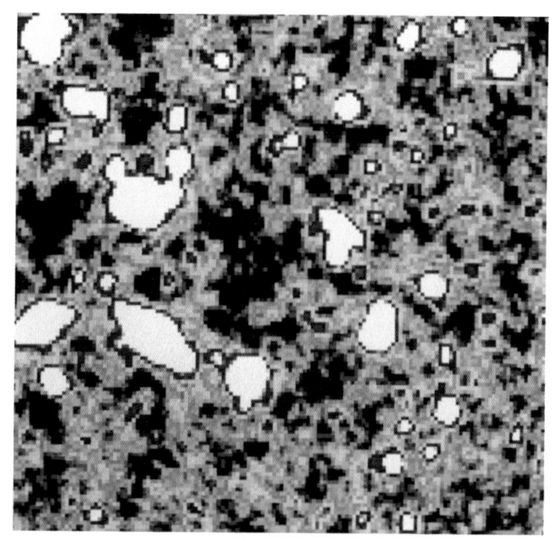

The NTT Deep Field in the constellation of Sextans. This picture shows a small part of the field and reveals galaxies down to the 29th magnitude — in the words of the ESO Press Release equalling "the faint light of a glow-worm in Garching, as seen from La Silla, 12 000 kilometres away".

Astronomy is a highly competitive science, and so, resting on one's laurels is not good: either for scientists or their telescopes. Having already yielded the sharpest image ever obtained with a ground-based optical telescope, at first light, the NTT soon broke another record. It was used to detect galaxies of the 29th magnitude, at least 2.5 times fainter than had ever been seen before by any telescope, be it on ground or in space. This was achieved in 1991 with an image of a small section of the sky called the NTT Deep Field. It was based on almost seven

[8] With a special instrument for high-resolution imaging in the near infrared, SHARP.

[9] After image stacking and appropriate image processing.

hours of observing time and the stacking of 41 individual images of the same field in the constellation of Sextans — selected for its apparent "emptiness", understood as meaning that it was free of bright objects. Of course, the field was anything but empty really: it was full of faint galaxies. *"As a matter of fact, it is not even certain that there is any place where we are able to see the sky background,"* wrote Bruce Peterson, the principal investigator, in *The ESO Messenger* (Peterson *et al.*, 1991). This was important, since what the team of scientists cautiously described as *"detecting and measuring extremely faint galaxies"* was in fact a bold attempt to see if the point could be reached beyond which we would see no galaxies anymore. Even if the NTT Deep Field image did not provide any spectroscopic information, from which redshifts could be inferred, it seemed clear that this point had not yet been reached. This notwithstanding, *"this NTT picture has given us a tantalising, first glimpse of what can be done with the new and improved observational means which are now at our disposal. It has given us a unique look into regions of the Universe, so remote in space and time that they have never before been explored. This is the type of work that will be at the frontline of optical observational cosmology during the coming years,"* wrote Peterson. There would indeed be more deep fields to come.

The first results from the NTT were presented at a press conference in connection with the inauguration celebrations on 6 February 1990. Seen from left to right: Manfred Ziebell, Daniel Enard, Ray Wilson, Harry van der Laan and Masssimo Tarenghi. (Photo: Hans Hermann Heyer)

The early success of the NTT also contributed to increasing the confidence of the Member States with respect to ESO's next project — the VLT. This was clearly pronounced during the festive inauguration of the telescope, held at the Garching Headquarters, on 6 February 1990 in the presence of, among others, science ministers from Belgium (Louis Tobback), France (Hubert Curien) and Germany (Heinz Riesenhuber). Italy was represented by Antonio Ruberti (at the time Italian Minister and later European Commissioner for Research), while Ambassador Jean-Pierre Keusch represented Switzerland. As already mentioned, the realisation of the NTT had become possible with the extra funding provided by the two new Member States, Italy and Switzerland. Indeed Ambassador Keusch put the early success of the NTT into the wider context of the VLT: *"With the commissioning of the NTT, its exciting career as one of the world's outstanding scientific instruments now officially begins. The other key role of our new telescope is of course that of a technological predecessor to the Very Large Telescope, decided by the Member States in December 1987.... ESO's budget will have to be increased considerably, but I am confident that Member States will honour their commitment and that the Executive, building on its experience with the NTT, will cope successfully with this new challenge."*

≈

The NTT brought new opportunities to the scientific community of Europe. La Silla now had two 4-metre-class telescopes and this enabled ESO to address a problem that had been present for a long time. As mentioned earlier, obtaining observing time is a highly competitive process. The telescopes had long been oversubscribed and thus many scientifically very highly-ranked research projects could not be pursued. In 1987, for example, ESO received more than twice the number of applications than the slightly over 300 that could be accommodated. At the same time, to satisfy as many users as possible, observing time, when granted, had been strictly apportioned. On the 3.6-metre telescope it could be one night (or less) and practically never exceeded a few nights. Indeed, of those proposals accepted, roughly half were granted less observing time than they had actually applied for (Breysacher, 1988). While this was often quite enough to secure a rich harvest of data, some research projects needed considerably more observing time, frequently spread over long time intervals. ESO's Director General described the problem as follows: *"The following scenario is ... a painful reality for many an ESO user. You request five nights in a judiciously balanced trade-off between minimal astronomical needs and your estimate of the OPC's*[10] *range. Then you get three nights, of which one is partly cloudy; your astrophysical goal shifts another year and the substance of your PhD student's thesis erodes precariously. The focus of your own scientific attention is blurred, you have to work in several areas at once and a rival/friend on another continent takes a decisive lead."* (van der Laan, 1988).

The solution, it appeared to van der Laan, was to introduce the idea of Key Programmes, and these were proposed in the spring of 1988. Successful Key Programme applications would be granted substantial observing time, thus enabling exciting new science to be done. Examples of observing projects from which this approach would benefit included optical identification of distant radio sources, the study of nearby clusters of galaxies as well as of high-redshift (i.e. exceedingly distant) galaxies, the search for quasars and gravitational lenses (as described in Chapter II-6). With a heavy bias towards extragalactic research, this clearly stimulated a shift in research priorities within the scientific community. It also helped the community to prepare for the much more powerful VLT, a telescope that with its larger light grasp would be ideally suited for exactly that kind of research.

The Key Programmes quickly became a hot topic among European astronomers[11]. Applying for massive amounts of telescope time would surely not be less competitive, and this had an interesting side effect, as can be seen in the description of one of the first successful proposals: *"At least three distinct groups ... were already involved in the study of gravitational lensing effects. Observations were being performed with ... the help of various telescopes on La Silla as well as at other observatories (Very Large Array [VLA], Canada France Hawaii Telescope [CFHT], Palomar, Kitt Peak, etc.). A general feeling existed that our individual work was progressing very slowly, the number of effective nights that were allocated to our programmes being modest.... We all knew that there was a total absence of coordination between our independent programmes and that, of course, we could not avoid duplicating observations of similar objects. Furthermore, each of our teams was hoping very much to broaden its observational interests: those studying highly luminous quasars wished to look also at distant radio galaxies, and vice versa, but this would remain a mere dream until ... it became really possible for our*

[10] As mentioned in Chapter I-5, OPC, the Observing Programme Committee, is the body that evaluates observing proposals and recommends time allotments to the Director General. It is the body that has to make the often painful choices between the many excellent proposals that the user community submit to ESO on a twice-yearly basis.

[11] In fact no less than 42 proposals were submitted in response to the first call in 1988. Twelve proposals were accepted.

present team to submit an ESO Key Programme entitled Gravitational lensing: quasars and radio galaxies. We do feel very fortunate today since our programme has been generously allocated 54 nights with the ESO 3.6-metre and/or NTT telescopes, 48 with the MPG/ESO 2.2-metre telescope and 9 effective nights with the Danish 1.54-metre telescope during the next three years, starting effectively during period 43 (1 April–1 October, 1989)." (Surdej *et al.*, 1989). Thus the advent of the NTT not only realised ESO's aspirations in terms of exploring new ground as regards telescope technologies, but also gave European astronomers a tool with which they could decisively advance their research. What the NTT did *not* do, was to reduce the pressure factor on the La Silla telescopes. It seemed that the more and the better observational facilities that were put at their disposal, the more ideas were developed in the increasingly vibrant European community of astronomers.

The Key Programmes were complemented by a new training programme for young astronomers. This allowed for doctoral students to spend up to two years at ESO, either at Garching or in Chile as part of their studies. This gave wonderful opportunities for promising young people and at the same time contributed to forging an early awareness of the European dimension in astronomy. We shall return to this in Chapter IV-8.

The initial success of the NTT, the Key Programmes and the Student Programme all pointed towards a bright future for European astronomy. But obviously the most important contribution to that future was the momentous decision to construct the VLT. Yet it would be while before that telescope would collect photons from distant celestial objects. In the coming chapters, we shall look at the arduous path towards the realisation of this project and the breakthrough for European astronomy that it signified.

Part III:
The Breakthrough

Artist's impression of the VLT in 1987. We see the four 8.2-metre telescopes in a straight line. The telescopes were supposed to be operated in the open air, but protected from the wind by a screen running along the entire array (not shown in this drawing). Inflatable enclosures would offer protection from bad weather. There is a huge gantry crane to service the telescopes. Seen to the left of the array are two auxiliary telescopes on tracks, one of them somewhat downhill. (Illustration by Jean Leclercq)

Chapter III-1

Back to the Drawing Board

"Whereof what's past is prologue,
what to come — In yours and my discharge."

William Shakespeare, from *The Tempest.*

In his memoirs Riccardo Giacconi, who was to become ESO's fifth Director General, writes *"VLT ... represented at least six times the organisation's annual budget"* (Giacconi, 2008). The VLT expressed the aspirations of Europe's astronomers, but it certainly also constituted a gigantic challenge for the engineers, scientists and project managers at ESO. If anyone was acutely aware of this fact it was Harry van der Laan, who took over as Director General from Lodewijk Woltjer in January 1988. Van der Laan was the third Dutchman in a row at the helm of the organisation. Before assuming the post, he had been in charge of the national Dutch radio observatories at Dwingeloo and Westerbork, as well as serving as scientific director of the Leiden Observatory, but he had also been involved in the discussions about the VLT project, notably as the first chairman of the Site Selection Committee. Van der Laan is extremely eloquent, perhaps — some might say — with an occasional hint of loftiness, yet always deeply human in his approach to his staff. Barely a month before, Council had taken its historic decision to embark on the VLT project, based on the Blue Book. Van der Laan's task was to turn the proposal into reality. The path towards this goal would be bumpy, which should not have come as a surprise to anyone. Pushing technology for science is about doing things that no one else has done before. How could this project *not* have entailed surprises? Indeed, as Adriaan Blaauw expressed it: *"... ESO again entered a phase of pioneering, but this time not just by European standards, but on a worldwide scale: whereas expertise for certain components of the VLT is available here and there, no ground-based observatory of the scope foreseen for VLT is in existence yet. Therefore ESO has had to build up in-house expertise, and in close collaboration with experts from outside, a set up qualified to do the job."* (Blaauw in Krige & Guzetti, 1995).

The main issues that would mark van der Laan's tenure were the detailed design of the VLT — what in contemporary ESO parlance is called a Phase B[1] — the site selection, the key decisions on interferometry, the awarding of the main industrial contracts (for the mirror blanks, figuring and the main mechanical structure). In parallel there was the development of an initial VLT instrumentation plan and the decision to devolve this programme to include national institutes, the introduction of the concept of Key Programmes and the student programme. In this chapter, we shall look at each of the VLT-related issues from a project management point of view, and discuss more technical isssues in the chapters that follow.

In reality, the Blue Book was effectively a conceptual description with many options and some top-level specifications. It could be nothing more, since much research and development (R&D) still had to be done. This applied, for example, to the primary mirrors, where there still was discussion about the possible use of metal mirrors instead of glass ceramics. Nobody had ever cast a blank of that size and while ideas existed about how to do it, they were still unproven. In the area of adaptive optics, development had also only just started. The Blue Book presented the Unit Telescopes as a linear array, but it also kept open the choice of site (though Paranal was prominently mentioned), a choice that might affect the final configuration of the array. And while there was no doubt that the interferometer would influence the layout, it was not yet clear how.

The VLT project also meant that ESO changed from being a single-mission programme organisation to a project-driven one with multiple activities running in parallel. Since the setting up of the TP Division, ESO had very much been a hands-on operation, in the sense that the people would take part at the most practical level in developing telescopes and equipment for La Silla. But with the VLT, ESO moved into a regime of heavy industrial interaction and cooperation, with stringent requirements for the upfront definition of detailed technical specifications, financial management and liability questions, especially in R&D-intensive areas. This in turn necessitated changes within the organisation and the recruitment of a number of new staff members. The first major change involved the splitting of the now venerable Telescope Project Division into two: a Project Division and a Technology Division. Two years later, the Project Division was transformed into a dedicated VLT Division[2].

---

[1] At the time this was called the Final Design.

[2] In January 1995, however, the VLT division was again split into two when a dedicated Instrumentation Division was established, moving some 40 staff from the VLT Division.

We have seen how Woltjer had prepared the ground for ESO's next steps, by navigating the political landscape, identifying promising technological solutions and very much by recruiting the right people. But his style had been fairly autocratic. This had hardly been an issue in the past. As a European organisation, ESO was steeped in a Franco-German culture, with a fairly strict hierarchical order. The academic world in the 1950s and early 1960s was hardly any different. But now it was nearly the 1990s, and van der Laan had different ideas about how ESO should be run, with a more inclusive, or participatory, management style. Furthermore, the constraints on budget and personnel[3], developing the VLT, completing the NTT and maintaining operations at La Silla, forced him to introduce a matrix organisational structure with about 20 individual VLT-related integrated task groups (covering optics, electronics and mechanics) — something new at ESO. The restructuring also meant changes at the upper management level. The early phase of the VLT project, which we would today call Phase A, had been skilfully led by Daniel Enard. With the transition in 1988 into the detailed design study, the Phase B, an initial triumvirate of Enard, Tarenghi and Robert Fischer led the project. However, in November 1991, after a brief interim period, during which time the project was managed by Joachim Becker, who had come from the aerospace industry, Massimo Tarenghi was appointed as project manager. Tarenghi brought both a profound scientific background and his project experience from the 2.2-metre and NTT projects. Ray Wilson, naturally, continued to play an important role as regards the VLT optics, although mainly as a consultant.

While the new organisational structure was being put into place, van der Laan set out to turn the VLT project into reality. First, this required a careful reassessment of the project, now from the point of view of a practical implementation, which meant taking firm decisions on a number of key issues. Work began on the detailed design and — for components with long lead times, such as the primary mirrors — with preparing the calls for tender. The outcome of this would reveal that the original project schedule had been too optimistic.

[3] The staffing level was set by the ESO Council and the Council had traditionally displayed considerable reluctance in increasing the staff complement. Realising the problems this created for the implementation of the VLT project, van der Laan pushed hard for a staff increase. In the past a solution, which had been widely used, was to hire personnel as auxiliary staff or in other categories intended for temporary functions. Towards the end of the 1980s this was no longer tenable and it was decided to convert many of these positions into regular international staff positions. Thus the limited staffing increase of nine positions, to which Council agreed under pressure from the Director General, mainly went into converting existing positions instead of increasing the personnel. It wasn't until 2006 that Council formally gave up its practice of determining the staff level and left the decisions to the Director General within the general limit of the budget.

≈

By 1992, the VLT had assumed the shape that we know today, seen here in this aerial photo: The four 8.2-metre Unit Telescopes in their enclosures, placed in an almost trapezoidal configuration, with the movable 1.8-metre Auxiliary Telescopes on tracks, the interferometric focus building at the centre of the photo and the control room building to the left, below the telescope deck. A comparison with the illustration on p. 200 shows how much it had changed. (Photo: Gerd Hüdepohl)

Meanwhile, over the next two years, the project would undergo significant changes and become the VLT that we know today: this was partly because a reconsideration of the scientific requirements led to the decision to add Cassegrain foci to all the telescopes. This was a costly proposition and had major repercussions for the overall telescope design. The changes included raising the altitude axis to gain free space underneath each telescope, significantly complicating the design of the primary mirror (M1) cell (among other things requiring a rotator and adding considerable weight at what is its weakest point) to the M2 unit as well as modifying the M3 tower. However, measured by the amount of science that has since been done with Cassegrain instruments on the VLT, there can be no doubt that this addition was important[4]. According to the Blue Book, the VLT was intended to be operated in the open air,

[4] During the period 1999–2009, the twin Cassegrain instruments FORS1 and FORS2 alone provided the basis for no less than 1359 research papers. In 2009, when X-shooter, the first second generation instrument for the Cassegrain focus, became operational, it instantly gained second position in terms of the amount of requested observing time.

but equipped with inflatable shelters to protect the telescopes in extreme weather conditions. Experiments with inflatable shelters began at La Silla, using a half-scale test setup in early 1988. However, when Paranal was finally chosen as the site, the features and conditions on this location firstly dictated that the idea of operating the telescopes in the open air should be abandoned[5] and secondly, it opened up options for a different configuration of the array than the one foreseen in the Blue Book. The idea of a linear array was therefore scrapped. But all of these changes took time. For example, the introduction of huge enclosures meant that extensive wind-tunnel tests were required to ensure that the seeing conditions on Paranal were not adversely influenced by the buildings or by their placement relative to each other. Finally, a major decision was that for the realisation of the project, ESO would act as its own main contractor. It would, in other words, not delegate the overall project management to an external company, but instead define, prepare and monitor a large number of individual contracts for individual items or services to be delivered by industry.

≈

The first major contract to be concluded concerned the primary mirrors, for which the Blue Book had presented two options — glass ceramics or metal blanks. ESO had carried out experiments with aluminium mirrors but eventually decided against this solution[6]. Two suppliers of glass blanks made offers to ESO — the Schott Glassworks in Mainz, Germany and Corning, the US glassmaker that had also supplied the blank for the 3.6-metre telescope. The Schott offer was for blanks produced in Zerodur, a patented glass-ceramic product with a zero thermal expansion coefficient. Corning, on the other hand, had proposed a traditional solution of fused silica. The Schott product was clearly superior from a technical point of view[7]. But it was

---

[5] Also, by this time, the NTT experience offered proof that a telescope enclosure could be designed in a way that avoided the dreaded dome seeing problem. The final decision to go for a scaled-up, simplified NTT building was taken in 1992 (Zago, 1992).

[6] For a while, an idea to form an industrial consortium under the name LAMA (Large Active Mirrors in Aluminum) with the participation of firms in France, Germany, Italy and Switzerland was floated. They would produce an 8-metre metal mirror for ESO with financial support by the European technology programme, EUREKA, which had been established in 1985 as a Franco-German initiative. The LAMA project was not an ESO suggestion, but the initiators were keen to exchange information with ESO, based on ESO's own experiments with metal mirrors which — among others — included the successful testing of two 1.8-metre nickel-coated aluminium mirrors (Dierickx & Zigmann, 1991), and for ESO to possibly procure an 8-metre blank from the consortium.

[7] The development of Zerodur dates back to the late 1960s.The first Zerodur blanks for astronomical purposes were for the 2.2-metre and the 3.6-metre Max Planck telescopes at Calar Alto in southern Spain in the 1970s. Zerodur made the Schott Glassworks a world leader in this most demanding market (Morian, 2003).

expensive and required a longer delivery time. Furthermore, Schott proposed to use a novel technique, spin casting, which had not been tested and entailed a risk that it was reluctant to carry itself[8]. Questions of liability became a major issue. However, after lengthy negotiations, a satisfactory solution was found. The VLT mirrors obviously required a huge mass of glass to be prepared and poured at a very high temperature into a mould. For this purpose, vital parts of the equipment had to be made out of platinum because of its unique non-corrosive characteristics. It was feared that the substantial amount necessary — about seven tonnes — might severely, if momentarily, impact the world market[9] and, in any case, involve costs that could not possibly be covered by the VLT project. Robert Fischer[10], however, came up with an innovative solution: Schott should *de facto* rent the amount of platinum for the duration of the casting, thus reducing the financial burden to acceptable levels[11]. In September 1988 a contract was signed with Schott for the delivery of four 8.2-metre blanks at a price of 55.5 million deutschmarks. As already mentioned, at that time no one had ever cast a glass mirror of that size, but Schott had firm ideas about how to do it. However, they also had to erect an entirely new production plant. In July 1989, the traditional ground breaking took place and the buildings, totalling 50 000 square metres, were completed in 1991 (Müller *et al.,* 1993). The second large VLT contract was also awarded in July 1989 — this time to REOSC for the polishing of the VLT primary mirrors at a price of 36 million deutschmarks. REOSC also needed a new facility, although it was more modest in size: a mere 1100 square metres, but equipped with a 32-metre high tower for interferometric tests. This plant was dedicated in April 1992 by Hubert Curien, then the French Minister for Research and Space. We shall look at the mirror production shortly. However, from the point of view of project management, the logistics of the mirror production, in combination with the expected construction of the telescopes, dictated a delay of several years[12]. It

---

[8] The necessary R&D was supported financially by the German Federal Government.

[9] The world production of platinum in 1987 was less than 130 tonnes, and the market price was in excess of 10 million US dollars/tonne.

[10] In projects like this, with their strong focus on technology and science, it often happens that rather little credit is given to the administrative side. However, in several conversations that the author has conducted with some of the key actors, Robert Fischer was recognised and praised for his contribution to the VLT programme, not the least for the superior way in which he guarded the integrity and credibility of the procurement activities.

[11] Accepting this solution, Schott bought the platinum at the London Commodity Exchange and sold it again after closing the 8-metre production line.

[12] The Blue Book foresaw the completion of the first blank in December 1993, but the new projection suggested November 1995. The delivery of the last blank, originally expected in December 1998 now slipped to August 2000, and possibly even into 2001, depending on the polishing capacity of REOSC. In the end, however, the delivery occurred earlier than feared.

appeared that the optics might not be completed before the year 2000. The news was presented to the Scientific Technical Committee (STC) in October 1988 and subsequently to Council. Among the major contracts still pending were the contracts for the M2 unit, the secondary mirrors with their complex control systems, and for the main mechanical structure. The contracts for the M1 mirror cells with their intricate support systems and, of course, the VLT enclosures were also pending.

To stimulate interest in industry and to give a fair chance to all, a four-day information meeting for European industry was organised at the City Hilton hotel in Munich in April 1989.

Almost 300 representatives from industry attended the VLT Industry Days at the Munich City Hilton. (Photo by the author)

≈

The sizes of the main VLT contracts were such that they whetted the appetite of industry, causing fierce competition not simply between individual companies[13] but also, to some extent, between countries. ESO has to act within strict procurement rules, aimed at securing the best deal for the organisation. Importantly, it does not

[13] In many cases, the number of potential bidders was small, due to the challenging technological requirements.

have to follow the principle of *juste retour*, i.e. awarding industrial contracts in a way that is proportional to the contributions of its individual Member States. However, even without the *juste retour* requirement, Member States eventually like to see a balanced return on their investment, at least within a time period of perhaps five or ten years. This is normally possible, but in a project with a few very large contracts the representatives of the Member States naturally keep a close watch on ESO's procurement activities. Occasionally, it may also cause situations where delegates must master the delicate balance between the interests of the organisation as a whole and the legitimate interests of the government and country that they represent.

In June 1990, the call for tender for the main mechanical structure went out to industry. This was the largest individual contract of the project, with a value of 76.5 million deutschmarks. Eight offers were received. In the end, two bids stood out as most promising, one from a consortium in Italy, another from a group of companies in France, Belgium and Switzerland. But both proposals raised points of concern. After evaluation by ESO's technical and administrative experts, their recommendation came out in favour of the Italian consortium. However, the Finance Committee, which normally deals with large contracts and must approve them before ESO can go ahead, could not reach an agreement. It eventually became an issue for the ESO Council and it was finally decided to organise individual meetings with bidders to review their proposals. These were carried out over the summer and, in September 1991, ESO finally signed the contract with the Italian consortium under the acronym AES, made up of three partners: Ansaldo Componenti (of the IRI Finmeccanica Group), European Industrial Engineering (EIE), and SOIMI (Società Impianti Industriali, a part of the Asea-Brown Boveri (ABB) Group). The contract covered all the steel structures, the hydrostatic bearings on which these heavy structures would rest, direct drive motors to move the telescopes and high-precision encoders to measure their exact position in order to ensure accurate pointing. A kick-off meeting followed in October of that year.

≈

In 1991, construction work started at Paranal. This was only nine months after Council had chosen the site. The work involved lowering the mountain by 28 metres, which meant removing 350 000 cubic metres of rock to create a telescope platform of the desired size. The blasting, which began on 23 September, was done by Interbeton, the Dutch company that had carried out the concrete work for the 3.6-metre telescope at La Silla thirty years earlier. ESO's men on the spot were two former

Blasting at Paranal.

employees, Peter de Jonge and Jörg Eschwey[14], brought back to ESO by Harry van der Laan. Blasting is hard and extremely dirty work — and quite obviously, potentially dangerous. Yet even in the rough atmosphere of the blasting operation, serene beauty could be found, in the person of Stuart Dunn, the blasting engineer. Also a trained opera singer, he treated his workers to sublime aria performances during the rest periods. But at an early stage, Dunn raised concerns about the stability of the rock, leading to an additional geological study by Fabien Bourlon, a French geologist. The study, however, confirmed that the VLT could safely be built there, but that it did require structural reinforcement of one of the telescope building foundations[15].

Then the news broke that a nitrate mine was being opened by the company Minera Yolanda at a distance of only 21 kilometres from the mountain. Readers will recall that when ESO chose La Silla it also secured the ownership of a substantial area around the mountain to avoid the potential adverse influence of mining activities in terms of light pollution and dust in the atmosphere. Mineral extraction in Chile is often done by open-cast mining. ESO had sought protection around Paranal in

[14] A hugely experienced construction engineer, Eschwey left ESO in 1980 for the USA. There he advanced to become Vice President of a large (billion-dollar turnover) construction company, before he was lured back to ESO. He would also stay involved with ESO after his formal retirement, helping both with the ALMA project and the Headquarters expansion in Garching.

[15] In order not to damage the granite base, the blasting left the last 1.5 metres unscathed. This was subsequently crushed with jackhammers and the rubble, basically pebbles, carefully removed in order to ensure a clean surface and sturdy connection between the base rock and the reinforced concrete.

Not really a happy lunch in the park — a thoughtful Jacques Beckers during a site-testing mission to the Paranal area.

a similar way as it had done for La Silla, but even if the piece of protected land around Paranal was not small, a major mine right on the outskirts was bound to create worries. To determine the effect on the observational qualities of the site, Jacques Beckers carried out a study of the atmospheric pollution from the mine. Fortunately, the mine was located southeast of Paranal and thus the prevailing winds could be assumed to offer some help. To everybody's relief, it turned out not to create any significant problem, and a few years later, the mining activities ceased.

The layout of the four VLT Unit Telescopes as well as the underground tunnels are clearly visible in this aerial photo, obtained in 1994. (Photo: Herbert Zodet)

≈

Gradually, despite the progress achieved — with the important project changes made and the detailed design of the project advancing, major contracts awarded, the site development underway etc. — a sense of crisis emerged. The shortage of staff had begun to bite and manifested itself in several ways: the internal structure was not functioning in an optimal manner and reorganising the top-level project management had proven to be fraught with problems. Compared to the Blue Book, the schedule was slipping. Key contracts were still pending and where work had begun in industry, especially on the main mechanical structures, the work was moving much more slowly than anticipated. And there were growing concerns about cost. In June 1992, the Director General had warned about an increase of 50 million deutschmarks, though this could only be an estimate.

Perhaps there was nothing unusual about the problems. It seems that any large project goes through distinct phases — from jubilation and excitement when it is approved, through a sobering realisation of what the project really requires, to a period of hard troubleshooting and, fortunately, in most cases, to a success that everyone can then acknowledge and enjoy. But by late 1992, troubleshooting was called for more than anything else. With the ending of Harry van der Laan's term of office, Council chose Riccardo Giacconi as ESO's next Director General. He would soon get a chance to demonstrate his awesome skills as a troubleshooter.

≈

Arriving in January 1993, Giacconi brought a breath of fresh air into ESO. Some would say a storm. For Riccardo Giacconi was not just a man with a mission, but an impatient one. His impatience was not only caused by his realisation that he would not have much time to accomplish his task, but also by his awareness that astronomers in the US were on track to step into a new age in ground-based optical astronomy. In fact, the very same year as he joined ESO, the first 10-metre Keck telescope obtained first light on its excellent site at Mauna Kea, in the Hawaiian Archipelago. Once again, US astronomers had been faster than their European colleagues and a new parameter space for discoveries had been opened up by them. Giaconni came to Europe from America to resurrect a project that needed to move quickly to an implementation phase, but soon encountered a string of further challenges. He is a tremendously energetic person with what sometimes appears to be tunnel vision in his search for uncompromising excellence, combined with an uncanny skill to strike at the heart of a subject. He had played a pivotal role in developing X-ray astronomy, for which he was later awarded the Nobel Prize in Physics (in 2002), had been in charge of the Uhuru satellite project, then the Einstein Observatory, and subsequently led the Space Telescope Science Institute in Baltimore before the launch of the telescope in 1990. He had also led the institute during the stormy time when the telescope's serious optical flaw was discovered, threatening to make Hubble the laughing stock of science — had the potential loss of the considerable, expected scientific exploits not been so dramatic. Riccardo Giacconi considers himself to be an adventurer rather than a scholar and his direct ways support that notion: an eloquent diplomat he is not. With his characteristic candour, he stated that the ESO he found had *"a wealth of engineering and scientific talent"*, yet — in his view — *"[the] organisation resembled a civil service agency rather than [a] research institution"*. Others would have disagreed. ESO, of course, operated within a European culture and had to respond to the ways and expectations of that culture. But, aside from that, ESO certainly was supposed to act as a service agency for the Member States' scientists[16]. But it is an undeniable fact that the result was a clash of cultures that led to tension between the staff and the Director General. That tension would prevail for all of Giacconi's tenure, but it is a tribute to both parties that it never impeded the strong desire to develop the VLT and to make it the success that it has become. Indeed it remains a fact that the wonderful VLT project, initiated by Lodewijk Woltjer and prepared by Harry van der Laan, was now infused by Giacconi's ideas, especially about operational aspects, and with his single-minded, sometimes ruthless, drive was reenergised and fully delivered. To paraphrase Giacconi, the situation caused by the uneasy relationship was "grave, but not serious". Serious challenges, however, were soon to come from elsewhere.

≈

In April 1993, Council convened for an extraordinary meeting, with several critical issues on the agenda. Most pressing were the state of the VLT project and the

[16] This was not simply a question of different opinions. Rather it is deeply rooted in the differences in the way research is funded, at least in astronomy, in Europe and in the USA. In the USA there is an integrated funding system that has a strong focus on the end result — the science — while in Europe funding is more diverse, with different funding regimes, each with their own area of responsibility. The problem has remained unresolved, as can be seen in the ASTRONET Roadmap from 2008: *"While European astronomers gain access to their major facilities as the result of peer-reviewed selection, they are generally unable to obtain dedicated funding to carry out the associated analysis and publication of results at a speed that is competitive with their non-European colleagues and competitors, the latter often being funded by substantial grants associated with the use of the facilities."* (ASTRONET, 2008).

relations with Chile, to which we shall soon return. The new Director General reported on his first impressions and expectations for the VLT project[17]. In June, he presented them in a more formal way in a written report to Council and since the assessment had not changed, we shall simply review the main findings and consequences. Compared to the original proposal, the VLT project was now suffering from a serious delay. The key reason for the delay was the lack of manpower. Furthermore, and partly as a result of this lack, the cost was going up. The cost increase was also due to the additional requirements on the telescope, not the least the inclusion of the Cassegrain foci. Some costs had not — or not in a sufficient manner — been considered in the Blue Book, e.g., for interferometry, various mirror handling tools and technical CCDs. In many cases, the budget contained in the Blue Book was based on a best estimate, while now, the real cost was becoming clear. The original price of 382 million deutschmarks had been set in 1986 prices. Giacconi's estimate, now based on a cost to completion calculation, was for 573 million deutschmarks, against 430 million deutschmarks, the original price, properly adjusted for inflation and contingency. But that was not enough. Giacconi also presented a plan for a significant increase in the personnel — 46 new staff to be added to the existing complement of 193 staff[18]. Finally he presented his ideas about a general reorganisation of ESO, with policy issues including project budget planning to be carried out within the Office of the Director General and a major staff increase in the VLT-related functions, such as a dedicated VLT system engineering group, more resources for software development and VLTI as well as data management, including data reduction and analysis, calibration and archiving. First light would now take place in January 1998 according to his projections. The plan laid the foundations for the operational concept that became an essential element of the ultimate success of the VLT, but at the time, the reaction was anything but ecstatic. It was shock treatment. At least one government representative speculated loudly about cancelling the entire project, but most Council members seemed to accept that there could be no going back. Giacconi got it his way: or almost. The implication of the cost increase was, of course, increased Member State contributions, possibly also involving a loan. At the meeting in December 1993, Council did not, in principle, reject an increase, but made it contingent on a financial audit, and in any case made it clear that the increase would have to be lower than that requested by the Director General. Since

[17] Some of the "first impressions" were based on a thorough project review that had uncovered several technical weaknesses — not unusual for a complex project like the VLT.

[18] This included an increase of 12 members of staff that Council had already approved at the preceding meeting in April.

neither the loan nor the full, increased contribution option found sufficient support, the only option left was a postponement of certain parts of the programme. Sensibly, but painfully, the VLT Interferometer fell victim to the hard financial realities. Furthermore the plans for creating adequate living quarters on Paranal in the form of a Residencia were provisionally shelved. Staff would continue to live in steel containers for several years to come.

The financial problems in the early 1990s caused considerable headaches both for Council, the ESO management and the staff. Here's a somewhat sarcastic comment by Philippe Dierickx, not only a master optical engineer, but also a skilled cartoonist.

Chapter III-2

Aux Instruments, Chercheurs!

"The 3.6-metre at La Silla had started operation in 1978.
In the following 15 years — before the VLT —
starting almost from scratch ESO succeeded to established itself
as a self standing, high quality instrument builder."

Sandro D'Odorico, retirement lecture,
31 May 2010.

In Chapter II-3, we reviewed the steps in instrumentation development pre-VLT. This period was characterised by a growing understanding of instrumentation not only as an issue of equal importance to the construction of a new telescope, but also as something that should be addressed early on and in conjunction with the telescope developers. Importantly, ESO had itself become a centre for the development of modern instrumentation. The VLT would benefit both from the experience gained and the capacity of ESO to conceive and orchestrate an almost all-encompassing instrumentation programme. Thus, almost ten years before first light, the first VLT instrumentation plan was presented to the scientific community. The trend was to develop permanently installed multi-mode instruments suitable for remote control, capitalising on the revolution in electronics and allowing for dramatically increased efficiency. But the VLT, conceptually different from any other large telescope, also gave ESO the opportunity to take the next step in instrumentation. With no less than 12 individual foci available (excluding the common foci), frequent instrument changes at the telescope, as had been necessary on the 3.6-metre telescope, would no longer be needed, thus enhancing the efficiency and the performance of the system. It would now be possible to develop a suite of permanently mounted instruments, each of them optimised for specific tasks[1]. In this chapter, we will look at these instruments. For the lay reader, this means diving into a world of technical

[1] During the early discussions, placing duplicates of the main instruments at each of the Unit Telescopes had been considered. The idea was never realised as the gain was found to be marginal and the trade-offs considerable. However, the legacy of these early considerations was the decision to build two nearly identical focal reducers, the FORS1 and 2 instruments.

terms and concepts and a possibly bewildering set of acronyms. For the astronomer it simply means looking at what has arguably been the world's most comprehensive and coherent instrumentation programme in astronomy.

In late 1988, two panels were set up to discuss the VLT instrumentation. One group, chaired by Jacques Beckers, dealt with interferometry, the other — led by Bernard Fort — with instrumentation for the individual Unit Telescopes. At ESO, the instrumentation programme was led by Jacques Beckers (diffraction-limited and interferometric instrumentation), Sandro D'Odorico (optical instrumentation) and Alan Moorwood (infrared instrumentation) (D'Odorico *et al.*, 1991). Both D'Odorico and Moorwood were old hands at ESO. Moorwood had joined ESO in 1978 after working at the European Space Research and Technology Centre in Noordwijk, the Netherlands. D'Odorico had arrived at ESO just two years later from the Osservatorio Astronomico di Padova. Together they would play decisive roles in ESO's instrumentation programme for more than 20 years. In June 1989, the first ideas were presented to the scientific community in a document entitled ESO VLT Instrumentation Plan: Preliminary Proposals and Call for Responses. Those days were clearly infused by great optimism and excitement. And ambition[2]. The introductory text concluded: *"There is time to get our act together, there is room for consultation, there is money for bright ideas, there are potential rewards for bold initiatives. We invite your response."*

The call resulted in close to 40 replies. Following their evaluation, in March 1990, the revised plan was presented to the Scientific and Technical Committee, followed by a new call, now for proposals for specific instrumentation. If the VLT was an ambitious project, the instrumentation plan was no less so.

≈

The instrumentation plan foresaw a total of ten different instruments, of which initially two were to be built by ESO and eight by consortia of national institutes. The rationale for retaining instrument-building activities at ESO was obviously to exploit and maintain existing expertise, but also to establish the necessary standardisation aspects for the full instrumentation programme. The rationale for devolving the

[2] At the time the VLT was to offer no less than 18 foci. In addition to the Nasmyth and the (newly added) Cassegrain foci as well as the coherent and incoherent common foci, it was also thought that each Unit Telescope would have a coudé focus. The technical installations necessary for the coudé were in fact implemented, but in the end no instrument was developed.

programme, on the other hand, was to exploit unique expertise in the scientific community, to strengthen the overall involvement and thus support for the VLT and, of course, to lessen the burden on the organisation. In any event, the instruments contained in the plan constituted the first generation suite of instruments for Europe's new telescope. The parameter space covered was exciting: high-resolution imaging in the optical and infrared wavelengths, low- and high-resolution spectroscopy in the optical and infrared domains and multi-object spectroscopy, an area that with time would become significantly more attractive. As a start, in the optical region, four instruments were planned, each with a more or less well-chosen acronym: FORS1 and 2[3] (FOcal Reducer/low dispersion Spectrograph) and UVES1 and 2[4] (UV-Visual Echelle Spectrograph). In the infrared domain, the plans were for ISAAC (Infrared Spectrometer And Array Camera) and CONICA (Coudé Near-Infrared Camera). Further instruments foreseen included a multi-fibre area spectrograph[5], a visible speckle camera, and an imager/spectrometer for the 8–14 μm range, a high-resolution visible spectrograph for the combined focus, a multichannel Fourier transform spectrometer[6] and finally, a high-resolution infrared echelle spectrograph.

It seems fair to say that with the VLT instrumentation programme ESO broke new ground also in terms of project management and implementation. Back in the early days, we have seen that the national scientific communities built specific instruments for the first telescopes at La Silla. But the VLT instrumentation programme was on a different scale and its realisation was also based on the experience from space activities[7] — with clearly defined requirements and specifications, in-depth design reviews and very tight project control, all following a clear project implementation path. Everything was governed by firm contracts between ESO and its various consortium partners, including penalty clauses for late delivery or other cases of non-compliance. If this was new for ESO, it was even more so for the national institutes with their more academic traditions. But mostly they lived up to the challenge, not least because it was a win-win deal. ESO would cover the hardware costs, the institutes the manpower costs. In return, the institutes would receive guaranteed observing time, arguably the most coveted commodity among scientists. Aside from the

[3] FORS1 and 2 were quite similar, but not identical. Orginally, FORS1 allowed for polarimetry, FORS2 not.

[4] UVES 2 was never built.

[5] MFAS, later known as FUEGOS.

[6] Later abandoned in favour of CRyogenic high-resolution InfraRed Echelle Spectrograph (CRIRES).

[7] The space experience that proved so invaluable was introduced forcefully by Joachim Becker, the VLT Project Manager, who came to ESO from the space industry.

immediate benefits it brought to ESO itself and indeed to the individual scientists in the Member States, ESO's decision to devolve its instrumentation programme contributed greatly to establishing a network of excellence that involved its user community fully in the project. In addition to the importance of harnessing expertise wherever it might be in Europe, came the strong feeling of ownership that is felt by the scientific community.

≈

The first science instrument to be mounted on the VLT was FORS, the Focal Reducer/Low dispersion Spectrograph, placed at the Cassegrain focus of the first Unit Telescope. FORS was developed under a contract between ESO and the Universitäts-Sternwarte München, the Universitäts-Sternwarte Göttingen and the Landessternwarte Heidelberg, with Immo Appenzeller as Principal Investigator[8] and Gero Rupprecht as the person at ESO responsible for the instrument. The contract was signed on 6 February 1992, together with a contract for the second external instrument foreseen, CONICA, an infrared camera and spectrometer built by a different consortium.

FORS was designed as an all-round instrument, but clearly aimed at extragalactic research projects, including the measurement of redshifts of distant galaxy clusters, counts of galaxies, the determination of their colours and shapes and the study of gravitational lenses.

On 15 September 1998, FORS1 was ready for action at Unit Telescope 1 (UT1). The very first technical image was stunningly sharp with a resolution of 0.6 arcseconds. *"Not bad,"* was the dry comment by Walter Seifert, the optical engineer. *"Let's see if we can do better!"* They had hit the focus, but in a sense they could do better still. For the formal first light image, the team had chosen a spiral galaxy, NGC 1232. This target was mainly picked because it would fill the entire field of view of 6.8 × 6.8 arcminutes, but it is also a truly beautiful sight. And the FORS image certainly did justice to that sight. In the year 2000, the popular US astronomy magazine *Sky and Telescope* included it among the ten space pictures with the highest impact in the 20th century. The picture became an icon of the VLT; it was used repeatedly and appeared everywhere, in books, magazines, on CD covers, on giant murals and in numerous exhibitions. It even featured as a giant back-lit reproduction on the ceiling of the

[8] With Rolf Kudritzki (Munich) and Klaus Fricke (Göttingen) as Co-Investigators.

The famous FORS image of NGC 1232, a beautiful spiral galaxy at a distance of 120 million light-years.

European Union pavilion at EXPO 2000, the World Exhibition in Hanover. Since their first days (or perhaps one should rather say nights) the two FORS instruments have been the workhorses of the VLT. This becomes evident from the publication statistics, often used to assess scientific productivity. Celebrating 20 years of FORS science operations in 2010 (i.e. roughly ten years for each instrument), Gero Rupprecht noted that FORS data had played a role in eight out of the ten most highly cited VLT science papers. Perhaps more impressively, its publication rate of one paper per observing night has made FORS the most publication-efficient instrument on the VLT ever, with UVES hard on its heels (Rupprecht *et al.*, 2010). During construction things had gone well and there was a remarkably smooth cooperation between the consortium partners. But this is not to say that there were no problems along the way. During the test phase in Europe, the collimator unit fell from a crane and was destroyed. Luckily, since FORS2 was under construction, the team simply exchanged the part and thus no delay resulted. During the first installation at Paranal, the instrument suffered from water damage. To the uninitiated, this may sound rather strange given that FORS was after all placed in the middle of the world's driest desert. It appeared that the source was coolant from a leaking hose linked to the VLT guide camera controller, in front of the instrument. The commissioning team was left with little choice but to dismantle the entire instrument and carefully clean every surface. But on the very next night, it happened again. In such cases, perhaps only dark humour can brighten the day. Jokes were cracked about FORS as Paranal's Yellow Submarine, referring to the yellow paint on its casing. More effective than the jokes, though, were different hoses that have been used ever since.

≈

The two instruments to be built by ESO were ISAAC and UVES. They would be available early on Unit Telescopes 1 and 2, respectively, and they too were unique instruments with a huge science potential. ISAAC, the infrared imaging spectrograph, provides for imaging and spectroscopy in the 1–5 µm wavelength domain as well as polarimetry in the non-thermal 1–2.5 µm domain.

ISAAC was the first VLT instrument developed by ESO and, aside from its specific characteristics, it is a testament to the progress achieved in instrumentation for infrared observations. As we remember, infrared instrumentation had from the outset been separate from instruments designed for visual wavelengths. This was partly because of the slower progress in the development of two-dimensional panoramic array detectors for this particular wavelength domain. But much had happened since the days, when infrared work was regarded as "a dark art" (Moorwood, 2009). ISAAC was the first instrument with a two-dimensional detector array in the near-infrared, making it essentially a near-infrared EMMI. As such it is an all-round instrument, suited for observing just about any kind of object in the infrared Universe, from star-forming regions in the Milky Way and brown dwarf stars to distant clusters of galaxies. ISAAC was mounted at the Nasmyth focus of the first Unit Telescope in November 1998.

As mentioned, the second ESO instrument built in-house was UVES[9], the two channel high-resolution UV-Visual Echelle Spectrograph, led by Hans Dekker. UVES was designed to be a direct competitor to HIRES, a spectrograph at the 10-metre Keck

[9] Software was supplied by the Osservatorio Astronomico di Trieste.

A look into the interior of ISAAC: inside the vacuum vessel. (Photo: Hans Hermann Heyer)

Telescope on Hawaii. Since this had been in operation for several years, the Keck astronomers had gained a considerable advantage and ESO would have to work both hard and fast to produce an instrument that could outperform HIRES.

Jean-Louis Lizon enjoys a relaxed moment during the installation of UVES. (Photo: Peter Gray)

With first light on 27 September 1999, UVES became the first instrument to be installed at Unit Telescope 2, Kueyen. Weighing no less than 8 tonnes, UVES was conceived for placement at the Nasmyth focus with its large, stable platform at the side of the telescope main structure. Like FORS, the early plans for UVES foresaw two instruments (Dekker & D'Odorico, 1992) with slightly different characteristics. However, in 1994, the idea was dropped. Instead, it was decided to build a single instrument with two arms (covering UV to blue, and visual to red light, respectively). The heart of this elaborate instrument was a mosaic of two optical echelle gratings, 84 cm × 21 cm × 12 cm in size — at the time the largest in the world. The two-arm solution meant that the performance both in the UV and the red could be optimised, providing exactly the desired competitive edge *vis-à-vis* its Hawaiian counterpart[10].

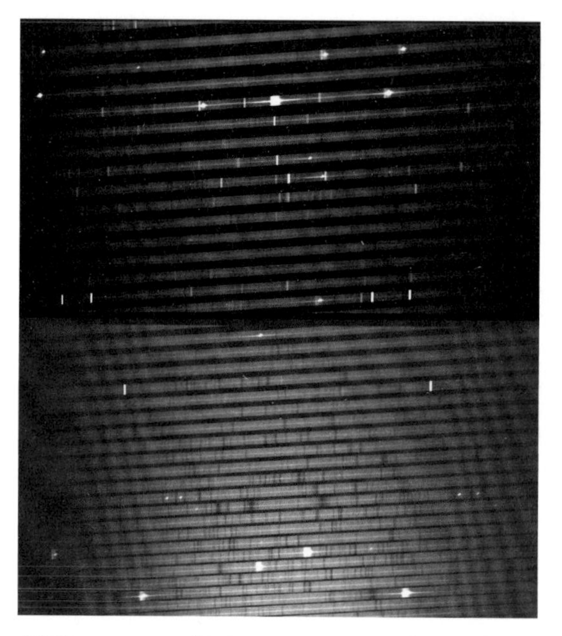

UVES spectrum of SN 1987A, covering the wavelength range 479–682 nm. The spectrum is divided into many individual parts, here seen as "steps", which are characteristic of an echelle spectrograph. The spectrum reveals both information about the chemical composition and state of the matter from the supernova.

The combination of very high resolution (potentially up to 110 000) and the light grasp of the 8.2-metre Unit Telescope made UVES ideal for studying the presence and composition of intergalactic gas (using distant quasars as background light sources), as well as investigating the physical conditions and element abundances in the faintest stars in the Milky Way galaxy and beyond. It is no wonder that UVES became a heavily sought-after instrument and has played such an important role in many of the discoveries made with the VLT. Indeed over the years, it has maintained its position as one of the top five VLT instruments in terms of

[10] UVES owes its success to much more than just the two-arm concept. Other features of the instrument include the field rotator and the atmospheric dispersion corrector. The end-to-end observing concept and its spin off, the centralised open access science archive with calibrated data also constitute essential elements of the story.

popularity (quantified, perhaps a bit negatively, in terms of oversubscription) and in the publication statistics.

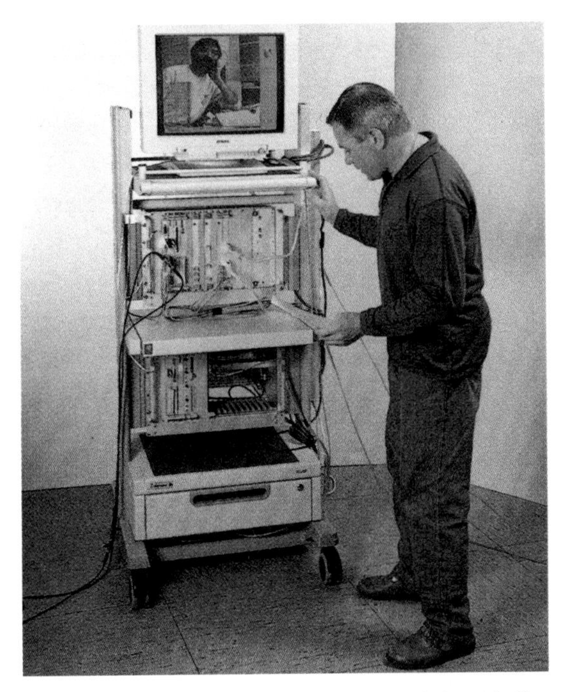

In the course of the 1990s, ESO undertook a dedicated effort to boost its activities in the field of detectors within its two detector teams, one for optical and one for infrared detectors. Among other things, the groups developed new and highly efficient detector controllers, FIERA and IRACE that for more than a decade became the standard for all instruments at ESO's telescopes. The photo, from 1996, shows Manfred Meyer from the infrared detector team with an IRACE controller.

Another of the instruments planned for early deployment was CONICA, built by the Max-Planck-Institut für Astronomie (Heidelberg), the Max-Planck-Institut für extraterrestrische Physik (Garching) and the Osservatorio Astronomico di Torino under an ESO contract. CONICA stands for Coudé Near-Infrared Camera, but due to programme changes linked to the financial problems that hit ESO in 1993, the individual coudé foci were abandoned and it was instead decided to move CONICA to a Nasmyth focus and fit it with an adaptive optics unit, renaming it NAOS-CONICA to which we shall return in the following chapter. The combined instrument only saw first light on 25 November 2001, at Unit Telescope 4, Yepun. The instrument may have been delayed, but it became very successful: over the first decade of operations it provided the basis for about 240 scientific papers (Fusco & Rousset, 2011), including spectacular images of the Galactic Centre and the first direct image of the brown dwarf 2M1207 with its giant planet companion (see also Chapter IV-5).

The science instruments at the VLT had changed considerably in comparison with the specifications originally set in the earlier days of the VLT planning. Not only had their performance improved, but their complexity had also increased dramatically.

The sheer physical dimensions of instruments had also grown, as well as their construction times. For some instruments, the lead time approached ten years and development happened in parallel with the telescope construction itself. Involving national institutes entailed challenges in terms of project management to secure delivery on time, within specifications and budget. So it comes as no surprise that the instrumentation programme underwent changes and modifications along the way — from the 1989 document to the Instrumentation Plan itself and then through the period of realisation.

The last instrument of the first batch was FUEGOS, a multi-object fibre spectrograph and one of the external instruments to be built by the Observatoire de Paris-Meudon, the Observatoire de Genève, the Laboratoire d'Astrophysique de Toulouse and the Osservatorio Astronomico di Bologna. Progress towards its realisation was slow and in 1997 ESO saw itself forced to cancel the original contract and involve additional partners to build a successor instrument, with ESO in charge of the project management. The two main components were the fibre optics and the mechanical part of the spectrograph. The contract for the fibre optics part was awarded to what became known under the name of AUSTRALIS, a consortium of institutes in Australia[11], and given the appropriate nickname of OzPoz, whereas the mechanical part, GIRAFFE, remained in the hands of the Observatoire de Paris-Meudon, but ultimately ten national institutes contributed to this instrument[12]. The final instrument, which was renamed FLAMES, would also feed the high-resolution spectrograph UVES. FLAMES was installed on the Nasmyth platform of Unit Telescope 2 between September 2001 and March 2002 (Pasquini *et al.*, 2002). An important feature of FLAMES was its large field of view, no less than 25 arcminutes in diameter. Combined with the fibre positioner and GIRAFFE, it allows up to 132 objects to be observed at the same time. As a novelty, FLAMES also contained a so-called integral field unit. We have discussed the unique role of spectroscopy for our understanding of the celestial objects that we observe and this chapter has illustrated how important this is to astrophysicists. Normally, astronomers will obtain spectra of an individual object, say a star or a galaxy, or of a number of objects using a multi-object spectrograph. Integral field spectroscopy marks the next step in sophistication, as it allows the simultaneous spectral analysis of multiple parts of extended objects, or for

---

[11] The institutes involved were the Anglo-Australian Observatory, the Australian National University and the University of New South Wales. The prospect of involving institutes in Australia came about in connection with Australia's bid to join ESO in 1995.

[12] Aside from the instituces already mentioned, the observatories of Geneva and Lausanne, Cagliari, Palermo and Trieste were involved.

example, of multiple objects in crowded fields such as a globular cluster of stars. The ability to study individual objects (or areas) with a minimum of angular separation is of course the outcome of much improved telescope performance, with larger and better telescope optics and, decisively, the application of adaptive optics.

Bernard Delabre and Bernard Buzzoni during the integration of the Nasmyth field corrector for FLAMES in September 2001. (Photo: Peter Gray)

≈

In 1993, a working group chaired by Laurent Vigroux and comprised of ESO staff and external experts reviewed and revised the instrumentation plan. UVES 2 was cancelled. On the other hand new instruments were proposed and discussed at a conference in Garching in May 1994. The new instruments that formed this second wave of the first generation instruments were VISIR, VIMOS and NIRMOS. In the same year, ESO decided to create a dedicated Instrument Division, from January 1995 headed by Guy Monnet. Before Monnet had been Director of the Canada France Hawaii Telescope. Together with D'Odorico and Moorwood, he would oversee the evolving instrumentation programme up to 2006.

Next in line, as regards ESO contracts, was VISIR (the VLT Imager and Spectrometer for the mid InfraRed). This instrument was designed to cover the 10–20 μm spectral domain, with the aim of studying, among other things, the interstellar medium, cool stars, and circumstellar and protoplanetary discs. VISIR would be the most powerful instrument of its kind in the world by far. The contract was signed in November 1996 with a French–Dutch consortium of institutes[13], with expected completion in early 2001. Realising the instrument turned out to be a bigger challenge than expected. In particular the wish to have maximum efficiency in the 20 μm wavelength range, meant that the instrument had to be cooled to 25 K (-248 degrees Celsius), not a trivial requirement for a complex opto-electronic and opto-mechanical piece of hardware. Non-instrument related problems, such as a fire and the subsequent flooding of the CEA building in Sarclay that housed the laboratory, added to the problems and so the development of the instrument was anything but quick. VISIR was shipped to Chile in March 2004. Upon arrival, it barely avoided a disaster. ESO's normal procedure is to subject new instruments to thorough testing in Europe and, upon successful completion of the tests, dismantle the instrument, pack and ship it to South America. However, to save time, VISIR was not dismantled and thus arrived in Chile in one piece and in a single box[14]. During customs clearance, however, an alert SAG inspector[15] noticed a wasp peering out of a hole in the wooden crate. Chile is understandably keen to protect its agriculture and enforces its vetenary border controls accordingly. The inspector therefore decided that the crate — with contents — should be destroyed. Luckily, Armin Silber from the ESO Instrumentation Division, was present to receive the consignment and he managed to convince the inspector that it would suffice to spray the outside of the crate with insecticide, arguing that the crate would be taken into the desert and any surviving bugs inside would perish there. And so, on 1 May 2004, VISIR saw first light (Lagage *et al.*, 2004), almost eight years after the signing of the contract. The scientists were happy and so was a SAG inspector who had travelled to Paranal to personally witness the burning of the crate — with or without any surviving bugs.

[13] The Service d'Astrophysique of the CEA, Saclay, the Netherlands Foundation for Research in Astronomy (NFRA), Dwingeloo, the Institut d'Astrophysique Spatiale (Orsay, France) and the Netherlands Foundation for Space Research (SRON).

[14] This was not the only instrument that arrived in one box. TIMMI (see Chapter II-3), simply occupied an economy-class airplane seat, with a ticket issued in the name of Mr TIMMI. Alas, in the meantime instuments had not only grown in capability but also in size and weight. Thus the total weight of the VISIR instrument amounts to no less than 2300 kg.

[15] Servicio Agrícola y Ganadero (Agricultural and Livestock Service).

Another instrument that went through a difficult gestation was VIMOS (VIsible MultiObject Spectrograph). The contract was signed in August 1997 between ESO and the VIRMOS consortium[16] and foresaw two instruments, VIMOS and NIRMOS, for the visible and infrared spectral bands, respectively. VIMOS is a multi-object instrument like FLAMES and can observe many objects (in fact, thousands) at the same time. The prime purpose for VIMOS and its near-infrared sister NIRMOS would be the simultaneous observations of a large sample of galaxies at different redshifts, i.e. cosmic epochs. As such the twin instruments were expected to become powerful tools for cosmological studies. VIMOS was installed at Unit Telescope 3, Melipal, in early 2002. Unfortunately it had a number of problems; the most conspicuous being that it exceeded its weight specifications by a third, or one tonne, which introduced serious problems of flexure and exceeded the allowed torque on the telescope. Further mechanical problems appeared and it took a long time before the instrument achieved its full performance. Yet, aside from its scientific importance, this instrument has also played an important role in the technical evolution process of advanced scientific instrumentation.

The problems with VIMOS, however, led to the decision to cancel its infrared sister, NIRMOS. Instead, the detectors acquired for NIRMOS found a good home in a new instrument, which, at least with respect to imaging, would act as a replacement, and with much improved performance — the High Acuity Wide field *K*-band Imager (HAWK-I).

To complete the overview of what finally constituted the first generation VLT instruments, we need to mention SINFONI (Spectrograph for INtegral Field Observations in the Near Infrared) and finally CRIRES.

Like NAOS–CONICA, SINFONI[17] combined a Multi Application Curvature Adaptive Optics (MACAO) module with a Spectrometer for Infrared Faint Field Imaging (SPIFFI) (Bonnet *et al.*, 2004). SINFONI was the second VLT instrument designed for integral field spectroscopy and the first for infrared work. It saw

[16] Laboratoire d'Astronomie Spatiale Marseille, Observatoire Midi-Pyrénées (Toulouse) and Observatoire de Haute-Provence in France, and, in Italy, the Istituto di Fisica Cosmica del CNR, Osservatorio Astronomico di Brera, Osservatorio Astronomico di Capodimonte, Istituto di Radioastronomia del CNR and Osservatorio Astronomico di Bologna.

[17] The SINFONI consortium comprised the Max-Planck-Institut für extraterrestrische Physik, the Department of Physics, University of California, Berkeley (US), Universität zu Köln, Max-Planck-Institut für Astronomie, Heidelberg, University of Oxford, Ingenieurbüro Weisz, Munich, Leiden Observatory, NOVA and the Netherlands Foundation for Research in Astronomy.

first light on 31 May 2004 on Unit Telescope 4, Yepun. In 2006, it demonstrated its awesome capabilities by enabling observations with a resolution of a stunning 0.15 arcseconds across a star-forming galaxy, BzK155043, at a distance of 11 billion light-years (Genzel, 2006). In February 2006, SINFONI was used for the first time with the laser guide star system on Yepun (see next chapter).

The last of the first generation VLT instruments, CRIRES, was developed as an ESO in-house project. Although it had been on the wish list since the earliest days of the VLT instrumentation plan, due to lack of resources it only started as a project in 1999 (Moorwood, 2003). Hans Ulrich Käufl was responsible for the project. CRIRES is a high-resolution infrared spectrograph with a resolving power of about 100 000, covering the spectral range from 1–5 μm. As the detector it uses a mosaic of Raytheon 1024 × 1024 pixel InSb CCDs, providing a focal plane array of 4096 × 512 pixels — reminding us once more of the incredible progress in the field of infrared area detectors over the last 15 years. The instrument saw first light in May 2006 on Unit Telescope 1, Antu. It was — and still is — a unique instrument. In terms of science it is particularly well suited to study a range of objects from the atmospheres of Solar System bodies, the chemistry and kinematics of the interstellar medium to the dynamics of the accretion discs around black holes. To some extent it can be seen as an infrared UVES, but the combination of nearly diffraction-limited spatial resolution (thanks to adaptive optics) and the high spectral resolution makes it extremely powerful, superior to any other similar instrument in the world.

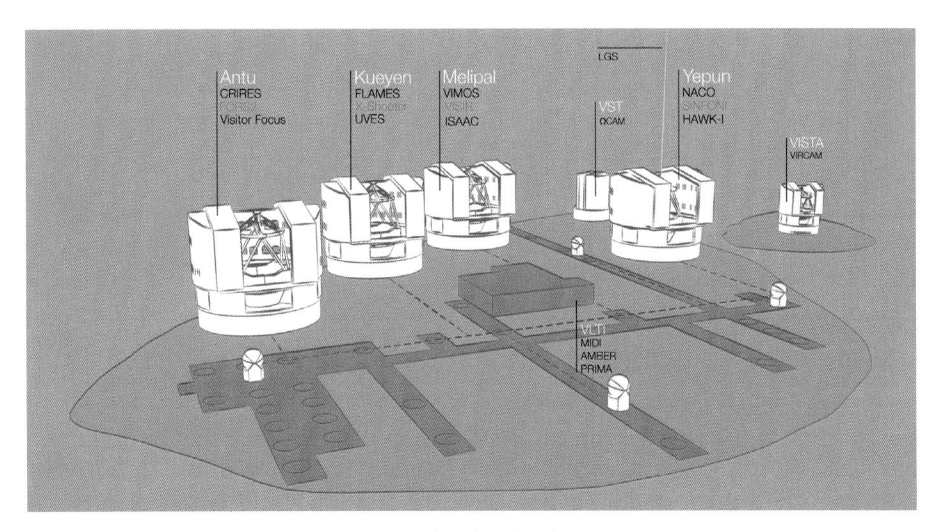

Overview of the VLT instruments by the end of the first decade.

Before we round up we must briefly return to HAWK-I. This is a high-resolution infrared (0.85–2.5 μm) camera with a field of view of 7.5 × 7.5 arcminutes. With a pixel size of 0.1 arcseconds it is able to exploit even the best seeing conditions at Paranal. As an imager, it partly replaces the cancelled NIRMOS instrument, though of course without the spectroscopic capabilities. HAWK-I was commissioned in 2007, seeing first light in August of that year. HAWK-I was chosen following the first call for proposals for second generation VLT instruments in November 2001. However, since it uses parts originally intended for an earlier instrument, it is often considered to be a generation 1.5 instrument.

≈

VLT instrumentation now moved towards the second generation to be implemented during the second decade of the century. The formidable point-and-shoot Cassegrain focus X-shooter — a single-object spectrograph covering the whole spectral range from the UV to the near-infrared *in a single exposure* — would be the first newcomer.

With a large fraction of first generation instruments in operation or about to be installed at the telescope, the process of identifying the first concepts of the second generation was initiated in 2001. ESO adopted the same approach of fully involving the user community through conferences on the scientific priorities, the first of which was held in June of that year, followed by feasibility studies and competitive calls for proposals. The advances in the scientific knowledge and understanding of the Universe, from the planets associated with stars in the relative vicinity of the Solar System to the primeval galaxies at high redshift, coupled with the technical development, especially in the area of the near-infrared detectors and adaptive optics, led to the identification of four new instrument concepts of highest priority, which were the subject of a call in November 2001. Following extended feasibility studies, four new instrument contracts were established with consortia of external institutes: the *K*-band Multi-Object Spectrograph (KMOS), X-shooter, the Multi Unit Spectroscopic Explorer (MUSE) and Spectro-Polarimetric High-contrast Exoplanet REsearch (SPHERE). ESO would contribute in many key areas like detector systems and AO and, in the case, of X-shooter, coordinating the full project including the final integration. As for the first generation instruments, each would constitute a major step forward in its scientific capability, but at the same time take its place in a coherent pattern of instruments with complementary and mutually enhancing contributions to astronomy.

KMOS, a near-infrared, multi-object spectrograph, will enable studies of a large sample of extremely distant galaxies detected with imagers such as the VISTA telescope or HAWK-I.

As already mentioned, X-shooter is a unique facility that allows *simultaneous* observations covering the spectral range from the UV to the near-infrared, in a "point-and-shoot" mode. It aims to provide comprehensive "instant" information about objects such as gamma-ray bursts, QSOs or X-ray binary stars, where time is a key factor in making successful observations. Mounted at the Cassegrain focus of UT 2, X-shooter was the first second generation instrument to go onto the VLT[18, 19].

Next in line is MUSE, an integral field spectrograph for the visual. It has a field of 1 × 1 arcminutes — significantly larger than that of VIMOS — and able to yield a resolution of 0.2 arcseconds. This instrument will rely on the full use of adaptive optics with four laser guide stars. SPHERE will be used for exoplanet studies, with extreme adaptive optics and all the tricks of the trade to characterise these objects and, if possible, search for biomarkers in their atmospheres. It is therefore the natural successor to NAOS–CONICA. However breathtaking the performance of these new instruments may be, it is beyond the scope of this book to deal with them, but in passing the author cannot but mention one additional, new instrument, ESPRESSO. This instrument is an ultra-stable[20] echelle spectrograph, designed for deployment at the incoherent, common focus of the VLT. With ESPRESSO, the VLT will finally (also) become what it was originally conceived to be — a 16-metre equivalent telescope, as it looks at the time of writing, around 30 years after the VLT project was approved.

But as impressive as the first generation instrument suite was, we still have to mention those designed for the common, coherent focus — i.e. for the VLT Interferometer. We shall look at these in Chapter III-15. In any event, the efforts and ingenuity going into scientific instrumentation have increased tremendously. So has, of course, the scientific output, but for that to happen, we need to look at an area that is of critical importance to ground-based astronomy. It is, in fact, decisive.

---

[18] X-shooter was commissioned in 2008 and made available to scientific users in 2009.

[19] The consortium behind X-shooter comprised the Niels Bohr Institute and the DARK Cosmology Centre of the University of Copenhagen, and the National Space Institute, the Observatoire de Paris and the Université Paris Diderot, with contributions from the CEA and the CNRS, the Observatories of Brera, Trieste, Palermo and Catania; the University of Amsterdam (Universiteit van Amsterdam), the Radboud University Nijmegen (Radboud Universiteit Nijmegen) and ASTRON.

[20] Expected to be able to measure radial velocity variations in the range of a few cm/s.

Chapter III-3

Breaking the Seeing Barrier

"An old dream of ground-based astronomers has finally come true."

Merkle *et al.*, 1989.

Astronomy is a constant struggle for photons. Astronomers build larger telescopes and improve their detector systems to gather more light. But increasing the size of the light-collecting surface, or rather the pupil, in theory also leads to better resolution. The pupil of the human eye will, when dark-adapted, achieve a diameter of roughly 8 millimetres. This enables us, in principle, to distinguish objects with an angular separation of 15 arcseconds (although physiological limitations of the eye normally lead to a much poorer value). A telescope with a diameter of 10 centimetres will allow us to distinguish objects with a smaller separation, roughly 1.2 arcseconds. This is the case for visible light, but the resolution that can be obtained with an optical system is also dependent on the wavelength of the radiation and it is important to remember the interplay between those two factors. But, for all practical purposes in astronomy, another factor is usually even more important: the terrestrial atmosphere. When we look at the stars from the ground, we see the starlight as it is after it has passed through the atmosphere. Since atmospheric turbulence has the same effect as an optical lens, and the atmosphere changes almost constantly, this leads to a continuously varying distortion of the wavefront of the light. This in turn degrades the image quality that can be obtained with the telescope. The sad outcome is that while larger telescopes will collect more light, and thus allow us to see fainter objects, telescopes beyond 50–60 centimetres in diameter will not provide us with better resolution, even if in theory they should do so[1] — unless one can do something about it. One of the first scientists to ponder this problem was the American astronomer Horace Babcock, who presented a paper in 1953 with preliminary thoughts on "The Possibility of Compensating Astronomical Seeing" (Babcock, 1953). At the time, no real practical solution was in sight,

[1] This applies to observations in the near-infrared. With shorter wavelengths, the maxium resolution is reached already with telescopes of about 15–20 cm.

but serious research began in the US in the 1970s, although not in astronomy: the first practical use was in satellite-tracking systems. Later, research in what had now been termed adaptive optics continued in the US in the framework of the Strategic Defence Initiative, popularly known as the Star Wars Programme, driven forward by the US President Ronald Reagan. For this reason, the development of adaptive optics was considered as classified information and could not be made available to astronomy. Still, some adaptive optics activities also began in astronomy, e.g., at the Kitt Peak National Observatory, where Jacques Beckers was working on the problem.

≈

In March 1981, ESO organised a conference on the Scientific Importance of High Angular Resolution at Infrared and Optical Wavelengths. We have mentioned this conference in Chapter II-1 in the context of the early VLT activities. Among the participants was John W. Hardy, who had been involved in classified US military research projects since the early 1970s, and therefore was not at liberty to reveal any details[2]. The French military was also undertaking work in this area, but this too was unknown to the participants of the conference. However, astronomers had clearly become active in the field, sensing the scientific gains that could be achieved by mitigating the effects of atmospheric turbulence. One of the participants, Fritz Merkle, had studied a system with electrostatic actuators. Another was Pierre Léna, a great promoter of interferometry (and, in general, high-resolution imaging), who realised that to make interferometry work the atmospheric distortion had to be overcome. It was one of those crucial moments that can happen at conferences, where the right people come together at the right moment. Woltjer realised the potential of this: perhaps this was the way towards the large optical interferometer that some were dreaming of in connection with the VLT. Later adaptive optics would be seen, not simply as the way towards the interferometer, but as a crucial technology in its own right, enabling a full exploitation of the powerful new telescopes — both to collect more light *and* to obtain sharper images.

As part of the preparations for the VLT, Pierre Léna led the interferometry study and this work was presented at the 1986 Venice conference. Ahead of the conference, on 21 November 1985, the group held a workshop at ESO in Garching. Several questions were still open and opinions varied widely. Within the global interferometry community some harboured doubts about whether it would make any sense at all

[2] Years later he described some of these activities in an article in *Scientific American* (Hardy, 1994). A fuller account of the efforts by the US military is given in Duffner (2009).

to use large telescopes as interferometers. Speckle imaging, on the other hand, no longer seemed to be the way forward and the option of adaptive optics, although at this time identified theoretically, had not been tested (or more precisely, used for civil rather than military applications). The issue was obviously of a technical and a scientific character, but it had a political side as well. Achieving consensus was important to secure the backing of the community for the VLT project. In this situation Léna, backed by Jean-Claude Fontanella (from ONERA[3]) and Michel Gaillard (CGE[4] research director), proposed that a prototype device be built to demonstrate the feasibility of adaptive optics.

The proposal was presented at the VLT meeting in Venice, leading to a contract with the Observatoire de Paris, ONERA and LASERDOT, the CGE successor company. The result became known as the VLT Adaptive Optics Prototype (though colloquially called Come-On[5]), which was first tested at the Observatoire de Haute-Provence during the nights of 12–23 October 1989, with a follow-up observing run in November of the same year. As so often is the case, the idea was stunningly simple, at least in theory: to carry out a real-time analysis of the image distortion of a star, using a Shack-Hartmann system of lenslets, just like in active optics. The lenslets created a series of individual star images on the detector, each different from each other. The difference, or to be more precise the relative displacement, of the individual star images from the theoretical location, would provide the necessary information for analysing the instantaneous distortion. This in turn would be used to control the shape of a small flexible mirror, inserted in the optical train and supported by piezoelectric actuators. Changing the shape of the mirror meant deforming it in a way that neutralised the original distortion of the wavefront of the light as it finally arrived at the detector after its bumpy ride through the terrestrial atmosphere and the telescope's surroundings. Put simply, the telescope would yield the same optical performance as if it were in space. Obviously, the correction would have to be conducted in real time, in the case of the prototype meaning about 100 times per second (100 Hz).

By today's standards the AO prototype was rather simple, with only 19 actuators[6], but it nonetheless provided full proof of the concept, delivering diffraction-limited

---

[3] Office National D'Etudes et de Recherche Aerospatiale.

[4] Compagnie Générale d'Électricité.

[5] Reminding us of the producers – CGE, the Observatoire de Meudon, and ONERA.

[6] NAOS-CONICA, the first adaptive optics instrument on the VLT used 185 actuators operating at 500 Hz. The next generation of extremely large telescope AO systems will have many thousands of actuators and work at frequencies beyond 1 kHz.

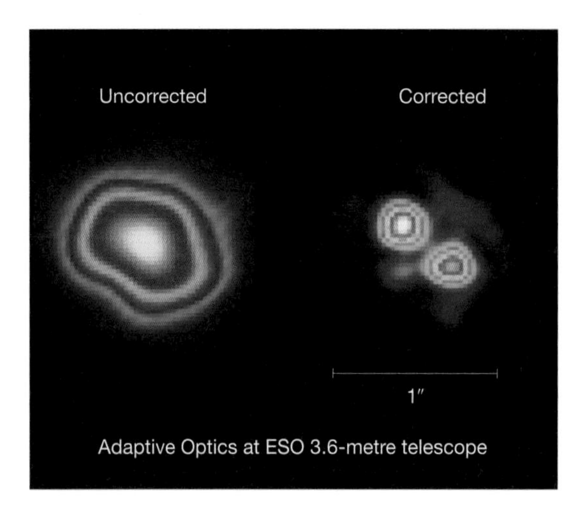

An early Come-On result at the ESO 3.6-metre telescope: A 5.5 magnitude star (HR 6658 located in NGC 6475), here observed at 3.5 μm. The left image shows a normal observation (without adaptive optics). The resolution is 0.8 arcseconds. The right image, with adaptive optics, features a dramatic increase in resolution to 0.22 arcseconds and reveals that the star is, in fact, a double star.

images[7] in the infrared wavelength domain (from 2.2 μm and beyond), and with noticeable improvements in resolution for shorter wavelengths. However, as simple as the idea was, realising it meant tackling a number of challenges. Firstly, the optical correction could only be applied to a small field around the optical axis, the so-called isoplanatic patch, typically only covering a few tens of arcseconds in the infrared. This was simply because light from stars located further afield on the celestial sphere would pass through different cells of air and thus needed to be corrected differently. Secondly, the adverse optical effects of the atmosphere are less pronounced in the infrared compared to the visible band and are thus easier to correct[8]. Thirdly, as we know the resolving power of the telescope is a function not only of the pupil size but also the wavelength, and this meant that AO would potentially yield much more impressive results in the visible, but that the system would require significantly more actuators and much faster correction. That in turn required dramatically increased computing power. Finally, the system needed a bright star to allow the image analyser to do its work. Since bright stars are not found everywhere in the sky — and could not be assumed to reside close to observational targets — much more sensitive detectors would be needed. But, despite these difficulties, a very promising start had been made, more than justifying the opening of a couple of champagne bottles at the ESO Headquarters. It was a real breakthrough. The ESO press release clearly found the right words: “Catching a Twinkling Star: Successful Tests of Adaptive Optics Herald New Era”[9]. It is interesting that within a period of only half a year,

[7] Meaning that the images reflected the maximum theoretical performance of the telescope and that the atmospheric disturbance had been eliminated.

[8] A further limitation was the rudimentary state of infrared detectors at the time.

[9] ESO press release, 26 October 1989

two technologies, albeit somewhat related, that would change telescope design forever and pave the way for giant advances in astronomy, became available. The first, servo-controlled active optics was, as already mentioned, invented at ESO, the other, adaptive optics, was first demonstrated in astronomy by a joint ESO/French team. If 1987 had been an *annus mirabilis* for ESO, 1989 was an *annus mirabilis* for the world of astronomy. And no less so for ESO. *Nature* expressed it this way: *"The gamble taken by ESO in 1985 when it decided to rely on adaptive optics for the VLT seems to have paid off."* (Dickman, 1989).

The following year, between 11 and 16 April, Come-On was mounted at the Cassegrain focus of the 3.6-metre telescope at La Silla. The original wavefront sensor used a Reticon array detector, able to work with a tenth magnitude star, and the camera was a 32 × 32 pixel infrared (IR) camera.

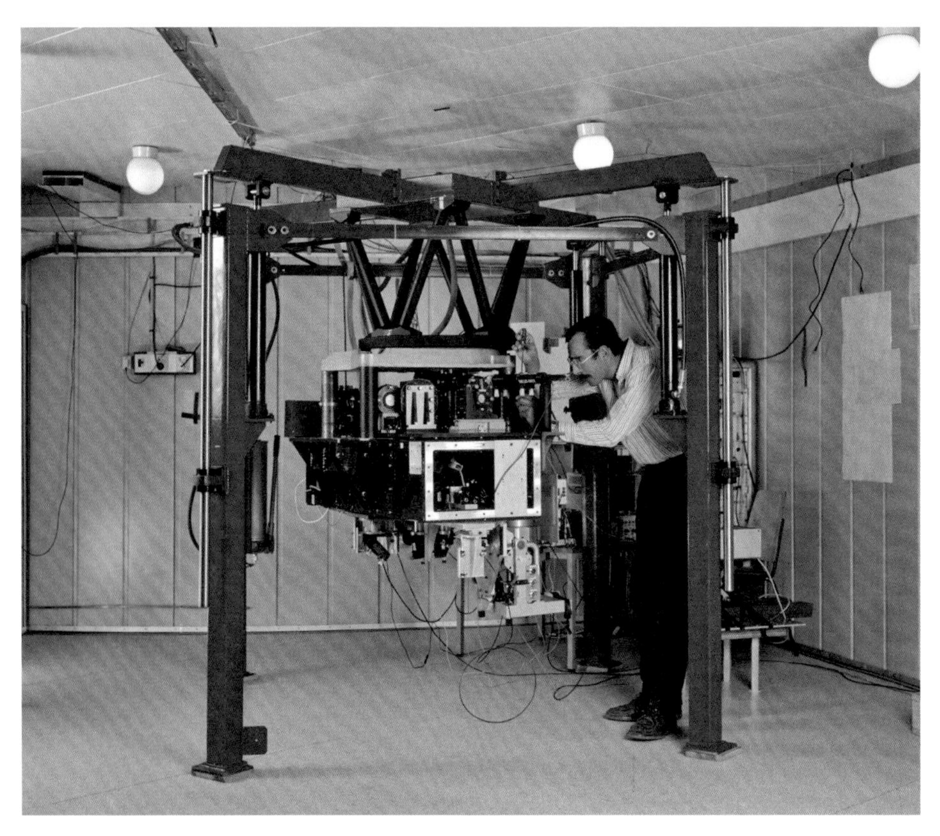

Michel Faucherre preparing Come-On+ for observations with the 3.6-metre telescope. (Photo: Herbert Zodet)

A second run at the 3.6-metre was carried out between 31 December 1990 and 8 January 1991 this time with a coronagraph[10] designed by the Observatoire de Paris-Meudon, and an improved detector as wavefront sensor, now allowing the use of 13th magnitude guide stars (Gehrig & Rigaut, 1991). Further observing runs followed, but in December 1992, the instrument was succeeded by Come-On+, now with 52 actuators and an electron-bombarded CCD as a photon-counting detector (sensitive enough to work with 15th magnitude stars) for the wavefront sensor, working up to 200 Hz — twice the speed of the first prototype. These early attempts, Come-On and Come-On+, provided the proof that it was possible to deliver the AO performance expected for the VLT, but as prototypes, they were not really scientist-friendly. Thus an AO run required a specialist team to be present at the telescope. ESO's philosophy, however, was standardisation and ease of use, allowing even the most complex systems to be used by scientists without any specialist knowledge. The term plug-and-play might be an exaggeration, but it would indicate the thrust of ESO's intentions for its users. Even so, impressive scientific observations were obtained with both prototypes, including images of the circumstellar disc around the star Beta Pictoris (see also Chapter III-7). The next step was to develop a user-friendly upgrade under the name of ADONIS (Beuzit & Hubin, 1993). Developed by ESO and the Observatoire de Paris, ADONIS featured improved control software and was be fitted with the 256 × 256 pixel SHARP II camera, developed at the Max-Planck-Institut für Extraterrestrische Physik, and also benefited from much improved mechanics. *"It worked!"* as Jean-Luc Beuzit, the person responsible for the instrument, later stated (Beuzit, 2009)[11]. *"Since 1996, the telescope operator has also been operating ADONIS!"*

With the successful demonstration by Come-On and Come-On+, it would appear that the time had come to address the AO requirements of the VLT. AO systems were foreseen at the coudé foci of each Unit Telescope, mainly to support the VLTI. In 1993 MATRA had begun a design study for a coudé AO system for the VLT[12]. Alas, the timing was far from optimal. As mentioned, in 1993 the financial difficulties had led to a postponement of the VLTI, but it also affected the AO coudé systems. As a result, CONICA, one of the VLT instruments that was already under development, was relocated to the Nasmyth focus. These changes ultimately led to a long delay, but eventually the contract for the AO system intended for CONICA

[10] The reason for the coronograph was that a main target was the circumstellar disc around Beta Pictoris. Only by avoiding the direct light from the star could the faint disc of matter be seen.

[11] At the time, half of the scientific papers based on AO observations worldwide were obtained by ADONIS.

[12] Despite the disturbance of the AO programme, the study later provided important input to the development of NAOS.

was signed in March 1997 between ONERA, the Laboratoire d'Astrophysique de l'Observatoire de Grenoble (LAOG) and the Observatoire de Paris, in collaboration with ESO. The Principal Investigator was Anne-Marie Lagrange, a former ESO astronomer now working in France. The combined instrument, NAOS–CONICA, or NACO for short, saw first light in November 2001.

≈

By 1998, ESO could again begin to seriously consider implementing the VLTI, including its AO component. The goal was a low cost, yet reliable and easy-to-operate system and the solution was the 60-actuator MACAO system, using wavefront curvature sensing and what are known as bi-morph mirrors, an idea developed by François Roddier. This system was simpler than the by now classical Shack-Hartmann-based AO system, but it nevertheless yielded medium-order correction and was able to work with systems with up to about 200 actuators. The principle which had been tested with great success at the 3.6-metre Canada France Hawaii Telescope, was taken over by ESO and found use in six AO systems at the VLT. Domenico Bonaccini Calia, who had joined ESO from the Osservatorio Astrofisico di Arcetri, in Florence, was responsible for the conceptual design. MACAO was approved in 1997 and installed between June 2003 and June 2004.

We have seen how, within a short space of time, the technology of adaptive optics improved — with more actuators, faster computers, more sensitive wavefront sensors and alternative ways to analyse the deformation of the wavefront itself. But some tricky problems remained. How, for example, could a larger field on the sky be corrected? Despite the increased sensitivity of the sensors, what was to be done when observing a target in a field void of sufficiently bright stars? And with the increasing number of instruments and AO systems, would it not make more sense to make the telescope itself adaptive? The answers were: multi-conjugate adaptive optics, laser guide star systems and, yes, adaptive telescopes.

The idea behind multi-conjugate adaptive optics is that if a comprehensive determination of the disturbances at different heights in the atmosphere and in different directions can be obtained, then it will be possible to obtain a fully corrected field significantly larger than that possible with traditional adaptive optics. For this to happen, a technique known to the astronomers as atmospheric tomography is necessary. This technique was developed in late 2000, in Europe mainly by Roberto Ragazzoni, Enrico Marchetti and Gianpaolo Valente (all from Padua), and François

Rigaut (Meudon), with financial support by the European Union Framework Programme. Atmospheric tomography implies probing the atmosphere in different directions, with the help of natural or laser guide star systems and thus constructing a three-dimensional map of the atmosphere at a given instant. Corrections can then be done by means of several deformable mirrors, and, as a result, a larger field of view can be covered with all the point spread functions (i.e. the rendition of the stellar images) subject to the same degree of correction over the field. Expressed in more simple terms, with conventional adaptive optics, the blurring of the stellar images would increase towards the edges of the field. With the new system this would not occur, despite the much fact that the field would be much larger. As an idea, multi-conjugate adaptive optics was relatively new. Jacques Beckers had proposed it in 1988 at the ESO Conference on The Very Large Telescopes and their Instrumentation (Beckers, 1988), but it was only after the technique of atmospheric tomography had come about that it became possible to test the idea. By the turn of the century, our understanding of the atmosphere, the ways to measure the continuous changes and the required technology had matured enough to take the next step: to build a demonstrator device. Among others, the demonstrator was strongly supported by Roberto Gilmozzi, who we shall meet later in the book. While strongly involved in the VLT project, he had also begun to dream of even larger telescopes, and it was clear to him that the larger the telescope, the more dependent it would be on adaptive optics. And so, in January 2002, ESO launched a demonstrator project[13]. Readers will by now have realised that astronomers, next to their ability to develop novel technological solutions in support of their research, are as astute when it comes to inventing acronyms. In this case, the acronym was truly emblematic: MAD, the Multi-conjugate Adaptive optics Demonstrator. Such an acronym would inevitably give rise to jokes. Hardly oblivious to this, the project manager, Enrico Marchetti (who had joined ESO), took the lead, putting a sign on the door to his office: Mad Manager, it said. But of course, there was nothing crazy about MAD. In October 2005, it was successfully tested in the laboratory and on 25 March 2007 on the sky, mounted at the VLT Unit Telescope 3, Melipal[14].

≈

[13] Aside from ESO, MAD involved a consortium led by the Universidade de Lisboa, providing the high-resolution infrared camera (CAMCAO), and the Observatories of Padua and Arcetri (Florence) developed the instrument control software and a new type of wavefront sensor.

[14] The honour of the first successful on-sky test of a multi-conjugate adaptive system, however, goes to solar astronomers, who applied their system in April 2005 on the German Solar Vacuum Tower Telescope at Tenerife (von der Lühe *et al.*, 2005).

It takes only a casual glance at the sky to realise that stars are not distributed evenly on the sky. We clearly recognise the Milky Way because here we find a much higher concentration of stars than anywhere else. And as we have seen in this chapter, adaptive optics is dependent on the presence of a guide star of sufficient brightness in the field of observation. But extragalactic research often involves studying parts of the sky far from the Milky Way's band of stars. If, in these seemingly dark parts of the sky, no star bright enough for adaptive optics can be found, perhaps one could be "made"? The idea was to shine a powerful laser beam with a wavelength of 5892 Å onto the sky. The laser would excite sodium atoms in the upper atmosphere, the so-called mesosphere, around 90 kilometres above ground[15]. These atoms would then re-emit the laser light, producing a point light source bright enough to be used for the wavefront sensor. This would be an artificial star so to speak — though astronomers with their particular phraseology would rather speak of a LGS, or laser guide star. The potential of this was mind-blowing, for the statement that "stars are not distributed in an even way on the sky" is, in the context of adaptive optics, a dramatic understatement. The sky coverage of conventional adaptive optics, i.e. the system that relies on the presence of a suitably bright star, is of the order of 1% only. Supplemented with a LGS, the sky coverage increases to perhaps 90%[16].

Once again, the US military[17] had been ahead of everyone else and undertaken significant efforts in this area, though — as could be expected — in absolute secrecy. Tests had been carried out as early as 1983 and in the period 1988–90, further tests had been performed from a US Air Force site in Hawaii (Hardy, *ibid.*). But, as in the case of adaptive optics, civilian astronomers were not far behind. In Europe, Renaud Foy (at the time working at the Observatoire de Paris and Centre d'Etudes et de Recherches Géodynamiques et Astronomiques [CERGA]) and Labeyrie — without knowledge of the US experiments — had presented their ideas about "adaptive telescopes with laser probes" in 1985 (Foy & Labeyrie, 1985). The gradual realisation, not the least because of the work in Europe, that knowledge about this technology could no longer be contained, led to the decision by the US military in 1991 to declassify information about American efforts in this area (Duffner, *ibid.*). In astronomy early attempts, notably at the Lick Observatory and at the W. M. Keck Observatory, appeared promising, and the Max-Planck-Institut für Extraterrestrische Physik in

---

[15] Weak radiation from the same layer of sodium atoms creates the "air glow".

[16] In the infrared *K*-band (2–2.4 μm) and using a visible-light wavefront sensor.

[17] Through the Defense Advanced Research Projects Agency (DARPA).

Garching (MPE) (through the German/Spanish observatory at Calar Alto in Spain) also began to develop a system, known as ALFA[18] (Quirrenbach, 2000).

In 1999, ESO set up a group to look at LGS systems, headed by Bonaccini. Among the recruits to the group was Wolfgang Hackenberg, who brought with him experience from the ALFA project. In 2003 work began on the first laser guide star facility at ESO, with a 4W continuous wave sodium laser called PARSEC[19], in a collaboration with the MPE (Garching) and the Max-Planck-Institut für Astronomie (Heidelberg). The system was installed at Unit Telescope 4, Yepun, in 2006 and became operational the year after. The main elements were a) a laser laboratory, of roughly the size of a small container[20], b) the laser itself, c) a relay system based on photonic crystal fibres to deliver the laser beam to d) a 50-centimetre launch telescope attached to the M2 unit of the telescope. Finally e) an LGS monitor, a 30-centimetre telescope about four kilometres away from the site, providing continuous information about the density of the sodium layer, the possible presence of clouds along the line of sight (of the laser) and the performance of the laser itself (Bonaccini *et al.*, 2001). This was only the beginning. The way forward lay in developing a simpler and better system. Simpler, in that it could be used as a routine feature during observations and did not require any specialised staff at all times. Better in terms of being lightweight (and thus imposing fewer demands on the telescope main structure) and by optimising the relationship between input, in terms of energy, and the output in terms of photons returned. This required an intensive R&D effort at the laboratory in Garching. LGS systems were spreading across the world, with different groups testing different technical solutions. To obtain a deeper understanding of the actual physical processes connected with laser-induced excitation of sodium atoms, ESO teamed up with a group of quantum physics experts of the University of California, Berkeley[21]. The result of these efforts was a new fibre-optics-based continuous-wave laser delivering a power of 50 W. The R&D effort also led to patents and licensing agreements with industry.

[18] Adaptive optics with Laser guide star For Astronomy.

[19] The Paranal Artificial Reference Source for Extended Coverage.

[20] The cabin was mounted underneath one of the Nasmyth platforms, thereby rotating with the telescope itself.

[21] Led by Dimitri Budker, but also involving William Happer at Princeton.

The laser guide star system at the Yepun telescope. The picture also provides a spectacular view of the southern Milky Way and the twin satellite galaxies, the Magellanic Clouds. (Photo: Gerd Hüdepohl)

≈

With the advances in adaptive optics, the dream of making the telescope itself adaptive had come closer to realisation. This could be done by turning the large secondary mirror into a fully deformable mirror: clearly a non-trivial task given its size and mass. Among the advantages of this are that it would offer *"fast wavefront correction without the addition of supplementary optics or mechanics. Moreover, with the two Nasmyth and Cassegrain foci this gain is threefold. The system gives better throughput to science instruments, lower emissivity for thermal IR instruments, large field of view accessible to all instruments and less complexity/crowding at the focal planes"* (Arsenault *et al.*, 2006). As in the case of the LGS systems, practical tests were started in the US, though based on seminal work at institutes in Italy, especially by Piero Salinari and his team at the Osservatorio Astrofisico di Arcetri. The first test was carried out in 2002 at the Multiple Mirror Telescope on Mount Hopkins in Arizona and later (in 2010) at the Large Binocular Telescope (LBT) on Mount Graham, also in Arizona. At the LBT, the 0.9-metre secondary mirror was controlled by no less than 672 actuators working at an incredible 1000 Hz, ten times faster than the first AO system. At ESO, at the suggestion of Norbert Hubin, who was in charge of ESO's AO activities, plans were initiated to develop the VLT Adaptive Optics Facility (AOF) in 2004.

This took the form of a feasibility study to be carried out by a consortium led by the Italian company Microgate[22]. The aim would be to turn Unit Telescope 4, Yepun, into an adaptive telescope, with the help of a new secondary mirror with 1172 actuators, operating at a frequency of 1200 Hz. It falls beyond the scope of this book to deal with the VLT AOF, which at the time of writing is still under development, but it nonetheless deserves to be mentioned in the historical context. Since the AOF is also a pathfinder project for the future E-ELT it is certainly not an end point in this development. It can also be seen as a powerful illustration of how telescope technology in its widest sense has developed over the last 30 years or so.

The development of ever more efficient adaptive optics systems has had a profound influence on contemporary astrophysics, notably in the field of extrasolar planets and high-resolution imagery of the Galactic Centre, i.e. studying the black hole at the centre of our stellar system as well as other areas that depend on high-resolution imaging. It is also a good illustration of ESO working at its best, with stable funding and the ability to mobilise a critical mass to "drive" industry. The alliance between national institutes, initially especially in France, but later also with German and Italian institutes, and ESO, as well as with Europe's high-tech industry, helped to catapult European astronomy to the forefront of this most competitive science. But our story has wider implications. Unsurprisingly, we see that in these extremely specialised fields of high technology, such as interferometry, adaptive optics and LGS systems, the group of experts in the world is very small. Many of them have found their way to ESO for a period of time, not least because ESO has been able to combine good working conditions with very interesting professional challenges. But many have also moved on to other projects elsewhere in the world. At a time when researchers' mobility and "brain circulation" are seen as a key to innovation, economic prosperity and our ability to tackle grand societal challenges with the help of science and technology, the examples from astronomy provide an interesting case to study. It also offers a nice illustration of the nature of hi-tech developments — the impetus of competition and the value of co-operation — that has led to the coining of the term "co-opetition".

However, in our story of the VLT, which basically spans the 1990s, we have taken a huge jump ahead in time. We shall therefore now turn to the other main elements of the project and return to the beginning of that decade.

[22] Together with the Osservatorio Astrofisico di Arcetri, Microgate built, assembled, delivered and tested the adaptive secondary of the MMT.

Chapter III-4

From double-sight to supersight: Interferometry

"I had always considered interferometry an interesting option for the VLT, but without much confidence in its early realisation."

Lodewijk Woltjer, 2006.

As Lodewijk Woltjer writes, one of the major elements behind the progress of astronomy has been the improvement in spatial resolution — the sharpness with which astronomical objects can be seen. In the previous chapter, we have discussed the relationship between resolving power and telescope size, one of the reasons why astronomers always dream of larger telescopes. But even in the best of worlds, there are limits on how big telescopes can become, if for no other reasons than hard financial ones. Could there be other ways to increase spatial resolution — to register the finest details in objects far away? The answer was found in a technique known as aperture synthesis using interferometry. It seems fitting to describe optical interferometry — the coherent combination of light beams from two or more separate telescopes — as the holy grail of astronomical imaging. The basic idea emerged from what had perhaps been one of the most fundamental challenges to physics — to understand the nature of light. The first experiment to demonstrate the wave property of light — known as the two-slit experiment — was carried out in the early 19th century by the Englishman Thomas Young, whose work was followed up by Augustin Fresnel of France. When light from a source passes through two separate slits, the waves emanating from each of the slits will interfere with each other, generating an interference pattern that can be analysed and thus provide information about the source. This led the French physicist Hippolyte Fizeau, in 1868, to suggest that it should be possible to measure fine details with interferometry, such as the diameters of stars (Monnier, 2003). Early examples, towards the end of the 19th century, demonstrated the validity of the idea. A first try at building such a telescope was undertaken in 1874 by Edouard Stephan at the Observatoire de Marseille (Léna, 2007) and in 1920/21 the diameter of the (bright) red giant star Betelgeuse (α Orionis) was determined with a stellar interferometer at the Mount Wilson Observatory. However, for practical astronomical purposes the concept was really developed in radio astronomy, where the much longer wavelengths made it somewhat easier. Optical interferometry, i.e. in the wavelength domains of visible and infrared light, remained a challenge. And a daunting one, indeed. In the early 1970s, the French astronomer Antoine Labeyrie breathed new life into the idea, with a set of interconnected optical telescopes on the plateau of Calern near Nice in France. The work of Labeyrie, i.e. the idea that ultra-high image resolution could be achieved by combining light from separate telescopes, came early enough to influence the ESO Conference at CERN in 1977. Yet the magnitude of the challenges that had to be met to make optical interferometry work in terms of controlling a complex dynamic system with nanometre precision was clear and many people remained highly sceptical. Woltjer's remark at the beginning of this chapter reveals the low degree of expectation. Pierre Léna recalls the substantial scepticism also expressed at the conference by Harry van der Laan and Robert Hanbury-Brown, both scientists with considerable knowledge in this field (Léna, 2008). Although everybody saw the promise held out by optical interferometry, their confidence in its feasibility was limited. But somehow, Fizeau's spirit prevailed: in the European context it does not seem unreasonable to see the technique of optical interferometry as a French preoccupation, but astronomers in other countries — notably in Germany — also thought that it was worth following up. Indeed it was.

But what is aperture synthesis? Recall that the larger the diameter of the primary mirror of a telescope (its entrance pupil) is, the sharper the image it yields. Expressed more technically, astronomers talk about the spatial resolution, rather than image sharpness, because spatial resolution describes more precisely how fine details can be discerned or how close objects can appear on the sky, and still be seen as distinct objects. The spatial resolution of an interferometer, on the other hand, is not limited by the size of the individual telescopes, but by the separation between them. However, in this case, the spatial information of an astronomical object is distributed over the entire area — known to astronomers as the (u,v) plane — whether it is covered by telescopes or not. Therefore, in order to obtain as much information as possible, it is necessary to locate the individual telescopes in a way that allows for an optimal sampling of the (u,v) plane, and then to fill the gaps by means of computational techniques generically known as image reconstruction. To improve the sampling, the telescopes can be physically moved in the course of an observational sequence, or simply, in time, the rotation of the Earth will do the job. Indeed, as the Earth rotates, the telescopes (as seen from the sky) probe different parts of the (u,v) plane. The Very

Large Telescope Interferometer makes use of both possibilities, by enabling the combination of light from the main 8.2-metre Unit Telescopes (in their fixed locations) and from a series of 1.8-metre Auxiliary Telescopes, mounted on tracks and thus movable over the entire telescope platform.

≈

As we have seen in the past chapters, the notion of doing optical interferometry on a large scale was around already in the early 1980s and the idea shaped quite a bit of the thinking behind the VLT, in spite of much scepticism. As the detailed planning of the VLT began in 1988, it was also time to address the issue of interferometry in a serious way. To lead the interferometry programme, ESO hired Jacques Beckers, who at the time worked at NOAO, the National Optical Astronomy Observatory in the US. Before joining ESO, he had been involved in an abandoned NOAO project known as the 16-metre National New Technology Telescope (NNTT), for a telescope with four 8-metre mirrors on a single mount[1]. We have briefly alluded to it in Chapter II-7. Beckers arrived in June 1988, charged not only with the interferometry programme, but also with the development of adaptive optics (which as we have seen in the past was very much linked to the needs of interferometry) as well as the site testing and thus, also to the selection of the site. This particular combination of tasks made sense, since *"it was completely clear that whether the interferometry was going to happen or not, the issue of interferometry had a major effect on the site choice and on the design of the telescopes,"* Jaques Beckers recalled in an interview with the author. At the time this *de facto* meant a choice between Cerro Paranal and Cerro Vizcachas, a peak close to La Silla and located within ESO's territory. Because of its natural shape Cerro Vizcachas would lend itself well to the linear telescope array that was foreseen in the 1987 Blue Book proposal. Paranal would not. ESO sought advice from a tiger team of internationally renowned experts from the interferometry community[2]. The team comprised Tim Cornwell (National Radio Astronomy Observatory [NRAO]/VLA), Chris Haniff (Cambridge/UK), Shri Kulkarni (Caltech), Jean-Marie Mariotti (Meudon), Jan Noordam (NFRA/Dwingeloo),

---

[1] In early 1987 an informal meeting took place between John Jefferies, director of the NOAO, accompanied by Jacques Beckers and Woltjer. The topic was possible cooperation between the NNTT, which was expected to be placed at a northern hemisphere site, and the VLT.

[2] The tiger team supplemented the VLT Interferometry Panel, mentioned earlier, which consisted of scientists from the ESO Member States: Robert Braun, Paolo di Benedetto, Dennis Downes, Renaud Foy, Reinhard Genzel, Laurent Koechlin, Antoine Labeyrie, Pierre Léna, Jean-Marie Mariotti and Gerd Weigelt, with Jacques Beckers in the Chair. The tiger team, however, earned its name by taking the critical, but at the time not universally supported, decision about the configuration of the VLTI (Beckers, *ibid.*).

Farrokh Vakilii (CERGA/Nice), Gerd Weigelt (Max-Planck-Institut für Radioastronomie [MPIfR]) and was chaired by Oskar von der Lühe (Beckers, 2003). The outcome was to abandon the linear array and instead choose a trapezoidal configuration, much like the one that was finally constructed. In the language of interferometry this was described as a "2D, non-redundant" solution. This pointed to Paranal, which had also proven to be superior on other accounts. Paranal had better seeing and the observing platform, which could be created, would allow for a change of the array to improve coverage in the *(u,v)* plane. The change could be achieved with small mobile telescopes, though at least one astronomer thought that the 8-metre telescopes should be movable (!). A further change that happened in this early phase was the decision to locate the point where the beams were combined in an underground system and not, as originally foreseen, above the surface.

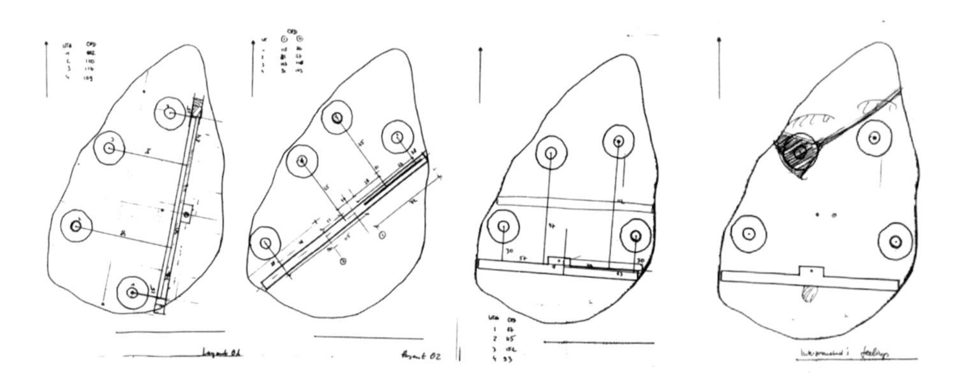

The three drawings on the left image illustrate some of options considered for the VLTI. The cartoon with the comment "interferometrist's feelings" to the right testifies to the agonising process that finally led to the current configuration. (Courtesy: Oskar von der Lühe)

While ESO solicited the advice of the tiger team, it also assembled a powerful — if small — in-house group of specialists, including Oskar von der Lühe, Bertrand Koehler, Fritz Merkle and Francesco Paresce, who later became VLTI project scientist.

≈

Work on the VLTI was progressing. Now the configuration had been determined — the location of the individual 8.2-metre telescopes had been fixed with two smaller auxiliary 1.8-metre telescopes moving on tracks and thus helping to

A happy moment at the ESO HQ: Daniel Enard (extreme left), Harry van der Laan, Fritz Merkle, Massimo Tarenghi and Oskar von der Lühe. The photo was obtained after the signing of the tripartite agreement between ESO, CNRS and MPG in 1992 for enhancing VLTI with a third auxiliary telescope. (Photo: Hans Hermann Heyer)

fill in the (u,v) plane. With additional funding from Germany and France, a third Auxiliary Telescope became possible[3]. But as already mentioned, the VLT project as a whole became engulfed in trouble, leading to the postponement of the interferometer as a result of the general project cost overrun. The March 1994 issue of *The ESO Messenger* summarises the decision as follows: *"In its October 4th and 5th 1993 meeting Council expressed its approval of the revised VLT/VLTI project ... for content, schedule and staff. Financial difficulties discussed in the Finance Committee meeting of November 8th and 9th 1993, and recent expression of concern in a diplomatic note from the French Government have led to reconsideration of this plan. Following the presentation and discussion of different alternatives for cost reduction, Council adopts further modifications to the VLT programme plan. This includes the postponement of*

[3] This was a voluntary contribution by the CNRS of France and the German Max-Planck-Gesellschaft signed on 18 December 1992. Originally the money had been set aside for an independent interferometry project, but when the VLTI began to appear as a realistic project, the funds were relocated towards this purpose.

the implementation of VLTI, VISA[4]*, Coudé Train and associated adaptive optics for all telescopes. In consultation with the Scientific Technical Committee a solution will be sought to introduce adaptive optics at the Nasmyth foci at the earliest possible time. Furthermore, the Executive will endeavour to reintroduce full Coudé and interferometric capabilities at the earliest possible date. This will include provisions for continuing technological research and development programmes devoted to this end."* This was a painful decision, indeed. Calls for tender, already issued, had to be cancelled. Key people, who in various ways were associated with the interferometer project, ultimately left the organisation — Jacques Beckers, Oskar von der Lühe and Fritz Merkle. In Council, the French science delegate, Pierre Léna, himself an ardent advocate of VLTI, resigned in the face of what he perceived to be dwindling support for interferometry, also in the community. The prospect for the realisation of this most ambitious part of the VLT project looked decidedly bleak. However, Giacconi, who had by then assumed the helm at ESO, decided to continue those activities that had a direct bearing on the construction work at Paranal. He clearly hoped that, with time, it would be possible to reinstate the VLTI programme one way or the other — as hinted in the Council declaration. And so, on Paranal the tunnel system was duly excavated, and in general the infrastructure prepared so as to be ready for better times. The loss of qualified staff was serious. The world of optical interferometry is a small one in terms of experts. Those who stayed did so in the hope of a revival of the project. One factor that contributed to maintaining a certain degree of hope was that, at the time, quite promising discussions were ongoing between Australia and ESO about membership. Australian membership would have brought financial relief, but it would also have meant the involvement of the Australian scientific community that possessed considerable knowledge about interferometry, albeit mostly from radio astronomy. Then, in early 1996, the talks with Australia collapsed.

On the other side of the Atlantic Ocean (or perhaps rather, the Pacific), things looked much brighter. Since the early 1980s, US astronomers in California had worked on plans for two 10-metre telescopes to be placed in Hawaii, what eventually became the Keck telescopes. We have referred to the Keck telescopes several times already. Like most other major projects, the original project (known as the Ten-Meter-Telescope, or TMT[5]) had had a somewhat bumpy time, but in 1993, Keck I had seen first light

[4] The VLT Subarray, VISA, was the array of auxiliary telescopes. At the June 1994 meeting, the ESO Council had already decided to continue with the development of the array, albeit with expected installation as late as 2003.

[5] Not to be confused with today's TMT, the Thirty Meter Telescope!

and since then it had pointed its huge mirror to the sky every night. Furthermore, at the beginning of the decade plans were being revived for building the second telescope, Keck II. It would then be possible to combine the light of the twin telescopes and also include what the Americans called smaller outrigger telescopes, the equivalent of ESO's Auxiliary Telescopes. In 1996, Keck II was ready and the interferometry programme could commence with support from NASA.

The VLT is a complex machine. In the previous chapters we have reviewed managerial and financial aspects. We have also looked at technical issues including scientific instrumentation. But the heart of any telescope is its mirrors. In the following chapter, we shall follow the production of these optical masterpieces and the associated equipment.

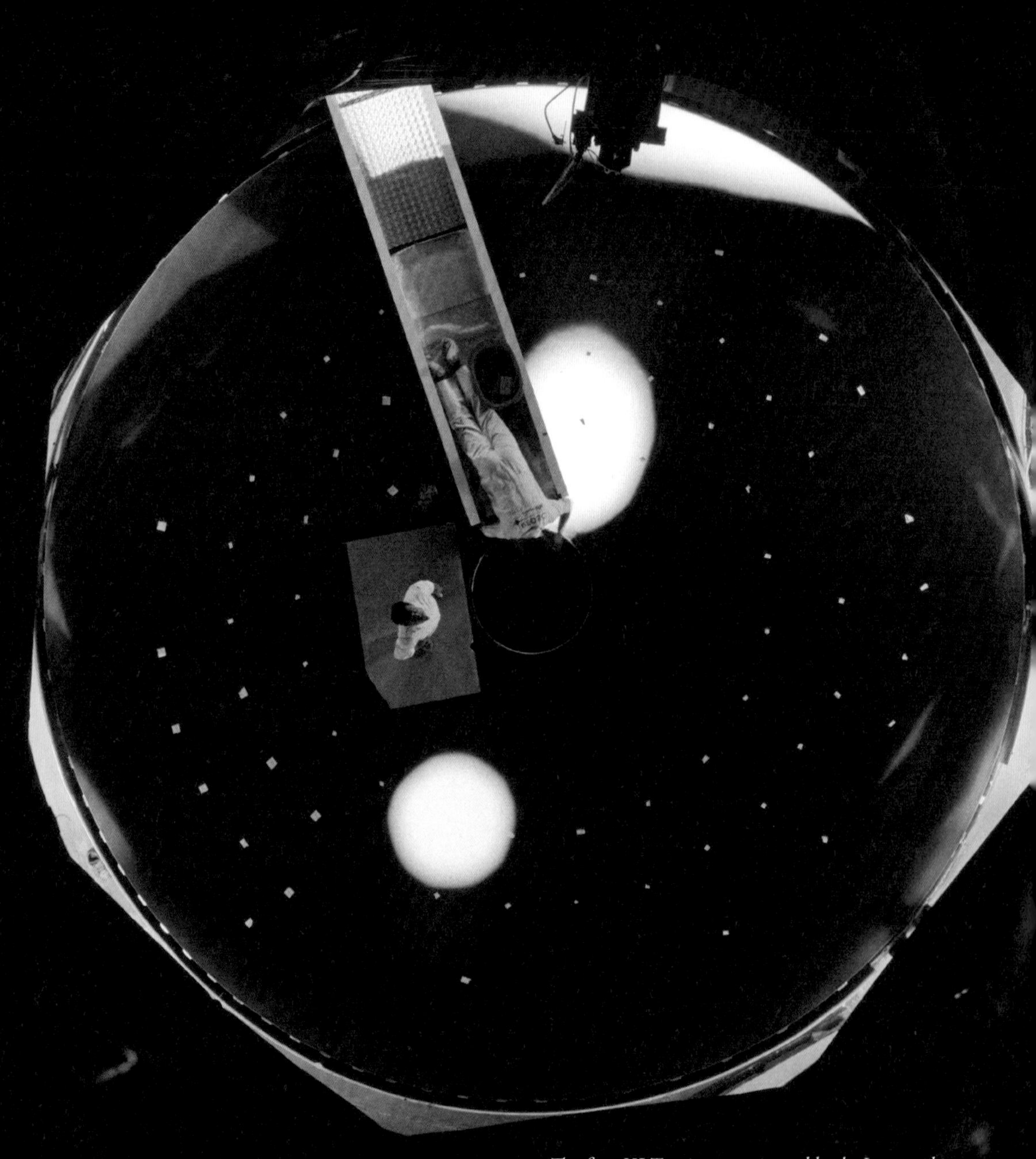

The first VLT primary mirror blank, Joe, on the polishing table at REOSC. The photo was taken from the top of the interferometric tower during the acceptance tests. (Photo: Hans Hermann Heyer)

Chapter III-5

The Return of the Dalton Brothers

"Joe Dalton was delivered in time and well within specifications... ."

Müller *et al.*, *The ESO Messenger*, 73.

Friends of late 20th century youth culture may associate the Dalton Brothers with a famous rock band. Others will perhaps think of the — infamous — members of the Dalton Gang who were killed in Kansas in 1887. Yet others will think of the well-known cartoon series known as Lucky Luke, created by the Belgian cartoonist Maurice De Bevere. With a Belgian optical engineer in charge of the VLT mirror contracts (and someone who knew his cartoon classics), it is perhaps less of a surprise that the Dalton Brothers — Joe, Jack, William and Averell — would also become associated with these spectacular mirrors. The engineer was Philippe Dierickx, who promptly assigned the names to the four blanks. He had joined ESO in 1986 after submitting an unsolicited application. At the time he worked at REOSC and his boss Jean Espiard, on learning that Dierickx might move to ESO, had commented: *"There you will meet the best telescope designer in the world — Ray Wilson!"* Before being accepted, Dierickx attended an interview at ESO, with Wilson as a member of the interview panel. When Wilson heard that the young candidate was working with Espiard, he remarked: *"Then you're working with the best telescope designer in the world — Jean Espiard!"*

In Chapter III-1, we mentioned that the contract for the provision of the four 8.2-metre primary mirrors for the VLT was the first major contract to be concluded. To produce the mirrors, the Schott Glassworks had to erect a new factory hall, and this would become the centre of attention, not just for the Schott experts but, obviously also for the ESO team, notably Massimo Tarenghi, Philippe Dierickx and Ray Wilson. A detailed technical account of the manufacturing process is provided in *The ESO Messenger*, 73 (Müller *et al.*, *ibid.*).

Schott had decided to use its new spin-casting technique, which meant spinning the mould while the liquid glass was cooling down. The mould had a curved bottom and with centrifugal forces at play, the blank would assume a shape more or less like the final mirror. In December 1990, the first pouring of glass — no less than 45 tonnes — occurred. Before the pouring, the glass was prepared in a huge vessel and heated. This process took 24 days. The temperature of the glass at the time of the pouring was 1300 degrees Celsius. The pouring itself took four hours, after which the mould, covered by an overhead heating system was moved to a different position and the spinning process began. Perhaps the most spectacular moment was the lifting of the cover with the heating system, exposing the glowing surface of the glass — 50 square metres — to ambient air and enveloping the whole factory hall in an intense heat wave that probably only those present will truly be able to imagine. The power radiated was 20 megawatt. Gradually the glass began to cool. When the temperature had dropped to 1100 degrees Celsius and the glass mass had assumed the desired concave meniscus form, the mould stopped spinning and was moved into a dedicated furnace for the annealing phase, which was scheduled to take four months. After that the first machining of the concave and convex sides was done with a diamond grinding tool, and then the glass again disappeared into a furnace, to undergo the ceramisation process, transforming the glass into the glass ceramic Zerodur. Then it was time for the drilling of the central hole and the fine grinding. With this completed, the blank was ready to undergo acceptance tests by ESO and subsequently to be transferred to REOSC for the final, but critical production stages. Put on paper many years hence, all of this seems like a fairly straightforward process, but at the time, though the individual steps were clear, there were complications — and that was hardly surprising! According to the contract, Schott provided a two-year warranty for the blanks. This *de facto* meant that Schott had to produce six blanks, of which ESO would acquire four. However, considering the probability of failures (we remind ourselves that such mirrors had never been produced before), Schott was actually prepared to cast up to 12 mirrors, although in the end this was not necessary. This could in principle have meant a production rate of one mirror per month, but the requirements for space for the subsequent production processes were such that this did not appear to be feasible. At the same time, the furnace could not simply be turned off — the production of the raw glass material had to go on, uninterrupted, with 60 tonnes being produced every month. Much of this glass eventually was used for other purposes. But then another problem occurred. After a few months of cooling in the annealing oven, the first blank broke. Soon afterwards, the second one broke too[1]. Schott mobilised some of its best engineers to find out what was going on. They carefully traced the history of each crack and managed to reproduce the beginning of the cracks. The cause of the problem, it turned out, was chemical contamination between the mould and the first millimetres of glass, leading to a crystalline layer with a different thermal expansion coefficient from the Zerodur. Stresses built up while cooling, resulting in cracks (Dierickx, 1993). The solution was to deploy a special alloy on the surface of the mould, reducing the crystalline layer dramatically. With this problem solved, the production resumed without any further serious mishaps. In July 1993, the first blank had completed the production cycle and was carefully placed in a specially designed container that would protect it during the subsequent journey, first to France and, four years later, on the long trip across the Atlantic Ocean to South America. In the afternoon of 19 July 1993, the journey began with a short drive to the Rhine port of Mainz. As if Heaven wanted to carry out a test of its own, it happened in torrential rain.

≈

Close to midnight, the river barge *Eldor* left the port with the first VLT mirror blank safely stored in its hold. The route foreseen took the vessel and its fragile cargo to Rotterdam, into the English Channel and up the Seine, passing through Paris during the night of 23 July, to arrive at Evry, its destination just south of Paris and just a few kilometres from the REOSC plant. Not surprisingly, the passage of the mirror led to considerable attention by the local population as roads were blocked by the slowly moving lorry carrying the 8.4-metre wide, and rather unusual-looking, mirror container. It may have caused some traffic delays and, perhaps, some irritation among motorists, but it also represented a chance to raise the visibility of the project in the public mass media. Of course, ESO's Information Service was on the spot, covering the mirror's journey with pictures and video footage, to be used for press releases and other publicity materials. A great deal of planning had gone into this effort, and it seemed that the passage through Paris would offer a unique chance to obtain a spectacular picture. The VLT represented the utmost in cutting-edge technology made-in-Europe. Now the mirror would pass close by the Eiffel Tower, itself a powerful symbol of Europe's technological prowess, albeit from bygone days. And so the team descended on Paris ahead of the vessel's arrival. The photographers first found the best location for their shot and then went out to the locks at Suresnes to meet the vessel. Here Herbert Zodet, one of the photographers, boarded the vessel to film the passage

[1] The exact number of pourings is still considered confidential by Schott.

from the bridge of the ship, while the others went back through the Bois de Boulogne to their chosen location near the Debilly footbridge to wait for the right moment, which would occur around 11 pm. Shortly before the boat arrived, however, the lights that had illuminated the Eiffel Tower and every night provide a spectacular view of this iconic building were abruptly turned off. To this very day the author of this book, who accompanied the team, feels the sense of despair as the *Eldor* passed quietly in front of the darkened Eiffel Tower and the photographers realised that they had just missed the shot of a lifetime!

The passage through Paris at night —the Eldor *in front of the Paris City Hall. (Photo: Hans Hermann Heyer)*

≈

Like Schott, REOSC had built a new production facility, which was located on a greenfield site at St. Pierre du Perray. The original REOSC plant was located in Ballainvilliers, a village some 20 kilometres south of Paris and to the west of the river Seine. Having signed the mirror contract with ESO, REOSC started to plan the necessary expansion of its original facility. But, completely unexpectedly, it was denied a building permit by the local town hall, which, according to the local press *"would rather lose [the company] than their rural*

Archeological excavations continued even during the construction of the new REOSC facility in Saint Pierre du Perray. (Courtesy: Roland Geyl, REOSC)

environment"[2]. Since the problem could not be resolved, REOSC erected its new plant at Saint Pierre du Perray, on the eastern side of the Seine and close to the harbour of Corbeil. Before construction started, however, the new site was inspected by archaeologists, who quickly made a find of Merovingian remains, causing another delay. Fortunately, after a few months, the exploration was completed and the construction work could continue. Despite these difficulties, the building was ready in time to receive the first VLT mirror.

On 26 July, the mirror container was unloaded from the ship and, at midnight commenced the last leg of its first journey, on the N 104 dual carriageway and finally on the local road leading to the plant. In the months to come, Joe — as the first blank had now been dubbed — would undergo fine grinding and polishing with extensive, continual testing, until the mirror had reached its final figure. The last, and most critical part, would be the polishing — from the point when the mirror blank was within a few micrometres from its specified shape. The polishing machine was placed directly underneath the test tower enabling *in situ* high-precision interferometric measurements of the surface. There can be no doubt that REOSC delivered a sterling performance and that the French engineers and technicians fully matched their colleagues in Germany who had produced the blanks. Nonetheless, the polishing of the first mirror did involve an incident. The grinding and polishing tables were fitted with a robot arm, under which the grinding or polishing tool was mounted. The tool itself comprised a set of carefully prepared ceramic tiles, glued onto a wooden mandrel. The tool would normally be checked every 24 hours, before a new polishing sequence began. As polishing involves sprays of water, the mandrel was protected by a coat of varnish, but at some point the wood had nevertheless absorbed some humidity, causing a minute displacement of the tiles. On that day, since the 24-hour period had not yet ended, the tool was used without further checking. The technician quickly discovered that something was wrong, but too late to prevent scratches. Fortunately, the scratches could be removed by a slight change in the radius of curvature of the mirror. As the mirror would remain within the stringent specifications, ESO agreed to this and the subsequent mirror blanks were also polished to the slightly modified figure. The acceptance tests, carried out by an ESO team involving Philippe Dierickx, Ray Wilson, Daniel Enard and Matthias Hess, gave testimony to the high quality of work carried out by the French opticians: the surface roughness was 2 nm. A precision of

[2] *Le Parisien*, 27 November 1990. The position of the town hall caused quite a stir, as can be seen in the 22 November 1990 Essonne edition of *Le Parisien: "Incroyable! La petite mairie bloque l'extension de l'enterprise de pointe et risqué de lui faire perdre un contrat de 140 millions."* (The amount indicated is given in French francs, the currency still used at that time).

21 nm was achieved over the whole optical surface of the mirror form during the polishing process, enabling a diffraction-limited resolution of 0.03 arcseconds in the visual part of the electromagnetic spectrum. *"For illustration, [it] corresponds to an accuracy of only 1 millimetre deviation over a surface with a diameter of 165 kilometres (equivalent to the entire Paris area),"* wrote Richard West in a press release on 13 November 1995. The incident had caused a delay of three months, yet as it turned out, this delay did not create problems for ESO. As other parts of the project were delayed as well, it was decided that the first mirror should remain in storage at REOSC until 30 October 1997. The delivery was thus in the first instance simply to the designated storage area. This was nonetheless marked with a celebration on 21 November 1995. This also provided for a photo opportunity with the key project managers lined up in front of the mirror: Giacconi, Tarenghi and Dierickx from ESO, Dominique de Ponteves, Jean Espiard, Marc Cayrel and Jacques Paseri from REOSC. The REOSC technicians watched quietly from behind the mirror. Dierickx noticed them, turned around and saluted the technicians. Giacconi immediately understood and clapped his hands, too. Then everybody joined in. *"It lasted quite a while — and was quite an emotional moment,"* Dierickx later recalled.

The REOSC team and key people involved in the contract. (Photo: Hans Hermann Heyer)

Jack, VLT mirror number 2, was delivered by Schott in October 1994, William on 21 September 1995 and Averell, the final mirror, was formally handed over to REOSC on 30 September 1996. REOSC, in turn, closed the order with the delivery of Averell on 14 December 1999. It was not only a milestone for the VLT project, but also for REOSC, whose technicians had outdone themselves. The fourth mirror far exceeded ESO's specifications by far. In fact, it was twice as good as the other mirrors[3], arguably making it the most accurate large telescope mirror in the world.

The first 8.2-metre primary mirror in its specially designed transport container being loaded onto MV Tarpon Santiago *at the port of Le Havre in November 1997. (Photo: Philippe Dierickx)*

≈

Much work obviously went into obtaining the highest possible optical quality of the mirrors, but an equal degree of attentiveness was required to ensure the safe handling and transport of these fragile pieces. A broken mirror means seven years of

[3] Between working on William and Averell, REOSC took time out from the VLT project to polish the twin 8-metre mirrors for the Gemini telescopes. ESO accepted this because of the delays at the construction site and ultimately the additional experience gained by REOSC benefitted ESO, as was clearly demonstrated by the quality of the fourth mirror, Averell.

bad luck, some say. Astronomers would likely concur, though for them, the loss of a mirror might have implications that could last much longer than seven years! The development of some of the mirror handling tools and of the transport equipment was also entrusted to REOSC[4] and tested in advance with a dummy mirror with the same dimensions and weight as the originals, but made of reinforced concrete. The first transport tests were carried out in the summer of 1992, i.e. one year before the first blank became ready. The concrete "mirror" underwent a sea voyage, road transport and positioning on the grinding table at REOSC. It was also used for testing the mirror cell, underwent a transatlantic voyage (again testing the container) and was finally used in connection with the integration of the telescope in Chile. Today, the concrete mirror rests on its laurels, in the shape of a support system, and is on permanent display outside the interim visitors centre at Paranal. But before it came that far, the disc had to endure elaborate tests. Aside from the importance of assuring the safety and proper functioning of the handling equipment, the key issue was of course to ensure the safety of the mirror during transport, whether exposed to irregular movements in rough seas or subjected to the inevitable shocks incurred as the river barge passed locks en route to its destination. Shocks might also occur on the road, possibly in Europe, but certainly in Chile during the last stretch through the desert. To test for this, the container was fitted with vibration sensors (accelerometers) and the data were carefully analysed. The tests included "high-speed" driving, at 25 km/h, full power acceleration followed by emergency braking, or acceleration while driving over a 5-centimetre-thick wooden beam. Reporting on the first tests in the *The ESO Messenger*, characteristically, Dierickx couldn't help remarking *"the beam is still O.K.!"* (Dierickx & Ansorge, 1992).

At times like these, when there is an increased focus on fostering close links between research and industry in support of innovation, it should be mentioned that the cooperation between ESO and its industrial contract partners for the VLT mirrors can only be described as exemplary. Philippe Dierickx said: *"Following the contracts was a fantastic experience.... We were very lucky to work with professional, but also profoundly honest people.... If they had a problem they were putting their cards on the table ... it was clear who was on which side, but they were genuine and saying exactly what they were doing."* Both Schott and REOSC also gained considerable experience in the field of large mirrors (casting, handling, polishing with *in situ* control, etc.) enabling them to attract customers from elsewhere, including the US.

---

[4] Supplied by SOCOFRAM under subcontract to REOSC.

Philippe Dierickx is not only well acquainted with cartoon classics. Here he gives his own account of the quality inspection of the first primary mirror. But, as with many cartoons, this lives from exaggeration. In fact, the tests showed that the mirror was fully within specifications. (Illustration: Philippe Dierickx and Ed Janssen)

≈

Fittingly, much effort went into the manufacture of the primary mirrors, giant discs 8.2 metres in diameter, but a with thickness of only 175 millimetres, i.e. an aspect ratio of about 47, an incredibly high value for such a large mirror. This meant that the support system would be at least as important as the mirrors themselves. The mirror cell is a complex structure, featuring active support elements as well as the interfaces to the tertiary mirror (M3) and the instruments at the Cassegrain focus. Importantly, since weight matters, it was necessary to develop a lightweight system that would possess the necessary stiffness and provided easy access to the 150 axial supports, with their hydraulics and computer-controlled active force actuators, as well as the 64 lateral supports located around the outer edge. The support structure also had to accommodate a mirror cooling system and various safety features for the primary mirror, including earthquake protection.

Stefano Stanghellini was responsible for this part of the design for the VLT Unit Telescopes. Originally a nuclear engineer, he had also worked in the aerospace industry and thus been exposed to the need for and challenges of lightweight construction.

This was valuable experience for telescope design as telescopes became ever larger. The call for tender for the M1 cell was issued in early 1993, with many of Europe's high-tech industries responding.

Before approving the final contract, at the instigation of Giacconi, ESO gave design contracts to two competing consortia, GIAT Industries together with SFIM (both based in France) and to a consortium of Carl Zeiss and MAN GmbH. of Germany. The companies were not only asked to deliver the complete design in the form of drawings but also in the form of 1:1 mock-ups that were assembled in a tent behind the ESO Headquarters as well as full size prototypes of the actuators. While the German consortium was highly experienced in telescope manufacture, GIAT were novices in the field. But GIAT brought valuable experience from the defence industry sector concerning lightweight products of high stiffness, based on laser cutting and laser welding manufacturing technologies using thin metal sheets, and the partial use of super-high grade Swedish steels. Whereas a cell using conventional technologies would have weighed some 18–20 tonnes, GIAT's proposed mirror cell had a mass of only 11 tonnes, yet was able to support not just the 23-tonne mirror, but also the complex equipment inside the mirror cell itself (including the support for

The first M1 cell, seen here at the GIAT factory in St. Etienne in 1997. The actuators for the active optics and the earthquake protection system are clearly visible. (Photo: Herbert Zodet)

the mirror as well as an earthquake protection system), adding another 8 tonnes, as well as the Cassegrain rotator and scientific instrumentation weighing 4.5 tonnes. GIAT/SFIM, with their team of young, highly dedicated engineers and the backing of its management (which was then diversifying away from purely defence-oriented projects) dispelled all doubts and secured the final contract for the delivery of the four M1 cells.

One of the problem areas revealed by the project review instigated by Giacconi in 1993 was with the 1.2-metre convex hyperbolic secondary mirror, M2, that is mounted high above the primary mirror and reflects the light to the Cassegrain focus behind the primary — or via a flat tertiary mirror to one of the Nasmyth foci. The secondary has other tasks as well, such as field stabilisation to compensate for wind buffeting. It must also be able to switch rapidly between neighbouring parts of the sky, a feature that is particularly important for infrared observations. It should be what astronomers call a "wobbling mirror" (for the stabilisation) and a "chopping secondary" for the continuous shift between target regions. It became clear that a conventional mirror would not be able perform as required at high frequencies and it was therefore necessary either to choose a light-weighted mirror or use new materials. The M2 would of also be an actively centred and focused mirror (in the sense of Wilson's active optics) and thus be housed in an M2 unit that would allow all the necessary functions. Initially, the contract for the M2 units was awarded to MATRA (of France), with MAN and REOSC as subcontractors. REOSC in turn chose a US-based company in Massachusetts, Carborundum, Inc. that offered a silicon carbide substrate for the M2 blank. In the end, however, the company could not deliver and the contract was cancelled. Since the initial solution could no longer be realised, in November 1994 the German company Dornier Satellitensysteme took over as the main contractor, but with REOSC still responsible for the optics. REOSC now turned to another US company, Brush Wellman, Inc., specialising in beryllium. Thanks to its combination of low density, high rigidity and advantageous thermal characteristics, beryllium is often used in highly demanding aerospace applications. In fact, beryllium had been considered by ESO, but initially thought to be too expensive. Furthermore, producing large beryllium mirrors is far from trivial. But it had been done. The 0.5-metre secondary mirror for the first Keck telescope was a beryllium mirror, for example. However, since this was intended for infrared observations, it was uncoated, made of "optical beryllium". ESO's requirements were tougher. They needed "space-technology" beryllium with super-high strength and characteristics in general. With the telescope covering both visible and infrared wavelengths, the mirror also had to be coated with nickel.

Despite early production mishaps, from ESO's perspective things went well initially. Brush Wellman successfully performed the isostatic pressing of the first blank and transferred it to Loral American Beryllium, Inc. in Florida for light-weighting and machining. Then, a few weeks before delivery, the company filed for bankruptcy. In a last-minute operation, ESO managed to secure the mirror blank and have it transferred to another company, Speedring, Inc. in Huntsville, Alabama. The asphérisation was then carried out by Tinsley, Inc. near San Francisco. The nickel coating was then applied before the mirror was finally shipped to REOSC for polishing[5]. Luckily the production of secondary mirrors for the other Unit Telescopes went more smoothly. At the time, the secondary mirrors for the VLT were the largest beryllium mirrors ever produced for civilian applications.

By early October 1997, the first of the secondary mirrors was delivered to Dornier for integration, and in December the first M2 unit arrived at Paranal just in time for the first light preparations.

As already mentioned, the contract for the main mechanical structure had been awarded to an Italian consortium under the name of AES.

≈

We will remember that the original VLT proposal envisaged operating the telescopes in the open air, with giant inflatable domes to provide protection only when adverse weather conditions interrupted observations. Open-air operation was a dream of telescope engineers and astronomers alike, as it would eliminate the need to combat the dreaded effects of dome seeing, the degradation of observing conditions due to local thermal effects. We shall revisit this topic later. However, once open-air operation was jettisoned, ESO began to design appropriate enclosures, under the leadership of Michael Schneermann, a German engineer. The shape of the proposed enclosures looked like something between a classical dome and the avant-garde NTT building, described in Chapter II-9. They comprised two cylindrical parts — a fixed part and a rotating structure. The fixed lower part, of concrete and rising to the Nasmyth level of the telescope, carried the large rotating part of the building. The upper part, a steel structure covered by insulating plates, had a 9.5-metre-wide observing slit equipped with a wind screen with variable wind protection and louvres to enable an optimal

[5] With its contractors ESO carried out pioneering work. The technologies developed under the ESO contract are now expected to be used for the James Webb Space Telescope.

flow of air through the enclosure. The rotating part moved together with the telescope, but independently of it, enabling an unhindered view of the sky. An essential element of the enclosure would be the elaborate active thermal control system to avoid dome-seeing effects.

The contract for the manufacture of the enclosures was awarded to a consortium named SEBIS. SEBIS brought together two of the AES partners (SOIMI S.p.A. and EIE s.r.l.) as well as other Italian companies.

Whether in Europe or on the site in Chile, work was going on to build the VLT. But meanwhile, many other things happened in the world of astronomy. Let us therefore take a break from the VLT tale and look at some important developments.

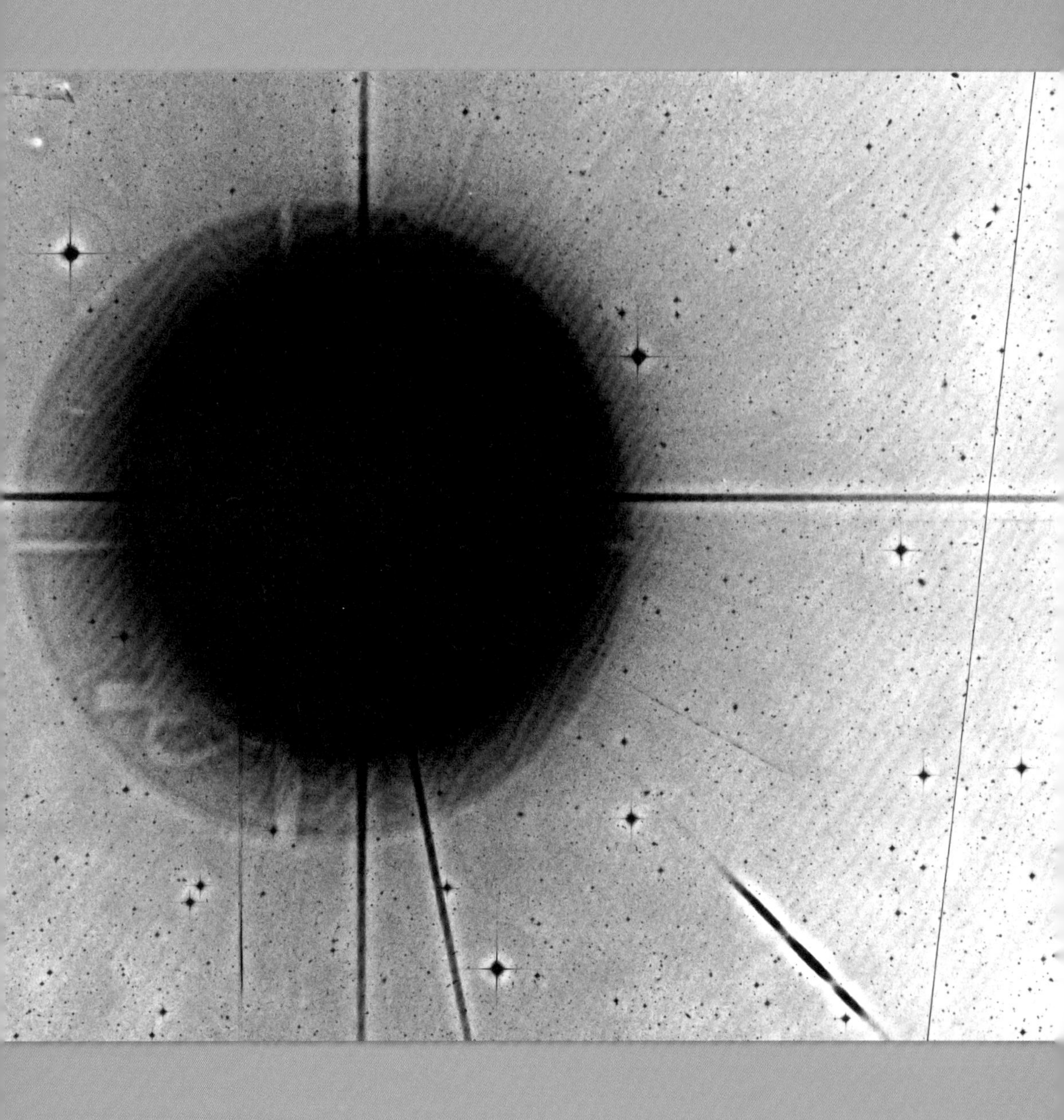

Racing towards their own destruction: like pearls on a string, the fragments of Comet Shoemaker-Levy 9, here observed with the ESO Schmidt telescope, can just be detected in the 4 o'clock position just outside the intense glare of Jupiter. The dark spikes are caused by reflections in the telescope.

Chapter III-6

Bang for the Buck

"... it was the celestial drama of the century."

Clark R. Chapman, *New Scientist*,
21 January 1995.

There is ample evidence that among the sciences, astronomy plays a special role in capturing the interest and the imagination of the general public. Even so, astronomical discoveries seldom reach the front pages of newspapers. However, when rare events occur in the sky, astronomers and the public alike follow them with great enthusiasm. The 1986 passage of Comet 1P/Halley is a good example, and eight years later, another event generated similar interest worldwide: the collision of Comet Shoemaker-Levy 9 with Jupiter. The periodic comet P/Shoemaker-Levy 9 (SL-9) was discovered by Carolyn and Eugene Shoemaker and David Levy on 24 March 1993. It was a very special comet as it seemed not to orbit the Sun, but the giant gas planet Jupiter. Furthermore, presumably in an earlier close approach, the gravitational pull from Jupiter had torn the comet's nucleus to pieces and now 21 large chunks — each estimated to be between a few hundred metres and a few kilometres in size — were moving like pearls on a string. Most interestingly, the orbit was such that slightly more than one year later, the fragments would collide with the planet. Since Jupiter is a gaseous planet, the collisions were likely to have a major, if unknown, effect on the planet, and it seemed reasonable that the observable outcome would provide important information, both about the comet itself and about the composition and structure of Jupiter, whose lower layers are shrouded by a dense atmosphere. The collisions were predicted to occur sequentially between 16 and 22 July 1994.

≈

In a first planning meeting in November 1993, cometary and planetary experts from Europe and the US met at ESO to discuss possible observations from La Silla and develop a coordinated set of observing proposals. Southern hemisphere sites such as La Silla were found to be favourable, since Jupiter would be 12 degrees south of the celestial equator at the time of the collision. But the observations would be difficult. Firstly, at least some of the impacts would happen during daylight. Secondly, the impact site itself was expected to be right on the limb of Jupiter or in fact slightly behind it. Also, the month of July is in the middle of the southern winter, with the increased associated risk of bad weather, underscoring the need to use telescopes in different locations, even on different continents. At an international meeting at the University of Maryland[1] in March 1994, it was agreed to establish a co-ordinated, worldwide observing campaign, similar to what had been done in connection with Halley's comet. At La Silla ten telescopes — including the 3.6-metre telescope, the NTT and the SEST — would be pointed towards Jupiter by the time of the impact. They would follow their target for as long as possible until the end of the observing campaign. Just about any observatory that could do so (a few at far northern latitudes were excluded), was planning to observe this unique event and this ensured that the entire sequence of impacts was well recorded. In addition, observations were planned with the NASA/ESA Hubble Space Telescope, NASA's Galileo spacecraft, itself en route towards Jupiter, and other spacecraft. Nonetheless, the observations at ESO constituted the largest coordinated effort of any observatory at the time.

As the time for the impacts approached, the excitement began to build, both among the observers and in the media, who had been extensively briefed by ESO. Its information service had, however, tried to temper expectations, as nobody knew how visible the event would turn out to be. Astrometric observations at La Silla began in May with the Danish 1.54-metre telescope and other instruments including the Schmidt telescope joined in during the last two weeks leading up to the impact. The significance of precise astrometric measurements lay in determining the precise exact impact locations and times, but the observations were far from easy as the faint cometary fragments had to be picked up against the increasingly disturbing glare of Jupiter. The Galileo spacecraft would be well placed to observe the event, but since it had problems with its antenna, allowing only limited data transmission, and the on-board recorders also had limited capacity, predicting the impact time as precisely as possible was important in order to maximise the scientific gain.

≈

[1] It was clear that the US would play a major role in the observing programme. Orbit calculations were done both by Don Yeomans at JPL and by Brian Marsden and his team at the IAU Minor Planet Center. Space satellites would also observe the event.

The first impact, "A", occurred on 16 July at 20:11 UTC — i.e. during daytime in Chile (but evening in Europe) — with the next some six hours later. At the 3.6-metre telescope, fitted with the infrared imager TIMMI[2], observations were possible right from the beginning — the observers using Venus (and observing through clouds) to initialise the telescope, before moving onto their target. Thus the first night (in Europe) offered two impacts, of which the fragment A — hitting the planet's atmosphere with a velocity of 60 km/s — left a spectacular imprint on the planet in the shape of a large 10 000-degree hot plume rising from the planet's troposphere. Based on prior observations of the fragments, most scientists had expected the first impact to be rather unspectacular, but they were in for a surprise. In fact, when the impact was initially observed, several observers at La Silla thought that the bright spot they saw was merely the Jovian moon Io crossing the face of Jupiter. These first results were, in a sense, simply too good to be true! The thorough training that scientists undergo to always be sceptical of first results, especially if they are spectacular, can at times also be a handicap. At La Silla, one member of the observing team had brought a laptop computer — a rarity in those days. It had a crude ephemeris programme that seemed to confirm that Io should be in this position. However, the object was too bright and when checking against the published high-precision ephemeris in the Nautical Almanac, it became clear that it could not be Io. It had to be impact A. (In fact the computer had been set at the wrong time — an oversight that caused embarrassment, but under the pressure of things is understandable). Yet the ESO observers were not alone in their scepticism. A very first report at the estimated time of the first impact, coming from Sutherland in South Africa, had also been disappointing, with a non-sighting being reported. But it soon changed. The German–Spanish observatory at Calar Alto in southern Spain was first to claim a sighting publicly (their result was also transmitted to ESO, and thus also to the media assembled at the ESO Headquarters) and soon thereafter, the observers at La Silla realised their initial mistake. Now with several confirmed sightings of the first impact, the adrenalin began to flow! Surprisingly, however, fragment B, which was expected to produce a much stronger visible impact, turned out to be a disappointment. Jupiter seemed to simply swallow the fragment. Nonetheless, it was clear that the time ahead would be highly exciting for everyone. During the following days and nights, some 120 000 individual frames were obtained with TIMMI. The NTT followed the event with its IRSPEC infrared spectrograph looking at the impact sites and with the SEST, astronomers looked for interesting molecules in the plumes generated by the falling cometary fragments. Unfortunately, in Chile bad weather set in during the last days

[2] TIMMI allowed both imaging and spectroscopy in the 10 μm atmospheric window.

and nights — after all no real surprise — but here the elaborate planning involving a global network of observatories paid off.

The largest impact recorded occurred on 18 July at 07:32 UTC when fragment G plunged into the Jovian atmosphere, leaving a giant dark pock mark of 12 000 kilometres in diameter, and it was estimated that the energy liberated in the atmosphere of Jupiter amounted to the equivalent to six million megatonnes of TNT — or several hundred times the world's nuclear arsenal. Indeed "bang for the buck".

≈

Richard West during the all-night media event on Saturday 16 July 1994 attended by journalists from the press, radio and ten television stations. Live broadcasts into the late evening news programmes were made. (Photo: Hans Hermann Heyer)

Meanwhile, at ESO Headquarters, a bustling press centre had been established, with daily news briefings — including a crowded all-night live event at the time of the first impacts — with observers at La Silla participating by video link from the control room of the MPG/ESO 2.2-metre telescope, and masterfully orchestrated by Richard West. Background material, including broadcast quality video footage, had been provided in advance, and during the "hot phase", daily media bulletins, produced overnight, were distributed each morning together with coffee and soft drinks *en masse* for the journalists and camera crew. It was the largest and longest lasting press event that ESO had ever organised and the media echo was overwhelming. So great was the public interest that ESO's web server momentarily gave up[3], causing some frustration as La Silla observers could not transfer their files to Garching. It also gave rise to a few witty comments by astute journalists. As one said to this author: *"Whoever claims that comets have no effect on the Earth, should see what it does to the ESO Information Service!"* But not all comments by journalists gave rise to smiles. One tabloid journalist, who felt that — in the middle of the media onslaught — not enough attention had been given to him, published the telephone number of an assistant at ESO, writing that ESO had installed an information hotline for the public. During one day, her office received almost 5000 telephone calls, while she was desperately trying to get on with her own work. The ESO Office in Santiago also organised daily press meetings, and as mentioned in Chapter III-9, this gave ESO a chance to present itself as an exciting place for science at a time when ESO was facing considerable difficulties in Chile.

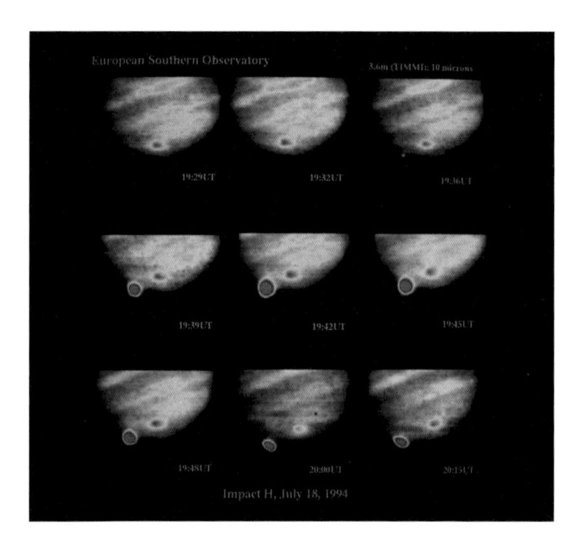

Time-sequence TIMMI observations of impact H. (Courtesy: Hans-Ulrich Käufl)

What was the outcome of all of this? Firstly, the scientific results were expected to provide new insights into the nature and composition of Jupiter's atmosphere. It appeared that the fragments disintegrated rather high up in the atmosphere of Jupiter causing ejecta to rise above the planet. It also became clear that the parent body of the cometary fragments must have been fairly inhomogeneous. This is based on the surprisingly different effects of

[3] In *The ESO Messenger*, Fionn Murtagh explained that on the morning of Monday, 18 July, the ESO *"www server collapsed with 250 simultaneous users, compared to a more usual figure of ten before that."* Alas, this was in the early days of the worldwide web! (Murtagh & Fendt, 1994). Indeed the first ESO web server was established at the beginning of 1994, just six months before the SL-9 event.

the individual impacts. Judged by the infrared emission, the fragments must have been fairly big — ranging from several hundreds of metres to over one kilometre. It also came as a surprise that the scars on Jupiter would remain visible for many weeks after the impact. As Richard West formulated it: *"Unique observations like these have led to important new knowledge, but at the same time they do not fail to raise a host of new and difficult questions."* As could be expected, spectroscopy showed the presence of many different elements, though it was difficult to determine their origin — from the comet or from Jupiter. The observations revealed the presence of substantial amounts of molecular sulphur, due to the abundance believed to come from the deeper layers of the Jovian atmosphere (West, 1994). Perhaps more mundanely, the event provided a stress test for the international community of astronomers in terms of the organisation and implementation of a global observing campaign, and in particular as regards the fast exchange of information between all observers. ESO, as the largest southern hemisphere observatory, and as on later occasions, such as the Deep Impact event in July 2005, clearly played a major role. Finally, ESO became known to the European public at large in way that had never been achieved before. At the same time, ESO provided both a good example to other organisiations and itself acquired ample experience regarding large public outreach activities. Such experience would stand ESO in good stead later on, e.g. on the occasion of the Deep Impact mission, already mentioned, and the International Year of Astronomy in 2009.

Aside from the scientific results and the widespread, if short-term, public attention, SL-9 occurred at a time of growing focus on the issue of near-Earth objects and the potential threat they may pose to the Earth. In 1991, the IAU had established a working group on the topic. Today, our awareness about impacts on the Earth is much improved. Our perception of the potential threat level may not have changed[4], but SL-9 provided a vivid illustration that the Solar System is not necessarily a tranquil place to reside in.

[4] Since 2007, ESO has participated in work of an Action Team on Near-Earth Objects established under the auspices of the United Nations Committee for the Peaceful Uses of Outer Space.

Chapter III-7

Paradigm Revisited

"Laßt uns ohne Vorurteil urteilen."

Immanuel Kant, *Allgemeine Naturgeschichte und Theorie des Himmels* (1755).

While engineers and construction workers toiled under the harsh Sun at Paranal, major discoveries had energised the scientific community and undoubtedly also sharpened the desire for new and more powerful instruments.

One of the most fundamental questions that humankind has ever asked is whether we are alone in the Universe. Is life, such as we know it, a unique feature associated only with our tiny planet or might it exist in other places as well? The equally fundamental questions of what, in the first place, life really is and how it might have come about on Earth are clearly associated. But there are also questions concerning the conditions that must be fulfilled for life to emerge. This, in turn, brings us straight to the issue of planets, and since our expectations of finding life on other planets (or their satellites) in the Solar System are not high (though hopes may still exist!), it is natural to ask whether planets might exist elsewhere in the vastness of the Universe. Indeed this question has fascinated people for centuries with widespread speculation, partly — for want of better information — assumptions based on statistical considerations, partly connected to our growing understanding of the formation processes of stars. Somehow, it seemed unlikely that there would not be planets beyond the Solar System. But nobody knew for sure.

In late 1995, however, all of this changed. In a paper in *Nature* with the title "A Jupiter-mass companion to a solar-type star", two Swiss astronomers, Michel Mayor and Didier Queloz of the Université de Genève, announced the first discovery of an exoplanet orbiting an ordinary star — an otherwise rather inconspicious main-sequence (G-type) star in the northern constellation of Pegasus, 51 Pegasi (Mayor & Queloz, 1995). As so often in science the discovery itself may have come as a surprise, yet this one somehow was already in the air. In 1986, Francesco Paresce, an

Italian astronomer working at STScI, and his colleague Chris Burrows published an image of the star Beta Pictoris, obtained with the MPG/ESO 2.2-metre telescope fitted with a coronograph built at STScI. The picture showed that this nearby and apparently normal main sequence A-type star featured a warped circumstellar disc of matter, 80 000 million kilometres across, and due to its peculiar shape it was suggested that we might be seeing a planet being formed. Beta Pictoris would soon also be observed with other telescopes including the ESO 3.6-metre telescope during one of the first test runs with the adaptive optics system and with the Hubble Space Telescope[1]. Planet hunting was becoming a hot topic in astronomy. What was new about the observations by Mayor and Queloz, however, was their method. The discovery was based on observations at the Observatoire de Haute-Provence, with the 1.9-metre telescope and a spectrograph, ELODIE, that was able to measure radial velocities with what at that time was a very high precision, *"better than 53 m/sec"* (Baranne, 1996)[2].

Exoplanets, as astronomers have dubbed planets outside of the Solar System, are elusive beasts. Firstly, they are small, compared to their parent stars. Secondly, they reside in orbits close to these stars, so that only telescopes with extremely high angular resolution stand any chance of seeing such a planet. What is more, the difference in brightness between the star and the planet can be of the order of a billion, so that the planet is normally lost in the glare of the star. With the telescopes available to astronomers at the time, the chance of ever directly seeing an exoplanet was close to zero. But astronomers are ingenious and they had worked out ways to detect the presence of planet-size objects by studying the effects they would have on the parent star. Basically this meant taking one of three different approaches. The first and most successful method, and the one used by Mayor and Queloz, was to look for tiny periodic variations in the velocity of the star. The idea was simply that the gravitational interaction between the star and the planet appear differently depending on where

[1] In 1996, observations with the Wide Field Planetary Camera 2 on Hubble strengthened this suggestion (Burrows, 2012). And in 2003, a French team led by Anne-Marie Lagrange carried out new observations of Beta Pictoris with the adaptive optics instrument NACO on the VLT. After careful analysis, the team detected an object, estimated to be eight times more massive than Jupiter (Lagrange *et al.*, 2008). Follow-up observations in 2008 and 2009 also with NACO, confirmed that the object was not simply a background star, but is in fact a planet in orbit around the star. This provides a nice illustration of how progress in observational astronomy is linked to progress in technology. The original observations simply showed a warped disc around the star. Moving beyond this stage required the resolving power of a much larger telescope and the application of adaptive optics.

[2] Over time the Observatoire de Haute-Provence has played an important role as a test observatory for advanced instrumentation. This was the case for the CORAVEL spectrometer, for ELODIE, its successor, and also — as we have seen — the test site for the first adaptive optics experiments.

the two objects are relative to us — e.g., the planet being either in front or behind the star, as seen from the ground. To measure these variations required very high-precision spectrographs and with time, the Swiss astronomers and their colleagues excelled in developing ever-more sophisticated equipment that brought about a true bonanza in exoplanet research. Other methods could be used in rarer cases, where a planet in its orbit passes in front of the surface of the star, as seen from the Earth. In this case, the shadowing effect of the planet leads to a miniscule drop in measured brightness of the star during the planet's passage. In even more exotic cases, the opposite might be the case: the gravitational effect of a system with a planet orbiting a star might enhance the light of background star, a phenomenon known as microlensing. Both of these phenomena have been studied from the ground, but the method of measuring velocity variations has yielded the most results by far during the first 10–15 years of exoplanet research[3].

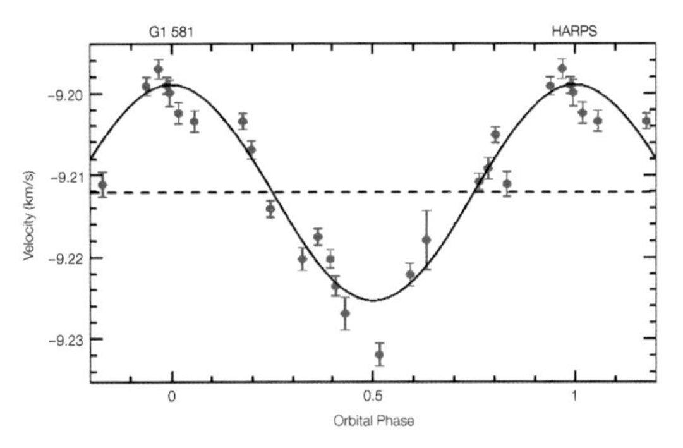

This HARPS result shows the radial velocity changes of the star Gliese 581, a red dwarf star at a distance of only 20 light-years in the constellation of Libra. The star is now believed to be surrounded by at least six planets, including one or more in the habitable zone, i.e. theoretically able to sustain some sort of life. Each of the planets exerts a gravitational effect on the star. The graph, produced in 2005 on the basis of HARPS observations, shows fluctuations in the radial velocity curve, induced by the planets as they orbit the star.

The Swiss 1.2-metre Leonhard Euler Telescope, named after the famous mathematician, was finished in 1998. It was fitted with CORALIE[4], an echelle spectrograph that increased the achievable precision in radial velocity measurements to better than 7 m/s. But this would not be the end of the story. The High Accuracy Radial Velocity Planetary Searcher (HARPS), also developed by the Swiss, initially achieved a precision of 1 m/s, which corresponds to the speed of a person walking. Today it offers a performance of a stunning 60 cm/s. HARPS is

[3] Beginning with the French COROT mission, launched in December 2006, telescopes in space now reap a rich exoplanet harvest by studying transit events.

[4] Unlike the names of most other instruments, CORALIE is not an acronym. It was named after a mechanical engineer and spectrograph specialist at the Observatoire de Genève (Mayor, 2012).

mounted on the 3.6-metre telescope, which not only has seen a second youth as a formidable planet-hunting telescope, but also given ESO leading position in this field of research.

And — how could it be any different? — plans now exist for even better instruments... .

HARPS was installed and commissioned at the 3.6-metre telescope in late 2003. Seen in the picture is Francesco Pepe, the HARPS System Engineer from the Observatoire de Genève. (Photo: Herbert Zodet)

≈

It has been said that astronomy is now experiencing a second golden age. This appears to be a reasonable statement, given the pace with which the science of astronomy progresses thanks to the application of ever more advanced technologies. Whereas, in the past, major discoveries might have occurred a few times during a century, today we see a flurry of new and exciting findings one after the other. Many seem to confirm our current worldview; others may challenge it. But, in 1998, astronomers were confronted with thoroughly unsettling set of observational data.

A group of 20 scientists, including ESO staff astronomers Bruno Leibundgut and Jason Spyromilio, published a meticulously prepared paper on spectral and photometric observations of ten Type Ia supernovae, which together with previous studies, meant that the sample was *"now large enough to yield interesting cosmological results of high statistical significance,"* as the group argued. The "interesting result", as the paper modestly said, however indicated nothing short of a revolution. Once more in the dry style of the scientific paper, the authors wrote that *"the results ... suggest an eternally expanding universe that is accelerated by energy in the vacuum."* (Riess *et al.*, 1998).

Since the 1929 discovery of the expansion of the Universe by the American astronomer Edwin Hubble — a discovery that underpins the Big Bang theory — the 64-million-dollar question had been how this expansion might change with time. According to conventional wisdom, the expansion would either continue forever or it would slow down, come to a halt and be reversed, with the Universe ending in a Big Crunch. The decisive factor, it was surmised, would be the amount of mass in the Universe — either enough for gravity to overcome the expansion — or not. Astronomers knew that the Universe contained much more mass than they could observe directly with their telescopes — nebulae, stars, galaxies, etc. — and the additional mass (which was by far the majority) was referred to as dark matter. The new observations, however, suggested a radically different scenario — that of a Universe with an *accelerating expansion.*

In a funny twist of history, Einstein had actually considered such an option when he developed his general theory of relativity. He had no knowledge of this phenomenon, but he had inserted a constant, which he called the cosmological constant, in his equations simply as a mathematical tool to prevent the Universe from collapsing. After Hubble's seminal discovery of the expanding Universe, he discarded the cosmological constant, which appeared "unaesthetic" to him, famously describing it as his biggest blunder. Prematurely, it would appear.

The new observations were of supernovae, the exploding stars that we discussed already in Chapter II-6, although of a different type from 1987A. This kind, Type Ia, is characterised by the fact that it has a specific intrinsic luminosity and can thus be used as standard candles, providing the observer with precise information about their relative distances. Furthermore supernovae are bright, so that they can be detected over vast distances. The idea of using Type I supernovae for this kind of research was not new. In fact it goes back to 1938 and is credited to Walter Baade, who, together with Fritz Zwicky, was engaged in supernova research. In the mid-1980s a group of

SN 1995K

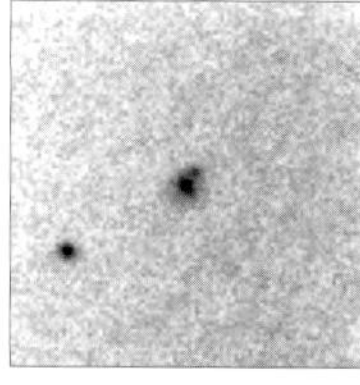

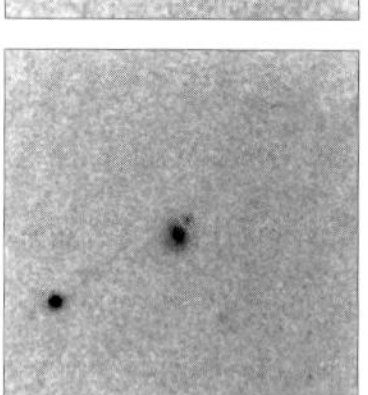

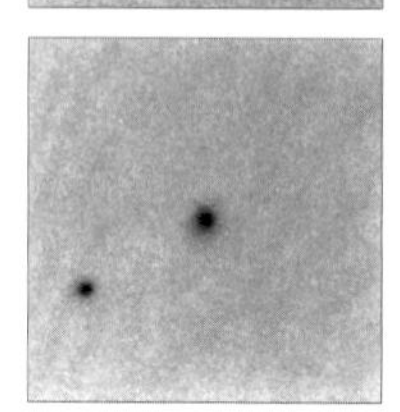

Supernova 1995K, a Type 1a supernova, observed with the NTT. It was one of the supernovae observations that led to the suggestion of the accelerating expansion of the Universe. The combination of observations with EMMI and SUSI, with its superb resolution, allowed the observers to quickly confirm the character of the target (Leibundgut et al.*, 1995).*

Danish astronomers had conducted a search over many years for these objects in distant clusters of galaxies, using a telescope at La Silla. However, they had netted only a few supernovae — enough to prove the concept, but not enough to justify its continuation amidst the ever-present struggle of prioritisation that individual researchers are confronted with. But the 1998 discovery of mysterious dark energy, a term invoked to "explain" the acceleration observed by this group (and also by a rival group), energised the worldwide astronomical community and set off speculation on how to verify the idea. In 2011, the work of both groups (the Supernova Cosmology Project and the High-*z* Supernova Search Team) was acknowledged with the award of the Nobel Prize in Physics to Saul Perlmutter, Brian Schmidt and Adam Riess. Dark energy raises dramatic questions. Where does this mysterious energy come from? If our ideas are correct and the Universe is continuously speeding up, will it eventually be ripped apart? Or are our present ideas about physics perhaps wrong, after all? If so, are we literally looking at completely new physics? Whatever the case, it once more showed us how little we actually know about the Universe and it provided at least a hint that many surprises lie in wait for us in our continuing quest to understand the world.

What was ESO's role in this effort? ESO telescopes had taken part in the programme and ESO astronomers had been members of both of the large and multinational teams. It was an example of how astronomy was gradually being transformed from the work of individual astronomers sitting patiently in their telescope domes and studying the heavens to elaborate and orchestrated undertakings requiring many different observations — either in different wavelength domains or with a number of different, but complementary telescopes and instruments — and involving large groups of scientists from many countries. Today it's not unusual for astrophysics papers in scientific journals to have a hundred or more authors, though the majority of articles still have a rather smaller number of authors behind them.

La Silla in 1991. At the very bottom of the picture, the office building of the maintenance group is seen. Above it, the administration and the scientific library building as well as the La Silla hotel. The buildings to the left are dormitories. Again from bottom to top, the Swiss 0.7-metre telescope is seen, followed by the Danish 0.5-metre telescope, the ESO 0.5-metre telescope, the Bochum 0.6-metre telescope and Dutch 0.9-metre telescope. In the middle: the ESO 1.52-metre, the ESO 1-metre, the GPO, the Danish 1.54-metre, the MPG/ESO 2.2-metre and the ESO 1-metre Schmidt telescopes. On the hill: the 3.5-metre NTT, the 1.4-metre Coudé Auxiliary Telescope and the ESO 3.6-metre telescopes. Finally, on the right-hand flank, the 15-metre SEST is seen. The small robotic telescopes are not visible in the picture but are located between the NTT and the 3.6-metre telescope. The large building below the NTT is the famous Astrotaller. (Photo by the author)

Chapter III-8

Upgrading, De-scoping

"In the VLT era, the functions and boundary conditions of the La Silla observatory will no doubt see drastic changes.... "

Johannes Andersen, *The ESO Messenger,* 78.

While ESO increasingly focused on the Very Large Telescope project, life at La Silla continued. The previous chapters have offered an impression of some of the scientific activities in which La Silla played a role. In the early 1990s a visitor to La Silla would have seen something close to a perfect setup. Sixteen telescopes were in operation every night, supported by an army of highly qualified technicians and engineers, organised into what was known as the Technical Research Support (TRS), headed by Daniel Hofstadt and with excellent optical and electronic test facilities. The huge mechanical workshop, the Astrotaller, was considered the best equipped workshop in all of Latin America. Astronomers would also find a large library and other facilities needed for their research. Other departments were devoted to civil construction and general maintenance of the infrastructure, from running a huge warehouse with supplies to operating the hotel and dormitories. There was also a self-service restaurant of some renown in the world of astronomy and leisure facilities for off-duty staff. At the foot of the mountain, an airstrip allowed for smooth and efficient transport between the observatory and Santiago, while an office in La Serena provided what one might call close support. La Silla was very much like a small buzzing town with white ESO cars everywhere, the pained sound of their oxygen-starved engines echoing among the telescope domes. There was lots of activity. It was a high-tech village in the middle of the desert. A self-contained world revolving around one topic: science. Could this be done any better?

In 1994, Jorge Melnick was appointed as Director of La Silla. Before that, he had been head of the astronomy department and a member of the management team in charge of the observatory. As an astronomer, Melnick focussed on the science output, which he thought could and should be increased. The prevailing approach to running the observatory, he found, was based on the (seemingly correct) assumption

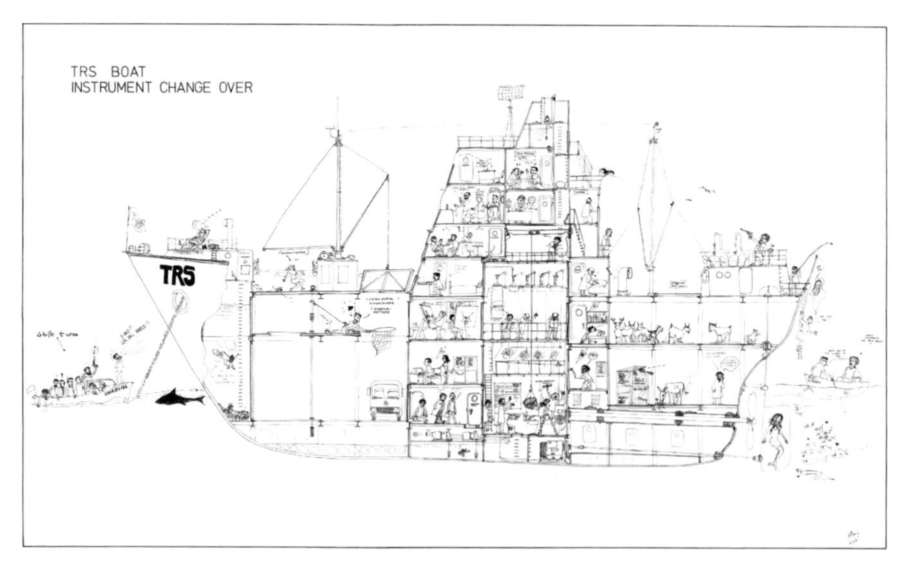

In this cartoon by Herman Nuñez, the Technical Research Support is depicted as an entire ship crew, providing an illustration of the versatile yet interconnected services provided. The original drawing was no less than two metres across, providing many details that cannot be properly reproduced here.

that because all telescopes are essentially prototypes, they break down from time to time. When they did, the TRS would move in and solve the problem. That occurred rather often. Melnick's philosophy was simple: in his view, telescopes should *not* normally break down. He therefore changed the local organisation and created telescope teams, each dedicated to — and responsible for — a particular telescope and each headed by a scientist. Of course, the telescopes did still break down, but the advantage of the dedicated teams was that they developed a very deep understanding of the peculiarities of the telescopes and instruments for which they were responsible. That, in turn, opened up opportunities for a thorough refurbishment of the telescopes, looking at the root causes for whatever problems were encountered rather than settling for quick fixes. This development was boosted firstly by the fact that the technical departments at Garching were increasingly busy with the VLT and could not devote as much attention to La Silla as in the past, and secondly, because there had been a growing suspicion that the telescopes, but especially the 3.6-metre telescope, were not performing as well as they should. Perhaps the first indication that the site offered better conditions than had long been assumed came with the commissioning of the Danish 1.54-metre telescope in late 1978. The MPG/ESO 2.2-metre telescope also yielded an image quality that was pleasing, but increased suspicion further. At the other end of the scale, the small DIMM telescopes developed by Marc Sarazin for site testing, also indicated that it should be possible to achieve better image quality. However, if anything, the stunning first light images obtained with the NTT confirmed the high quality of La Silla as a site for astronomical observations. Along with the NTT there were also much improved image analysis tools, based on a Shack-Hartmann set-up with a CCD detector. It was no longer necessary to use photographic plates, which had to be sent by the weekly diplomatic bag to Garching for analysis there, with the results communicated back by telex, at best. A portable wavefront sensor, called Antares, became available and gave the optical experts at La Silla the tool they needed to test the intrinsic quality of the telescopes. In a brief article in the *The ESO Messenger*, Alain Gilliotte presented a sobering result: *"Almost all telescopes tested suffered from spherical aberration.... Discrepancies between theoretical and real matching of primary and secondary mirrors produced the spherical errors.... "* (Gilliotte, 1992) In Chapters II-4 and II-9, we looked at the detrimental effects of spherical aberration, although in the cases of the Hubble and the NTT, the original problems were of a different magnitude. The analysis also revealed serious problems with dome seeing and mirror seeing caused by convection around the telescope, e.g., by a warm primary mirror heating the colder ambient air during the night or by insufficient dome ventilation. Moreover, classical problems such as tube flexure could now also be revisited, and so a targeted upgrade programme for the 3.6-metre telescope started in 1995. The problems related to thermal effects were addressed by removing heat sources in the dome, improving the ventilation and introducing fans at the primary mirror. The spherical aberration was corrected by lowering the focal plane by 166 millimetres. Certain aberrations had been found to vary with telescope elevation so changes to the M1 cell, including modifications to the astatic levers, were made. After improving the dome and the M1 mirror cell, the team took a closer look at the secondary mirror (M2) and the top ring. A new top ring was constructed by the Astrotaller as well as a new M2 unit, which was installed in August 2004. With further fine-tuning of the telescope, the image quality turned out to be better than 0.4 arcseconds, reaching values comparable to the external seeing monitors (Gilliotte *et al.*, 2005). The 3.6-metre was now close to its best possible performance. In 2007 the control system of the lateral pads was upgraded and the image quality of the telescope is now better than 0.2 arcseconds at zenith. Since the telescope remained in service during the upgrade[1], the project took considerable time, but the problems were solved one by one and thus, when the HARPS instrument with its ultra-high performance (and correspondingly high requirements on the

[1] In early October 2006, the telescope had to be taken out of service following a serious breakdown of the rotating dome. Preliminary repairs allowed restricted operations as of November of that year. However, it was clear that replacement of the bearings was necessary, a task that took half a year (Ihle *et al.*, 2007).

telescope) arrived at La Silla[2], the upgrade had almost been completed and the telescope was now able to deliver spectacular science results with the new spectrograph[3]. Since April 2008 HARPS is the only instrument available at the 3.6-metre, removing yet another potential source of image degradation: frequent instrument changes.

It may seem surprising that it took so long for ESO's long-standing primary telescope to reach its full potential. However, as we have seen, the understanding of the many technical aspects of telescope performance developed over time and was also subject to the development of better analytical tools. Also only with the advent of the VLT with its huge capacity, did the changing scientific demands on the 3.6-metre telescope allow for a drastic streamlining of its operation. Perhaps one of the most important lessons is the value to science that lies in allocating sufficient resources to maintenance and upgrades of telescopes over their entire lifetimes.

Not only the 3.6-metre telescope, also the other telescopes at La Silla were subject to thorough scrutiny and many improvements were carried out. Thus motors, hydraulic systems and encoders of the MPG/ESO 2.2-metre telescope were exchanged. The telescope was being prepared for the installation of the Wide-Field Imager, which would later deliver spectacular images of celestial objects. The guiding system of the Schmidt telescope was changed, the adaptor for the Danish 1.54-metre telescope was enhanced and many other upgrades were carried out.

≈

All of this work, together with activities that we shall describe below, yielded impressive results. La Silla became more of a scientific powerhouse than ever before. In fact, in 1998, when the world of astronomy marvelled at the VLT first light images, La Silla could take pride from the fact that in this year it had produced more scientific papers than any other ground-based observatory in the world (ESO Annual Report, 1998) — 400 papers in total. By comparison, the Hubble Space Telescope produced 514 papers, the mighty Keck telescopes, the largest in the world at that time, 94 articles (Grothkopf, 2010). However, the upgrade project and the reorganisation in general also contributed considerably to raising and maintaining the motivation levels of the La Silla staff at a time when it was clear that the observatory would

---

[2] HARPS saw first light on 11 February 2003.

[3] The last part of the upgrade was completed after the installation of HARPS. According to Gilliotte *et al.* (2005) the final modifications brought a considerable flux increase (40%), undoubtedly welcomed by the astronomers.

soon face substantial reductions and that the balance in terms of frontline science would increasingly shift towards Paranal. Rightly, the staff at La Silla took pride in their contribution to ESO and European astronomy in those years. And of course, during the 1990s, La Silla *was* Europe's observatory in the southern hemisphere.

We have already touched upon the importance of the NTT. We shall soon return to this telescope in the context of the VLT development, but it is also worth mentioning the NTT's role in supporting, in a sense fortifying the shift towards extragalactic research by European astronomers. The 3.6-metre provided the first step — it had whetted astronomers' appetites, but maybe more than anything, it was the NTT 's superior optical performance that opened the door to the deep Universe. Thus in 1998, 60% of the observing time on the NTT was dedicated to that kind of research (in addition to about 40% for the 3.6-metre). Many individual objects in the Magellanic Clouds were observed, but its programme also included spectral classification of galaxies, studies of Seyfert galaxies[4], clusters of galaxies, the search for UV-bright quasars, and studies of gravitationally-lensed quasars, etc. — the whole gamut of extragalactic studies. And of course La Silla also played its role in the orchestrated observing campaign that led to the seminal discovery of the accelerating expansion of the Universe and dark energy, discussed in the previous chapter.

Studies of objects closer to home, i.e. in our own galaxy, also prospered. For example a major investigation aiming at understanding the evolutionary history of our cosmic home, the spiral galaxy that we call the Milky Way, by studying a large number of individual solar-type stars, their composition, ages and motions in space. The study took 20 years and involved more than 1000 observing nights by a Danish/Swedish/Swiss team of astronomers (Nordström *et al.*, 2004). The idea was that detailed studies of our galaxy would allow us to test the ideas about the evolution of spiral galaxies in general. To use contemporary language, one could say that our current knowledge in this area was subjected to a stress test. A necessary one, it would seem. *"Classical disc evolution models spectacularly fail every classical test when confronted with our data,"* the scientists concluded (Nordström *et al.*, 2004b). Science is about constructing new knowledge by constantly challenging conventional assumptions! The main instruments used by the group were two high-precision CORAVEL spectrometers, mounted on the Danish 1.54-metre telescope at La Silla and on the Swiss 1-metre telescope at the Observatoire de Haute-Provence. Aside from the scientific merit of the project, we also note the Swiss preoccupation with high-precision

[4] A particular type of galaxy with a very bright core and hosting an active black hole at its centre.

spectrographs, which subsequently led to spectacular results in the search for exoplanets.

≈

An area in which European astronomy and observations at La Silla played a major role was that of star formation, which had evolved from a research niche to a major field of activity. One of the leading scientists in this field was Bo Reipurth, who worked as a staff astronomer at La Silla for eleven years. Stars are formed by the contraction of matter. Unsurprisingly, in a Universe with a mean density of a few atoms per cubic metre, stars form where matter is concentrated in clumps, dense clouds of dust and gas, such as we see them in our own Milky Way. But this means that the early phases of star formation happen behind a veil of dust, impenetrable by light, although increasingly transparent to longer wavelengths[5]. However, from a certain point in its lifecycle, a new-born star may reveal its presence by ejecting streams of matter into interstellar space. These streams or jets may become visible as they break free of the parental cloud, or when they collide with other interstellar matter, ionising it in what are known as bowshocks. Astronomers call these phenomena Herbig–Haro objects[6] (HH). Finding HH objects

HH-111 in Orion, observed with the NTT. This is one of the most prominent examples of a jet streaming out from a new star (located at the bottom of the image, but still hidden in its parental cloud). Further observations showed a somewhat fainter jet in the opposite direction. The bowshock is also clearly seen as a mushroom-shaped nebula near the upper edge of the picture.

[5] The main observational problem, however, is that at long wavelengths, the resolution is insufficient to resolve objects of small angular extension.

[6] After their discoverers George Herbig of the Lick Observatory (Mt. Hamilton, USA) and Guillermo Haro of the Observatorio de Tonantzintla (Puebla, Mexico).

means finding the signposts to stellar nurseries, which can then be studied with large telescopes and their full complement of infrared instruments. But detectors at that time were small; hence precise pointing of the telescope to the appropriate position was necessary. Reipurth therefore used the ESO Schmidt telescope with its large field to first identify candidate targets (Reipurth & Madsen, 1989). Follow-up observations were then made with the NTT, among other telescopes, to confirm the nature of the objects and, over time, to study their evolution and motion in space.

Holger Pedersen checking one of the GMS telescopes. (Photo by the author)

These studies provided examples of important scientific work carried out with relatively small telescopes (at least seen in the perspective of the VLT) — studies requiring massive amounts of observing time but of relatively bright sources, thus not in need of the light-collecting capacity of the new large facilities. Important work was done with even smaller telescopes. Thus, since 1982, La Silla has seen the arrival of a succession of robotic telescopes with names like GMS[7], TAROT-South[8] and REM[9], all relatively small telescopes aimed at detecting the optical components of gamma-ray bursts.

≈

Despite the impressive scientific achievements, the heyday of La Silla was coming to an end. The financial pressure of maintaining two sites, La Silla and Paranal, at

[7] Gamma-ray burst Monitoring System.

[8] Télescope à Action Rapide pour les Objets Transitoires.

[9] Rapid Eye Mount.

this level could not be sustained, a fact that had been clear at least since the decision to develop Paranal had been taken. Furthermore, with the 8–10-metre class telescopes coming into service, it was obvious that the science at La Silla would change, as Johannes Andersen expressed it: *"... the work [at La Silla and other observatories with 1-4-metre-class telescopes] will be largely conditioned by the research done at the 8-metre giants, and their tasks then will be different from now."* (Andersen, 1996). Andersen presented a report by a working group set up in 1994 by Giacconi to consider the future of La Silla. Aside from himself (as Chairman), the working group consisted of Sergio Ortolani, both members of the Scientific Technical Committee, Michel Dennefeld and Hans Schild, representing the Users Committee and Jacqueline Bergeron, Jim Crocker, and Jorge Melnick from ESO. The report recommended that ESO should concentrate its efforts at La Silla on maintaining a limited number of four telescopes, obviously including the 3.6-metre and the NTT with upgraded instrumentation, but that in general the observation programmes carried out at La Silla should be seen as complementary to what would be done with the VLT. An important follow-up report, prepared under the leadership of Birgitta Nordström and entitled La Silla 2000+ appeared in 1999, listing the needs of ESO's users in the light of the first experience of the VLT operations [10]. The STC followed the recommendations, giving top priority to the operation of the 3.6-metre and NTT, followed by the 2.2-metre and, as third priority, the continued ESO use of the Danish 1.54-metre telescope. Whilst it endorsed an upgrade of SEST, it also requested that ESO *"investigate the possibility to shift submm activities to Chajnantor as soon as possible,"* [11] noting that were that to happen, SEST operation could cease.

La Silla was thus set to experience a period of decline, with telescopes closing [12] as well as a reduction in staff and budget. This posed a huge challenge to the management who had to keep up staff motivation levels and secure the necessary expertise to keep the observatory running as a first-class scientific facility. Melnick and his lieutenant Gaetano Andreoni were well aware of the potential problems and they introduced a number of measures aimed at alleviating the situation. One was the introduction

[10] Further reductions were agreed in 2007 with the La Silla 2010 plan, leaving the 3.6-metre telescope with HARPS as its only instrument and the NTT with EFOSC 2 and SOFI at the Nasmyth foci and finally continuing support for the MPG/ESO 2.2-metre telescope following an extension of the agreement with the Max-Planck-Gesellschaft (Saviane *et al.*, 2009).

[11] STC recommendations, 10 November 1999.

[12] Some telescopes continued to be operated by national groups or through sharing arrangements. Thus already in 1996, Brazil became a 50% partner in the ESO 1.52-metre telescope. This lasted until 2002. Between 2007 and 2010 Brazil also had access to the 2.2-metre telescope.

of the "La Silla University" as *"an in-house institution devoted to find ways of embedding learning in a range of meaningful contexts where ESO staff members can learn to use their knowledge and skills creatively, to make an impact on the world around them."*[13] The University offered courses in a wide number of topics ranging from science and management training to social and arts activities, such as drama, music and even the martial arts. The impact of this was considerable since La Silla was not simply an observatory, but a small and isolated society, where people stayed together day and night for periods of various lengths, but seldom less than a week and often much longer. A further tool both to retain motivation and optimise operational processes, necessary as staff reductions became the order of the day, was the introduction of the ISO 9000 quality management system. The ISO 9000 family of standards is designed to assist systemic changes in enterprises. It is based on well-defined service standards, with processes monitored by internal auditors and thus, in a sense, devolves responsibilities to the staff at large. In 2003, La Silla was formally certified as an ISO 9001 undertaking, quite possibly the only observatory in the world that has been granted this status.

≈

As we return to the VLT story, we will still for a moment stay at La Silla and revisit the NTT. We will remember that this telescope had delivered spectacular results during its first period of operations, but later, things changed. Dietrich Baade formulated it this way: *"Sufficiently many excellent observations have been obtained to raise the expectations of the observers' community substantially beyond the traditional level. However these hopes have often been disappointed."* (Baade *et al.*, 1994). There may have been several reasons for this, as indicated by Baade and his colleagues: *"It has often been remarked that after the commissioning period the NTT apparently never fully repeated this early performance. However, it deserves to be mentioned that since the end of 1990 the average seeing recorded with DIMM2 kept deteriorating until 1993 when a dramatic improvement started… ."* It also became evident that, under the pressure from the VLT project, ESO had moved resources away from the NTT too quickly and had not given itself time to eradicate the "childhood diseases" that most new high-tech projects will display for some time. These problems, it would appear, came as a blessing in disguise. ESO was of course focused on achieving first light for the VLT according to schedule. Both the complexity of the telescope and its software and the necessity of developing a plan for the science operations for the

[13] La Silla University Mission Statement.

VLT warranted substantial "live" testing ahead of the magical moment of first light. In the summer of 1993 Giacconi had put Baade in charge of both developing the science operations plan and of looking into the problems still plaguing the NTT, which were mainly issues of reliability. Baade, in turn, put together a team, hand-picking staff both from Garching and from La Silla. The idea of having a dedicated NTT team of experts, which would operate with a high degree of autonomy, would soon be taken over at La Silla in general. The original NTT team included La Silla staff astronomer Gautier Mathys, Paul Le Saux, Anders Wallander, Domingo Gojak, Riccardo Parra and Percy Glaves. Inside ESO they were seen as the Dream Team. Roland Gredel, at the time also La Silla staff astronomer, worked with the team as well, mostly on infrared instrumentation. One of the first problems that the group saw was linked to the control software. There were issues of training and expertise in using the software, but it also became clear that upgrades would be difficult and it was decided to install the VLT control software, instead. However, to be compatible with the VLT, it would also be necessary to exchange the control electronics, and all of this would effectively mean taking the telescope out of service for a while. The NTT had always been seen as a testbed for VLT technologies, but in the meantime — and despite its problems — it had become a user instrument and a highly coveted one at that. Decommissioning it for a period of almost one year was not easy, but it would prove to be a good investment. In October 1995, Baade passed the baton to Jason Spyromilio, who carried out what became known as the NTT Big Bang, which occurred between July 1996 and July 1997 — in time to gain experience with the new operational environment before VLT first light.

One of the first observations with the upgraded telescope was — how could it be any different? — another deep field, produced by combining a large number of relatively short exposures (amounting to 32 hours of observing time in total) with the Super Seeing Instrument and reaching objects of about 27th magnitude. *"The image illustrates well the capability in deep imaging at good angular resolution at a 4-metre-class ground-based telescope,"* wrote the principal investigator, Sandro D'Odorico, in the *The ESO Messenger* (D'Odorico, 1997). A few years before, a *"4-metre-class"* would have meant one of the largest telescopes available. But his formulation lets us sense that the astronomical community had now set its sights on greater things.

However, while astronomy was blossoming during the 1990s, boosted by new technologies and larger telescopes coming into service, more mundane issues began to take centre stage at ESO.

Chapter III-9

Clouds on the Horizon

"It seems to me that ESO's overall situation in Chile has become more difficult than in the past...."

Hans Peter Kunz-Hallstein, fax to Riccardo Giacconi (undated, but apparently June 1994).

The change of Director General in January 1993 — in the middle of the VLT project — signalled a crisis. But if ESO found itself in rough waters, it was nothing compared to what was still in store for the organisation.

As already mentioned, the original *Convenio* between ESO and Chile dated back to 1963. By the time the VLT project began to assume a firm form, much had changed both at ESO and in Chile. Chile had implemented a peaceful transition from the military regime of Augusto Pinochet to a democracy, led by President Patricio Aylwin and backed by a broad coalition of political parties. The country was seeing the beginning of a strong economic upturn, and many well-educated people who had lived in exile for many years had returned to their native country, bringing with them valuable experience from the outside world. The issue of installing the VLT in Chile, at a new site, had of course not been foreseen in the 1963 accord and the wording of the text could be considered somewhat ambiguous as regards the legal status of new facilities in different locations from the original one. For that reason, but in fact rather more out of a fundamental wish to seek what it considered to be a more equitable relationship, on 11 November 1991, the Government of Chile requested a modification to the 1963 agreement[1]. ESO was not unwilling to consider this, but was struggling with finding ways to do so that would be compatible with the principles upon which the organisation rested. We shall return to the details below. Shortly afterwards, in December 1991, Council met in Chile — its first official visit since 1972. A few

[1] Through its diplomatic channels, the Chilean government had begun to press for this as early as 1988, with the first direct exchange with ESO occurring in 1989. At the same time, several Council members had also expressed the view that the transition of government would constitute a good opportunity to put the relationship on a new footing.

days before Council's arrival, the then Director General van der Laan, accompanied by Gerhard Bachmann and ESO's lawyer in Chile, Antonio Urrutia[2], met the new Chilean government's minister for science and education, Ricardo Lagos, for a first exchange regarding the future relationship, including the growing labour issues and the country's wish to obtain guaranteed access for Chile's astronomers. At the meeting, van der Laan held his ground, referring to the principles for granting observing time and to the status of international organisations as regards national legislation. However, after the meeting a worried Bachmann remarked to him: *"You won the argument, but [logical] arguments in politics don't work!"* Further meetings followed with Edgardo Boeninger, Secretary General of the President's Office and finally, on 9 December, with the President of the Republic, Patricio Aylwin. For the latter meeting, van der Laan was accompanied by Franco Pacini, President of the ESO Council, and Per-Olof Lindblad, member of the Council for Sweden.

At the suggestion of Minister Lagos it was decided to form a joint Chile–ESO working group to consider possible solutions. In parallel, ESO established its own working group, chaired by Henrik Grage and with participation by Peter Creola, Pierre Léna and Gerhard Bachmann[3]. The deliberations of the internal working group led to an invitation in June 1992 from ESO to Chile to join the organisation as a regular member.

On 19–20 April of the following year, an ESO delegation visited Santiago to discuss the situation with the Chilean Government, represented by Ambassador Mario Ataza. The delay was partly due to the fact that in the meantime, ESO had a new Director General (Riccardo Giacconi) and also because of forthcoming congressional elections in Chile. The delay, however, had not brought calm into the situation, which was becoming increasingly political. Given the outside pressure, the discussions were not easy, with quite a bit of brinkmanship being exercised. It was evident that in the prevailing political climate, the proposal for Chilean membership of ESO could not be pursued further. On the contrary: later in April, the Chamber of Deputies raised the stakes by adopting a resolution, which formally requested the

---

[2] Urrutia was a well-connected lawyer who had provided legal counsel to ESO since the early days. He had also acted as a go-between *vis-à-vis* the military government. But with the changes in Chile at the beginning of the decade, it became clear that ESO needed a more broadly based independent legal support to deal with a wide range of problems. Following the visit of the negotiating team in 1994 (see p. 297), ESO therefore engaged the prestigious Philippi law firm in Santiago. Laura Novoa, from this firm, became ESO's new Counsel in legal matters.

[3] From 1994 this group was comprised of Catherine Cesarsky, Franco Pacini, Gerhard Bachmann and Henrik Grage (Chair).

Government of Chile to re-negotiate the Treaty *"and also mentioned the possibility that the Chilean government could make use of its right to revoke the Treaty"*[4], a dramatic measure in international relations, should it finally be taken. However, negotiations continued in June, this time in a much more positive atmosphere and leading to a shared understanding in principle of how the wishes expressed by the Chilean government could be accommodated. A document was now on the table and, in a joint press release by the Chilean Ministry of Foreign Affairs and ESO on 23 June 1993, it was stated that *"the Agreement will be submitted for ratification to the Congress of Chile and to the Council of the ESO."* It seemed that the problems had been solved.

The labour problems had built up over many years. When ESO came to Chile in 1963, finding qualified Chilean staff was difficult, and in many cases impossible. Local staff were therefore only hired for menial work, leaving the specialised tasks to ESO's international, i.e. European, staff. But things had changed and gradually, qualified Chilean staff began to work side by side with their European colleagues. For Europeans, in turn, working in Chile meant uprooting families, settling far from their home countries, accepting the undisputed hardship of working in the desert, including the long periods of separation from spouses and children. This had to be compensated by a premium on the salary and thus, with time, the situation led to dissatisfaction among the Chilean staff, who felt discriminated against[5]. On the other hand, introducing equal pay at a level that could still attract European staff, it was feared, might have had a disruptive effect on the Chilean labour market. La Silla in particular constituted a close-knit society with the staff living and working together for weeks, while at the same time being in an isolated place. Given these conditions, it is remarkable that both at the personal and professional level, relations between local and international staff always remained cordial, but at the institutional level the situation became increasingly tense.

≈

ESO is an international, intergovernmental (treaty-based) organisation in the sense of Article 102 of the United Nations' Charter. Such organisations are not subordinate to national legislation and therefore are obliged to maintain their own regulations

[4] ESO Press Release, 14 May 1993.

[5] This view was shared by some Europeans, as Blaauw readily acknowledges in an article in the *Annual Review of Astronomy & Astrophysics* (Blaauw, 2004), which also reveals that the problem had basically remained unresolved for 20 years.

including methodologies to determine staff pay, social security and pension schemes. It was widely assumed that such systems were not suitable for trade union involvement, since trade unions normally work within, and use, national legal frameworks, a view also held by the International Labour Organization. But ESO local staff saw a larger role for their staff association, including its transformation into a formal trade union and the introduction of collective bargaining, as a way to improve their conditions. Neither was foreseen within the ESO system.

The Chilean staff sought allies within the political parties of Chile, arguing that they *ought* to work under Chilean law, and thus began to exert pressure on the organisation through the country's political system, a system that was still trying to find its feet after more than one and a half decades of forced inaction. They sent an enquiry to the International Labour Organization and later also petitioned the European Parliament, accusing ESO of violating Article 23 of the Universal Declaration of Human Rights by not allowing trade unions[6]. For ESO, however, the situation was less simple. It could not go against the basic tenets of its legal status as an international organisation. It was also clear to ESO that the demands of the union activists, should they be met, might *de facto* mean less favourable conditions for its local staff than those under which they were already working[7]. But national sentiments were in play, and in any case, the conflict soon became one element of a potentially lethal cocktail that was brewing.

≈

Chilean astronomers also aired grievances. During the 1970s and 1980s Spain had established a large astronomical observatory in the Canary Islands with telescopes on the islands of Tenerife and La Palma (Observatorio del Roque de los Muchachos). Here a number of universities and national research organisations from many European countries had placed telescopes and in compensation — also for Spanish investments in infrastructure — provided Spanish astronomers with 20% of the available observing time. This had been very important in developing Spanish astronomy to its present high level. In Hawaii, observatories had made a similar concession to the local astronomers. In Chile, AURA had ceded 10% of the observing time to Chilean

[6] ESO never objected to staff members being members of national trade unions, but for the reasons mentioned could not agree to accord such unions with negotiating rights.

[7] Studies confirmed that ESO was among the top five best-paying employers in the country. Together with other benefits working for the organisation meant enjoying a very attractive employment package.

scientists. ESO, however, had not done so — and, at least for a while, did not consider it part of its remit to deal specifically with Chilean astronomy, as Woltjer's remark on p. 94 makes clear. It is true that when ESO came to Chile, Chilean astronomy was not sufficiently developed to take advantage of the ESO facilities, but meanwhile, a lot had happened in the country. There was, however, a more fundamental issue in that ESO found it difficult to give guaranteed time to scientists from one country, which was not even a Member State, while denying a similar, privileged status to astronomers from its own Member States, i.e. those countries that actually funded the organisation. This touched upon a basic principle of ESO: telescope time is granted on the basis of scientific quality, not passport. Astronomers wishing to use ESO's facilities apply for telescope time in competition with their colleagues. Their proposals are subject to a peer review process, their applications being considered by the Observing Programmes Committee and the best are selected. Competition is tough since ESO receives many more applications than it can accommodate. Hence, while ESO had certainly always welcomed Chilean observers[8], there was considerable reluctance to grant them guaranteed time, which might be seen to subvert the principle of scientific excellence based on peer review. The Chilean community of astronomers saw this quite differently. The presence of the foreign observatories with their continued traffic of observers from around the world had had an impact on Chilean astronomy. The Chilean scientific community had grown in numbers and the quality of their science had improved, too. Understandably, so had the ambitions of Chile's astronomers. In 1966, the Universidad de Chile had established an astronomy department. Graduates had often continued their studies overseas and, upon their return, bred new life into the small community of astronomers. Over time, the university had expanded its astronomy programme and other universities had or were about to follow suit, including the Pontificia Universidad Católica de Chile and Universidad de Concepción. The universities in the north of the country, in La Serena and in Antofagasta, also aspired to engage in astronomy. The example from Spain had shown how beneficial a privileged status for a national group could be, at least in the phase of development in which Chilean astronomy then found itself. In February 1988, when it was decided to build the VLT and it was thought that it would most likely be placed at Cerro Paranal, the Chilean government sought formal access rights for Chilean astronomers, a suggestion that ESO turned down for the above reasons. They therefore began to press their case via the International Council of Scientific Unions (ICSU) and also through the Chilean Congress, where

[8] ESO had, in fact, also employed astronomers of Chilean nationality at its Headquarters in Europe since the earliest days of the Science Division.

some members discussed a motion demanding a halt to the VLT project on Paranal. Evidently, the issue could not be contained and solved within the boundaries of the scientific community, and — at least in the public sphere — it became fully intertwined with the labour problems. And if that were not enough, yet another conflict emerged — of a very different nature and, this time, as something of a surprise.

Chapter III-10

At the Brink

"Latorre issue a real problem!"

From a slide presentation by Riccardo Giacconi
to the ESO Council, 1 December 1993.

In October 1988, as it became increasingly clear that Cerro Paranal would be the site of choice for the VLT, the Chilean Government donated 725 square kilometres of land around the mountain to ESO, on the condition that ESO would commence construction work within five years[1]. The government had considered itself the lawful owner of the land since 1987, following the implementation of a 1977 Government Act, which had the aim of clarifying land claims and ownership in Northern Chile, a vexed issue with a history that goes back to the time before the Pacific War in the late 19th century[2].

Following the decision by the ESO Council to place the VLT on Paranal and, in early 1990, the beginning of the actual construction work, ESO had formally fulfilled the conditions of the donation, and hence the ownership, and expected no difficulties in this respect. However, in April 1993, the Chilean Andrés Santa Cruz family sued the Chilean government and ESO for damages, claiming that they were the rightful owners of Paranal[3]. The mountain, said the family, had been donated to their

[1] Law no. 643, published on 19 October 1988.

[2] The situation regarding land deeds had been complicated partly because the relevant archive, located in Copiapó, had been destroyed in a fire in the early part of the 20th century. With the 1977 Act, the Chilean Treasury had regained the ownership of the land north of Taltal and incorporated it into its National Properties. The Act requested possible claimants to land titles to register and present appropriate documentation of ownership to the authorities within a period of ten years. By failing to register, claimants would forfeit their ownership. By registering, however, the owners would be liable to property taxation — a consequence that may have prevented some owners from coming forward. The Act was introduced in 1977, meaning that with the expiry of the ten-year grace period, the ownership of Paranal had irreversibly gone to the Chilean State.

[3] Long forgotten, and unknown to ESO, was the fact that Paranal had been the subject of a land dispute before — back in 1905 — involving the Santa Cruz family and mineral prospectors wishing to open a nitrate mine in the vicinity of the mountain.

ancestor, Vice-Admiral Juan José Latorre Benavente, in gratitude for his services to the nation during the Pacific War. Allegedly, the heirs now "wanted their mountain back" — or sought financial compensation in the form of several tens of million of US dollars, an absurd amount from any reasonable point of view. Yet, ESO was ill prepared for this situation.

Before the court, ESO claimed legal immunity in full accordance with the 1963 *Convenio*. Surprisingly, the judiciary decided first to hear the case and only afterwards to consider the question of immunity. According to the judge, the fact that ESO had appeared before the court (to claim immunity) meant that ESO acknowledged the jurisdiction of the court and thus, in a sense, had forfeited its protection. Furthermore, in Parliament the argument was also made that the immunity might not extend beyond the geographical locations where ESO had originally established its facilities, disregarding an exchange of notes in 1983/1984 between the Government of Chile and ESO according the privileges and immunities which derive from the 1963 *Convenio* to all future astronomical observatories which ESO would install in Chile with the agreement of the Government. To the uninitiated reader all of this may appear strange, but it must be understood that it happened in the aftermath of the demise of the military dictatorship. Perhaps understandably, guarantees from the former government seemed not to carry the same weight anymore — even in matters of international relations.

In March 1994 the lower chamber of the Parliament stepped in again, requesting the Government to order ESO to halt construction at Paranal, underscoring that this was fundamentally a political issue. This notwithstanding, on 17 March 1994, the Latorre party filed a request with the civil judge of Taltal (the provincial town nearest Paranal) aiming at a court injunction against ESO's contractor Skanska-Belfi Ltd. — charged with the construction work — for a prohibition to "effect new works" on its alleged property. The conflict escalated quickly with court hearings and injunctions being issued, overturned and reissued. Some involved ESO directly, others the main contractor. *"Two weeks ago,"* reported Allison Abott in *Nature* on 21 April 1994, *"different court members issued four contradictory judgments within days of each other."* Following a court ruling, construction work ceased for a few weeks in what ESO — maintaining its immunity — said was a gesture of goodwill. After that, work resumed. Halts to construction were very costly. At the time they were estimated to cost one million deutschmarks per month.

≈

While the court battles were fought, negotiations about changes to the *Convenio* continued. From the outset Riccardo Giacconi sympathised with the wishes of the Chilean astronomers. He thought that their demands were fair and he argued in their favour in the ESO Council. Also, despite the tough rhetoric[4], he had much understanding for the Chilean staff. Given his position, however, he could not possibly condone some of the actions by the union activists, but he was committed to finding a mutually agreeable solution. In late May 1994, he led a small delegation including Peter Creola (President of the ESO Council), Catherine Cesarsky (Vice-President of the Council) and Henrik Grage (Chairman of the Council ESO-Chile Working Group) in a meeting with representatives of the Chilean Government, including a visit to La Moneda to see the President of the Republic. To hedge his bets, Giacconi also began to consider alternative sites for the VLT. To this end, in April he ordered that a site-testing campaign in Namibia should be conducted, which began in August and lasted for a full year. Nonetheless, the discussions seemed to have gone well and by the autumn of 1994 a full understanding was reached between the Chilean government and ESO, leading to the initialling of an agreement by Ambassador Carlos Ducci on behalf of the Chilean government and Peter Creola for ESO on 22 November. Despite this, pressure continued to build up in Chile. Some members of Congress remained highly critical of the Agreement and the Chilean Minister of Foreign Affairs and his collaborator, Ducci, both trying to resolve the conflict, even had to fend off the threat of criminal proceedings against themselves for having initialled the Agreement with ESO on the accusation that it "violated the Chilean constitution". At the same time, imported items for the Paranal site were blocked by the customs authorities.

≈

The court cases over the land ownership also proceeded and ended at the Chilean Supreme Court, which issued contradictory rulings on five occasions. The principal topic of legal dispute was the precedence of property rights over an international treaty or vice versa. Since no court ruling had been issued so far regarding the property rights of the plaintiffs the Supreme Court decided in its last ruling to wait

[4] In one instant in 1993, however, he went beyond rhetoric, suspending (with pay) two local staff members for gross disloyalty to the organisation. However justifiable the suspensions were, it temporarily poured oil on the fire, since they were also elected as staff representatives. They were later reinstated.

until these rights were clarified before issuing its judgement. In late January 1995 a Supreme Court ruling could be interpreted as an injunction to stop the construction at Paranal. The situation was becoming explosive. Not surprisingly, ESO was the focus of intense media attention. Daniel Hofstadt, who had been appointed ESO Representative to Chile, had his hands full. Luckily, he had good help from his lieutenant, Rodrigo de Castro, a Chilean journalist, who had returned from exile in Europe to his homeland to work for ESO. Daniel Hofstadt had joined ESO in 1972 and, as already mentioned, had headed the TRS Department at La Silla and later served as head of the management team there. Perhaps aided by his Alsatian origin, which had made him sensitive to the ways of different cultures, perhaps also by his wide range of interests, from technology and science to the arts and philosophy, after his appointment as ESO Representative he quickly developed into a highly successful diplomat, earning the respect of his Chilean counterparts who rightly came to see him as an honest and dedicated interlocutor[5]. Yet, if anything, this was his battle test. In a dramatic press conference in February 1995 Hofstadt explained that ESO *"was not in Chile to handle court cases but to do science. It was for the Chilean government and its State Defence Council to settle the issue. ESO could only work in good faith and with the commitment of the Government to safeguard its projects"*[6,7]. Luckily, at Foreign Affairs, (then) Undersecretary José Miguel Insulza began to take an interest in the case[8]. Hofstadt and Insulza went on to develop a close relationship, which helped to solve the problems.

But then things went from bad to worse: a ruling by an Antofagasta court on 20 March again ordered ESO to halt construction work at Paranal and sent a court official to inspect the site. Referring to its immunity, ESO denied him access, while, at the same time, the Chilean government — in support of ESO — had launched an appeal against the Antofagasta verdict in the Supreme Court. But on 28 March, the Supreme Court in Chile upheld the injunction to stop construction and decided to send the visiting judge back again to Paranal. The conflict was spinning out of

---

[5] Hofstadt was later awarded the Bernardo O'Higgins Medal, on honour bestowed on selected foreign citizens in recognition of their outstanding services and contributions to Chile.

[6] *El Mercurio* sharpened the statement somewhat by writing *"La entidad está en Chile para hacer ciencia, no para participar en polémicas o litigios"* (*El Mercurio*, 14 February 1995).

[7] Hofstadt had to face the press on many occasions in this period, but at least in one case, science came to his rescue. The collision between Jupiter and Comet Shoemaker-Levy 9 (see Chapter III-6), which also became a media event in Chile, gave ESO a welcome opportunity to present a very different image in the Chilean press than the negative one that had dominated for a while.

[8] Insulza was appointed Minister of Foreign Affairs in September 1994.

control. By then the various issues had become completely interlocked, involving not just ESO, its staff and the Chilean astronomical community, but also the three branches of state in Chile: the judicial, the legislative and the executive, and this at a time when the democratic system was still trying to find its way within a new and untested constitution. Of course the media covered the conflict extensively[9]. Thus, on 30 March, it came to a showdown as the Supreme Court decided to show its teeth with the arrival of the visiting magistrate. Immediately before, Italy's foreign minister Susanna Agnelli, who was on an official visit to Chile, had discussed the matter with the Chilean president[10]. The president had assured Agnelli that the Government supported ESO. However, it could not defy a ruling by the Supreme Court, and so events took their course. ESO's official press statement, issued on the following day, makes no bones about it: "*... at 12:40 h Chilean time, a Chilean court official, Mr. Javier Jimenez,* Receptor Judicial, *accompanied by Chilean* Carabineros, *forced an entry to the premises of ESO on Cerro Paranal without permission of ESO and without agreement between the Chilean Government and ESO, although warned that this act was in violation of the status of ESO as an international organisation on the basis of the 1963 Convention and subsequent agreements between the Government of Chile and ESO.*" The ESO area around Paranal is a desert area without fences or any other clearly visible demarcation. But where the lone road from the B-70 desert road — the old *Panamericana* — leads to the observatory camp, there is a gate with a barrier. This was where, on 30 March, the magistrate accompanied by police and the ESO representatives met. The barrier was fixed with a metal chain. ESO had received advance warning of the impending court action and key representatives of the organisation including Daniel Hofstadt and the Head of Administration, Willy Buschmeier, were present. Given the legal nature of the event it followed rather strict rules; one could almost describe it as being carefully scripted. The magistrate was expected to arrive at the gate and demand access. ESO would not grant this. On the other side of the barrier, the ESO representatives were then supposed to read out an official protest when the magistrate crossed the gate after cutting the chain in a symbol of forced entry. Of course ESO staff had strict orders not to use any physical force. But for a moment the magistrate, surrounded by running television cameras (and almost certainly without experience in handling court actions of such magnitude and with clear international implications), seemed to have forgotten the script. Arriving at the

[9] *El Mercurio* alone published 49 articles between 1 February and 25 April 1995. Other newspapers in Chile were equally active.

[10] Shortly before the President had paid a visit to Europe, where the topic had been raised in political discussions in several ESO Member States.

chained barrier, he did what any sensible person would do: he simply went around it. A young ESO staff member panicked, fearing that the carefully foreseen sequence — an adherence to which was deemed of legal importance — would be endangered. He laid his hands on the magistrate's shoulders, exclaiming *"No, no! You can't simply do this!"* For a moment, everybody froze. Did this completely unintended incident constitute an act of violence? Luckily, a sober-minded police officer intervened, calmly explaining to the magistrate that he needed to retreat to the other side of the barrier and first read out his court order. He did. Upon making his demand, he cut the chain and walked in, unhindered, to inspect the site. The visit lasted longer than planned. When the magistrate had completed his job and returned to his car, which an assistant in the meantime had parked on the site he saw that the car had a flat tyre — the effect of driving into the construction site with its rugged terrain and many sharp pieces of metal lying around. To his relief ESO staff came to help and changed the wheel. In this way, the day ended on a more conciliatory note — at least at the human level, and in spite of the somewhat farcical air, it symbolised that now the time had come to defuse the conflict.

Daniel Hofstadt (to the right) reading out his declaration facing the magistrate (to the left) and surrounded by media representatives. (Photo: from El Mercurio*)*

In the media, it also seemed that the gravity of the situation had begun to sink in. Thus, *Las Ultimas Noticias* used the headline: *"The Incident at Paranal Affects the Image of Chile."*[11] The daily *La Tercera* was equally concerned, stating that *"When cutting the chains of the observatory, we have undermined our respect for an institution consecrated by an international agreement to which Chile subscribed."*[12] At the same time, in an organisation-wide video

[11] *"Lo ocurrido en Paranal afecta la imagen de Chile"* (*Las Ultimas Noticias*, 1 April 1995).

[12] *"Al cortar las cadenas del observatorio, hemos puesto en entredicho nuestro respeto a una institución consagrada por un convenio international que Chile suscrito."* (*La Tercera*, 2 April 1995).

conference for all ESO staff, Riccardo Giacconi unequivocally declared that *"if ESO cannot continue on Paranal, it can operate nowhere in Chile."* The stakes could be raised no higher.

≈

What had now become a major diplomatic incident of course also involved the representations of ESO's Member States, and on the following day, 31 March, the ambassadors[13] met with Chile's Minister of Foreign Affairs. Since the President of the ESO Council, Peter Creola, was Swiss, the Swiss ambassador, Bernard Freymond, acted as an unofficial spokesperson for the Member States. On 1 April Daniel Hofstadt visited the Chilean Ministry of Foreign Affairs to deliver an official protest.

Shortly thereafter, Giacconi arrived in Santiago. In a breach of normal customs, his arrival had not been announced in advance in order to enable the necessary talks to take place without public attention — and heavy press coverage.

Teitelboim, Giacconi and Hofstadt leaving the meeting at the Centro de Estudios Científicos. The meeting was supposed to have been confidential, but the press had caught wind of it and the picture was taken by a photographer from the Chilean broadsheet newspaper El Mercurio. *Though smiling to the press, the three realised the seriousness of the situation.*

On Friday 7 April, Giacconi, accompanied by Hofstadt and de Castro, met with Insulza and Claudio Teitelboim (Bunster), the science advisor to the President, at a lunch at the Centro de Estudios Científicos, Teitelboim's private institute. However, the press had found out and was waiting. Giacconi was uneasy, fearing that too much media attention might complicate the search for a solution. Nonetheless over the weekend, the final details of the

[13] Michel Delfosse (Belgium), Ole Wøhlers (Denmark), Gerard Cros (France), Werner Reichenbaum (Germany), Emanuelle Costa (Italy), Steven Ramondt (the Netherlands), Bernard Freymond (Switzerland) and Teppo Tauriainem (first secretary, Sweden).

agreement were hammered out between Hofstadt and Ducci, with the text being translated on the Monday. Everybody understood that speed was essential to avoid further complications. But the Chilean Government still needed more time to secure enough political support. On the Wednesday morning, 12 April, Hofstadt and Giacconi were invited to breakfast at the home of President Frei, together with Insulza. *"It was helpful that both [Chilean Minister of Foreign Affairs] Insulza and [President] Frei had been refugees in Italy — Frei working at the Politecnico di Milano and Insulza in Rome,"* Giacconi later recalled. He felt that it helped them to establish common ground. After the encounter, both Frei and Giacconi spoke to the waiting journalists. The President assured Giacconi in public of his full support in resolving the crisis. This was an important gesture and a signal to the ESO Member States.

A few days later, Giacconi's apprehension about the media turned out to be justified. A newspaper published an interview with him, obtained before the agreement was reached, in which he had aired a certain degree of frustration under the enduring pressure. Given the developments in the meantime, his remarks had, in a sense, become outdated, but the article made for a good story, though it also threatened to re-ignite the conflict. Quick, concerted action on the part of ESO, the government and Teitelboim, quenched the fire.

Finally the way lay open for the finalisation of the last version of the new agreement which would be termed the "Interpretative, Supplementary and Amending Agreement" to the 1963 Convention between the Government of Chile and the European Southern Observatory. Although much haggling had taken place over the fine print, the broad lines of the agreement remained as in the 1994 draft accord. It set out to *"confirm and regulate the application"* of the existing Convention. With the agreement ESO committed itself to *"harmonise its [staff rules for local staff] with the principles and objectives of Chilean labour law"* including freedom of association for workers and collective bargaining. This notwithstanding, it would be implemented in a way which was compatible with international law and thus respecting ESO's immunities and privileges as an international organisation. ESO also agreed to guarantee 10% of the available telescope time to scientifically meritorious proposals from Chilean astronomers. This would apply to all telescopes, with the stipulation that for the VLT, half of that time would require projects involving astronomers from ESO's Member States. Applications for telescope time would be subject to the same evaluation procedure and criteria as those from Member State astronomers, i.e. dealt with by the Observing Programmes Committee. Finally, ESO agreed to financially support the development of astronomy in Chile with a Joint ESO/Chile Committee to manage the funds. Chile, in turn, reconfirmed its commitment to ESO and to respecting its international status and extended it to possible future sites in the country. The government also agreed to increase its own investment in Chilean astronomy *"with the objective of promoting the efficient use of the installations of ESO by Chilean scientists."* Finally, Chile accepted an obligation to compensate ESO with respect to immovable installations should Chile terminate the agreement leading to a cessation of ESO activities in that country, thus providing a reasonable reassurance to ESO that it could safely proceed with its investments.

On 18 April 1995, Roberto Cifuentes, an emissary of the Chilean Government, and the ESO Director General met at the ESO Headquarters to sign the new agreement, which as is customary for such accords, would subsequently be submitted to the Chilean Congress (and the ESO Council) for ratification. But the crisis was not yet over, as the ESO press release, issued the following day, made clear: *"There still exist a number of problems which impede ESO's activities in Chile, including the issue of ESO's immunity from national jurisdiction, financial damage incurred to ESO and its VLT project, and the practical implementation of custom clearance and accreditation procedures to permit the continuation of work. Consequently, Council has directed the ESO Management to continue direct negotiations with the Chilean Government to resolve all pending issues, to continue work on Paranal, and also to convene an* ad hoc *Working Group to propose alternatives for the possible placement of one VLT telescope or of the entire VLT/VLTI."*

Towards the end of the year, however, closure was in sight. The Chilean government had acknowledged its responsibility in dealing with the Latorre claim and had agreed to pay a sum of money to settle the case out of court[14]. Nonetheless, the government was wary of a claim for damages from ESO. After all, the work stoppages had led to a significant financial loss on ESO's side — estimated at the time to be of the order of eight million deutschmarks (or roughly four million euros). At the Council meeting in Milan in November, it was decided, though, not to press any claims and with

[14] The resolution came in the form of an out of court settlement between the Government and the plaintiffs. As the Paranal case became more explosive the claim for the land prize had escalated to 130 million US dollars. For a piece of desert without any mineral riches this was totally out of scale with commercial land prices. An assessment report from the Universidad Catolica had determined that Paranal was a unique place for astronomy, and therefore it should be valued accordingly. Probably for the first time in human history the price of a piece of land was valued according to the quality of its atmosphere for astronomical observations. In any event, the negotiations came down quickly to 33 million US dollars but this amount was still not acceptable for the Government. It took some time to reach an agreement for the 8.8 million US dollars, which was paid by the Chilean Treasury to the Santa Cruz family to close the court cases.

this, communicated in a *Nota Verbal* to the Chilean Foreign Ministry on 4 December 1995, the final hurdle was cleared. On 16 January 1996, the government submitted the *Acuerdo* to the parliamentary ratification process.

In Chile, the Chamber of Deputies approved the agreement on 4 June 1996 with a huge majority, followed by the Senate on 5 September, clearing the way for a meeting of the ESO Council in Santiago in early December of that year, on which occasion the documents, or instruments as they are properly called, of ratification could be exchanged. A six-year conflict, which had thrown ESO into its gravest crisis, had been brought to a satisfactory close[15].

In the end, level-headed people had prevailed. There were many and they were found at all levels. Nonetheless, a few names stand out and deserve to be particularly mentioned. The President of the Republic, Frei, and his Foreign Minister, Insulza, had given their full support to ESO. More than that: Insulza had gone to great lengths to enable the final outcome. As mentioned, the science advisor to the President, Claudio Teitelboim, had played an important role in defusing the conflict. In Congress, the senator for Antofagasta, Arturo Allessandri — the nephew of President Jorge Allesandri with whom Heckmann had negotiated — defended ESO and the agreement against opposition and helped its passage through the legislative assembly. On ESO's side, the Council President Peter Creola and the incoming Council President Henrik Grage had worked hard together with the Director General. Still, there is no doubt about the decisive role played by Daniel Hofstadt, who became ESO's front man in this diplomatic, political and public imbroglio. We have also mentioned his close collaborator Rodrigo de Castro, who thanks to his background was uniquely placed to bridge the gap, clear up the misunderstandings and rebuild trust and good will between the parties. He later assumed high posts in Chile, including editor-in-chief of *La Nación*, the government newspaper. The diplomatic representatives of ESO's Member States in Chile were involved and the Swiss ambassador, Bernard Freymond, in particular played a constructive role.

[15] The full implementation of the agreement, proved to be complicated, not the least as regards the labour issues. Thus this part was only completed during the term of the next ESO Director General, Catherine Cesarsky.

Chapter III-11

Tranquillity in Chile, Icy Winds in Europe

"The deliberations, notably in the financial domain, were not always easy as most of the member countries were striving to stabilise their national budgets and had to prioritise their expenses."

Riccardo Giacconi, *ESO Annual Report 1997.*

With the formal exchange of the ratification documents on 2 December 1996 and the subsequent Foundation Ceremony at Paranal two days later, the door was finally closed on this difficult period[1]. The Chilean President used the ceremony to reinforce his message that good relations had been fully restored. In his introduction, welcoming the participants (including the Royal Couple of Sweden[2], Foreign Minister Insulza and the ESO Council), he explicitly added: *"Profesor Riccardo Giacconi, Director General,* estimado amigo." An important gesture and one that did not go unnoticed. He said: *"... I would like to take this occasion to publicly express our appreciation to the ESO Member States, to the Organisation itself, its executives and to all those in Chile who have contributed in this task and to say that when I took office there were not only stones, but rocks on the path, but there was also the political will to overcome them, since we could not allow hindrances in the construction of this great enterprise."*

The foundation ceremony had been intended to happen back in 1994, but — understandably — the conflict had prevented it. Finally, the time had come to celebrate the VLT project. To mark this unique moment, it was decided to insert a time capsule in a cavity in the outer concrete wall at ground level of Unit Telescope 1. The

[1] Ten years later in a joint publication, the fruits of the new cooperation in terms of astronomy projects in Chile were presented at a press conference in Santiago.

[2] The royal visit, although to some extent coincidental, was certainly welcome, for at the time, the Swedish government was critically reviewing Sweden's participation in European research projects, looking for savings. The press, however, looked at ESO with different eyes, with almost poetic press headlines such as *"The Royal Couple looked into eternity" ("Kungaparet såg in i evigheten").*

time capsule was an aluminium cylinder 15 centimetres in diameter and 45 centimetres in length, sealed hermetically and containing a document signed by President Frei on the occasion, a copy of the *Acuerdo* between ESO and Chile, one outstanding scientific paper from each of ESO's Member States and Chile, a copy of the day's issues of the Chilean newspapers *El Mercurio* and *La Época*, the texts of the official speeches delivered on the occasion, and various other pieces of information about the VLT project.

The President of the Republic of Chile, Frei, with the time capsule. On the left, Peter Creola (President of the ESO Council) is seen, to the right the Archbishop of Antofagasta, Monsignore Patricio Infante Alfonso. (Photo: Hans Hermann Heyer)

The President and his guests, the King and Queen of Sweden, had flown into a specially prepared airstrip at the foot of the mountain and the empty enclosure of Unit Telescope 1 had been transformed into a festive environment. ESO staff had worked hand in hand with government officials, security forces, technicians, caterers and many others to create a successful high-level event for the 250 invited guests in what was still a rough building site in the middle of the desert. In his memoirs, Riccardo Giacconi describes the feeling of hosting a formal luncheon with such an audience in these unusual surroundings as *"eerie"*, but everybody was happy. Of course, much work remained to be done before the VLT could begin to observe, but now the road lay open. Or so they thought. A few months later, and this time in Europe, a chill would set in. And it was not because of the impending winter.

Preserved for the future — The Time Capsule

The Time Capsule, inserted into the wall of UT1, contained a collection of outstanding papers by scientists from ESO's Member States and Chile, as well as by the scientific group at ESO itself. They were:

Belgium: Magain P., Surdej J., Swings J.-P., Borgeest U., Kayser R., Kuehr H., Refsdal S. & Remy M. 1988, *Discovery of a quadruply lensed quasar: the "clover leaf"* H1413+117, Nature, 334, 325

Chile: Ruiz M. T., Maza J., Wischnjewski M. & Gonzales L.E. 1986, *ER 8: A very low luminosity degenerate star,* ApJ, 304, L25

Denmark: Nørgaard-Nielsen H. U., Hansen L., Jørgensen H. E., Aragón Salamanca A., Ellis R. S., & Couch W. J. 1989, *The discovery of a type Ia supernova at a redshift of 0.31,* Nature, 339, 523

France: Aubourg E., Bareyre P., Brahin S., Gros M., Lachieze-Rey M., Laurent B., Lesquoy E., Magneville C., Milsztajn A., Moscoso L., Queinnec F., Rich J., Spiro M., Vigroux L., Zylberajch S., Ansari R., Cavalier F., Moniez M., Beaulieu J.-P., Ferlet R., Grison Ph., Vidal-Madjar A., Guibert J., Moreau O., Tajahmady F., Maurice E., Prevot L. & Gry C. 1993, *Evidence for gravitational microlensing by dark objects in the Galactic halo,* Nature, 365, 623

Germany: Eckart A. & Genzel R. 1996, *Observations of stellar proper motions near the Galactic Centre,* Nature, 383, 415

Italy: Vaiana G. S., Casinelli J. P., Fabbiano G., Giacconi R., Golub L., Gorenstein P., Haisch B. M., Harnden Jr. F. R., Johnson H. M., Linsky J. L., Maxson C. W., Mewe R., Rosner R., Seward F., Topka K.& Zwaan C. 1981, *Results from an Extensive Einstein Stellar Survey,* ApJ, 245, 163

The Netherlands: Mathewson D. S., van der Kruit P. C. & Brouw W. N. 1972, *A High Resolution Radio Continuum Survey of M51 and NGC 5195 at 1415 MHz,* A&A, 17, 468

Sweden: Edvardsson B., Andersen J., Gustafsson B., Lambert D. L., Nissen P. E. & Tomkin J. 1993, *The chemical evolution of the galactic disk,* A&A, 275, 101

Switzerland: Mayor M. & Queloz D. 1995, *A Jupiter-mass companion to a solar-type star,* Nature, 378, 355

ESO: Danziger I. J., Bouchet P., Fosbury R. A. E., Gouiffes C., Lucy, Moorwood A. F. M., Oliva E. & Rufener F. 1988, *SN 1987A: Observational Results Obtained at ESO,* in Kafatos M. & Michalitsianos A. G. (eds.), *Supernova 1987A in the Large Magellanic Cloud,* Proceedings of the 4th George Mason Astrophysics Workshop, Cambridge University Press;
Danziger I. J., Gouiffes C., Bouchet P. & Lucy L. B. 1989, *Supernova 1987A in the Large Magellanic Cloud,* IAU Circular No. 4746

≈

An icy wind was blowing from Paris. In June 1997, Lionel Jospin, a French socialist, had assumed the position of Prime Minister under President Jacques Chirac. As Minister of Education he had chosen Claude Allègre, an accomplished geophysicist and also an energetic, but strident politician. At a meeting among the Research Ministers of the EU on 10 November of the same year, Allègre proposed a considerable cut in the budgets of the major intergovernmental organisations such as ESA, CERN and ESO, *"to obtain a 'right' balance between research carried out in national laboratories and in the international organisations"*. His proposal was reiterated in a formal letter to some of his counterparts in the other countries, shortly before the ESO Council Meeting on 1–2 December 1997[3]. The proposal, which implied a budget cut of about 25%, could hardly have hit ESO at a worse moment. The VLT was coming close to completion, which also meant that the organisation had to shoulder a peak in construction costs as well as mustering all its reserves to adhere to the timeline set for the project. Delays would only increase the cost and other savings, such as those that were believed to be feasible by closing La Silla (as the Minister seemed to suggest) were clearly unacceptable to Europe's community of astronomers and would in any case be associated with increased costs for a while. Aware of the political initiative, Riccardo Giacconi had travelled to Paris ahead of the Council meeting in the

[3] One of the most important, recurrent tasks for the December Council meetings is to decide on the budget for the coming year.

hope of establishing a dialogue with the Minister. However, Allègre had excused himself, leaving a representative to talk to the ESO Director General. Still, Giacconi had gone along with the meeting anyway and argued that ESO was already in the process of reducing costs and underlined the advantages for France stemming from its membership. Furthermore, a unilateral request to reduce the French contribution was difficult to reconcile with the obligations set out in the Convention. But the meeting ended with the assistant simply declaring that whilst he *"showed understanding for [the] situation [he] confirmed that Minister Allègre would expect that his policy guidelines would be followed…."* Despite the considerable problems this would cause ESO, Allègre's demand dovetailed with the overall obligation to reduce government spending deficits in preparation for Phase 3 of European Monetary Union — commonly known as the introduction of the euro — in which five of ESO's Member States participated, including its largest contributors.

In his arguments for reducing the contributions to the European organisations, Allègre had invoked the principle of subsidiarity. Subsidiarity basically means that tasks should be dealt with at the lowest appropriate level — be it local, regional, national or international. This is often described as "a central principle" in the EU context and it was enshrined in the Maastricht Treaty signed in 1992. It was thus part of the dominant political thinking and vocabulary of the time. However, it also reflected the traditional French position that ESO should function as an *Observatoire de Mission*, a view that has emerged again and again[4]. But it undoubtedly threw ESO into a serious crisis. The immediate effect was that Council, caught by surprise, postponed the approval of the budget for 1998.

In the end, on 18 February 1998, the budget was approved. It foresaw a much smaller cut, but to maintain the balance between the Member States, their contributions were reduced equally. It hurt, but ESO could live with this solution. The Great Financial Scare was over. Fairness dictates that we stress that the situation that had arisen was exceptional. Of course, ESO does not exist in isolation. It is dependent on its Member States and their economies. And whilst there will always be a certain tension between the budgetary interests of each of the Member States and the interests of an organisation like ESO, the Member States have indeed maintained a decidedly benevolent attitude towards the organisation over the years, supported its financial requests, paid their dues — sometimes ahead of time — and occasionally even made additional voluntary contributions to enable specific projects to proceed.

[4] The model of the *observatoire de mission* had been applied in France for the Observatoire de Haute-Provence.

≈

Giacconi reacted swiftly and forcefully. In February 1998 he presented a paper to Council entitled "The Role of ESO in European Astronomy". Having summarised the development of the organisation so far, he drew parallels with the situation in the US, showing that ESO's role as provider of observational facilities was crucial for European astronomy. Including the VLT, he stated that *"approximately three quarters of all telescope area available to European[5] astronomers will be provided through ESO."*[6] In the US, only some 20% came from public observatories, such as those operated by AURA. The majority was operated by what he called independent observatories, mainly private universities and institutions.

Citing the recommendations of a recent visiting committee, he argued cogently why the Member States should continue to support ESO at a level commensurate with its tasks. He also demonstrated that a strengthening of ESO did not imply that national institutes would be disadvantaged or lose influence, thus addressing the subsidiarity argument head on[7]. But he went a step further, presenting an exciting perspective for the further evolution of the organisation, including some quite bold suggestions to which we shall return. Among the key points he made was the need for *"a strategic plan for European astronomy, which unfortunately does not yet exist"*, an analogue to the US Decadal Survey prepared by the National Academy of Sciences. For want of an overall plan in Europe, he implied, ESO had developed its own mid-term and long-range plan, fully supported by its various committees with representatives of the scientific community[8]. As for the future, and using the ESA/ESO ST-ECF as an example, he emphasised close cooperation with ESA to exploit obvious synergies between ground-based and space-borne observational facilities. In the field of

[5] In the paper, the term European was used as being congruent with the ESO Member States. Since at the time neither the UK nor Spain, both important actors, were not yet members, this was certainly a bit of a stretch at that time. Today, however, it would have been more appropriate.

[6] According to Giacconi's calculation, professional astronomers in the ESO Member States had access to a total light-collecting area of about 306 square metres compared to the 511 square metres available to to US astronomers.

[7] He also addressed the issue of cost, showing that from a perspective of effectiveness and in considering its tasks, ESO did not need to shy away from comparison with other institutions.

[8] He did not, however, draw a clear distinction between a general strategic plan and ESO's plan. Later, his successor as Director General, Catherine Cesarsky, again pressed for an overall plan. In the end, in 2007–8, it was realised by the ASTRONET consortium — ten years after the Giacconi paper. In an ESO context, the first considerations about a European plan, go back to discussions in the ESO Scientific Technical Committee in the mid-1980s.

international relations, he thought that ESO should *"represent European interests in negotiation with the US and Japan"*. He continued: *"It is clear that in the future even larger facilities that will be needed in astronomy will require transcontinental scientific collaborations and ESO can play a unique role in these developments."* For an organisation under immediate threat of a significant budget cut this may have been seen simply as a bold attempt of forward defence. Yet it was hard to deny the correctness of his analysis. The document was a true *tour de force* leading logically to a new mission statement — simple, yet sweeping — that ESO should be *"an institution which will provide an important, continued contribution to European astronomy in providing technology and ground-based facilities beyond the reach of national groups."* The inescapable conclusion, then, was that *"ESO's activities must reach the critical mass required to carry out those programmes which are beyond the capabilities of single nations, but are essential to maintain a competitive research level in the international context."*[9] Having been endorsed by the ESO Council in June 1998, this document has informed the thinking and the organisational ethos up to the present day.

≈

In following these important political developments, we have moved forward in time, but many other things happened during those years — both at ESO and within the world of astronomy. We shall consider some of them individually.

After the ambassadors' meeting in Santiago, construction work continued. Precious time had been lost, but things looked brighter now. In Europe, the first 8.2-metre primary mirror was undergoing acceptance tests at the REOSC plant just south of Paris, while the test assembly of the first main mechanical structure had begun in Milan. In Chile, the difficulties with imports had been removed and the construction of the enclosures was proceeding. By the end of July 1995 the concrete foundations and support for the rotating enclosure of UT1 had been completed, and the other telescope buildings were also progressing, although the steel structures carrying the ring upon which the enclosures would rotate had not yet been mounted at the other UTs. The concrete works for the interferometric tunnel had largely been finished, as they had for the interferometric laboratory building.

[9] The politically oriented reader will note that this was not only logical, but in fact fully concordant with the subsidiarity principle.

Then, on Sunday 30 July 1995, at 01:15, a major earthquake struck the region. The earthquake was measured at 7.8[10] on the Richter scale. The epicentre lay 30 kilometres west of Tocopilla, i.e. in the Pacific Ocean, about 250 kilometres from Paranal. *"I woke up as my bed seemed to move sideways, back and forth, in a funny, yet soft movement,"* explained Jörg Eschwey, who was in charge of the construction on site. *"Then it changed into a series of hard shocks, as if one was driving over a poor dirt track. Everything was moving horizontally. This was clearly a major earthquake. The generators, however, kept working, but my first worries were about gas. We used gas for the kitchen and also for heating the water. Every container had a gas heater, and my worry was that sparks from a severed electricity cable might cause a gas explosion. When the quake was over — it felt like a long time, although it probably lasted no more than a minute or so — I first went into the office. Everything had come down. Shelves had fallen over; monitors were lying on the floor — smashed. The telephones were still working, so I called Massimo Tarenghi, who was in Europe and told him about it. Then I went up to the platform expecting to see cranes turned over and collapsed scaffolding. The air was filled with dust. You couldn't see it for it was of course dark, but you sensed the sweet smell of it everywhere. The road was partly blocked by a two-metre boulder that had fallen down from the top, but I managed to get around it. I first stopped at the interferometric tunnel, where workers doing the night shift were busy. To my surprise, they were still working, seemingly unaffected. Also on the platform, as far as I could tell, no serious damage had occurred. It was moonlight, so that at least a first assessment was possible."* He went back to the base camp, 3.5 kilometres away and took a look at the situation there. Fortunately, no one had been hurt[11] and the damage in the camp was minimal. As daylight returned, a thorough inspection of the site was carried out. Practically no damage could be seen, no cracks whatsoever could be found in the concrete walls and, as they measured the foundations, they too were fully intact and had not even moved. Naturally, everybody was relieved, but for the engineers it was also a confirmation that their structural calculations had been correct.

≈

[10] This was a local estimate. The US Geological Survey lists this earthquake as a grade 8 on the Richter scale, and thus among the top 30 on the list of large earthquakes in the 20th century.

[11] Others had been less fortunate. In the region three people had perished and 130 buildings had been seriously damaged, including the cathedral in Antofagasta. ESO subsequently donated a sum of approximately 100 000 deutschmarks towards the reconstruction effort. ESO has several times made special donations in connection with natural disasters in Chile.

To anyone who is familiar with Chile, the fact that strong earthquakes occur in that country comes as no surprise, and had been taken into consideration when ESO decided in favour of Chile in the first place. In fact the largest earthquake ever recorded had happened on 22 May 1960 — with the epicentre between Concepción and Temuco — only two and a half years before ESO began to look more seriously in the direction of this South American country. The South American continent sits on a giant tectonic plate, which moves westwards, away from the Mid-Atlantic Ridge, and thus from Europe. However, as it moves, it collides with the Pacific Nazca Plate, which is moving eastwards. This creates what geologists call a convergent boundary and, thus, a subduction zone: the Nazca Plate, so to speak, "crawls under" the South American Plate, creating enormous tension in the lithosphere. This is the reason for the existence of the Andes mountain range and the many volcanoes found in the western part of South America, and also for the recurrent earthquakes, some of which can be very severe. Satellite imagery of northern Chile reveals a number of fault lines. One such line runs in north–south, 500 metres east of Paranal. A small fault line even runs directly under one of the Unit Telescopes of the VLT. How on Earth could one decide to put an astronomical telescope — such a hugely expensive high-precision instrument — in a place like this? *"No problem!"* Jörg Eschwey would say. Who could provide a better answer than Eschwey, the immensely experienced engineer, who had worked for ESO back in the 1970s, then spent a long time in the US, only to be called back at the start of the VLT project? *"No problem. You simply have to build accordingly,"* was his straightforward reply to the author in 1991 during a visit to Paranal. Indeed, the VLT specifications were written with an eye to the geological conditions found in Chile. The requirements are that a VLT Unit Telescope (and its enclosure) should be able to *continue* operations, without any damage at all, following earthquakes up to 7.75 on the Richter scale. This is what ESO calls an OBE — an Operating Basis Earthquake. Should an MLE, a Maximum Likely Earthquake, occur (defined as 8.5 on the Richter scale, at a distance of 150 kilometres), the observatory must be able to survive this too. In this case, however, minor cosmetic damage (but no structural damage) to the buildings is considered acceptable. Statistically, an MLE can be expected with a frequency of one in a hundred years, twice the foreseen lifetime of the VLT. This performance is achieved through damping and shock-absorbing elements in the telescope foundations as well as an intricate system to protect components, and especially the primary mirror, against shocks, among other things.

≈

As already mentioned, the political problems had repercussions on the construction work. But the delays were not simply a result of the difficulties arising from those problems. A scathing report to the ESO STC by the VLT Project Manager in 1995 describes serious difficulties in the area of civil engineering. The contractor seemed to have underestimated the work and was losing money. In the end, ESO relieved the company of its contractual responsibilities prematurely and invited the company that had originally come second in the bidding process, the French firm SPIE-Batignolles, to complete the work. SPIE-Batignolles, however, had lost interest in the project and so, SOIMI was charged with the completion of the work. Alas, further delays occurred and so, ESO decided also to end this contract and to take the remaining work into its own hands.

In other technical areas, however, the progress was very noticeable. In Milan, the test assembly of the main mechanical structure of one of the telescopes had begun. At GIAT in France, the work on the mirror cells was also going according to plan.

The Council and the telescope: the ESO Council members during their visit to the Ansaldo plant in Milan on 8 November 1995: (From left to right): Philippe Brosser, Bernard Fort, Bengt Gustafsson, Johannes Andersen (STC Chairman), Joachim Krautter (OPC Chairman), Edwin van Dessel, Stephane Berthet, Jean-Pierre Swings (Council Vice-President), Peter Creola (Council President), Emil Broesterhuizen, Riccardo Giacconi (ESO Director General), Dieter Reimers, Guglielmo Castro, Poul Erik Nissen, Gerhard Bachmann (Head of the ESO Administration), Gustav Tammann, Arno Freytag, Franco Pacini and Francesco Bello (observer for Portugal). The original picture caption from the 1995 Annual Report notes "Johannes Andersen (on his heels...)" but no pun was intended. The photo provides a good impression of the size of the new telescope, standing 24 metres high and with a moving mass of 430 tonnes. (Photo: Hans Hermann Heyer)

Chapter III-12

Tuning a Formidable Science Machine

"ESO has revolutionized the operations of ground-based astronomical observatories with a new end-to-end data flow system, designed to improve the transmission and management of astronomical observations and data over transcontinental distances."

Computer World, 2005.

The VLT was gradually taking shape. But, as ESO developed this wonderful new tool, the VLT also transformed the organisation itself. New terminology had entered into the institutional language, such as cost to completion, full-time equivalents and work packages. Another term that became prevalent was end to end, which was used in several contexts, but basically described a holistic view of a subject. For the VLT it could mean an integrated view of design, construction and operation. It could also mean an integrated view of the scientific process of proposal formulation, observation and utilisation of data. The VLT lent itself well to this kind of thinking because the entire facility was designed as a whole, rather than the organic growth exhibited by most other observatories that had expanded over time. As an operational concept, integrated thinking was introduced at ESO by Giacconi who was used to such an approach from his background in space projects. However, it was not customary among ground-based facilities and projects and ESO's embrace of these ideas marked a radical break from many long-established traditions.

The traditional operational concept for ground-based observatories roughly consisted of the following steps. Having been granted the coveted telescope time, the astronomer would travel to the observatory, take charge of the telescope for the allocated time, carry out the observations (which often involved making photoelectric measurements or taking images, either on photographic plates or, increasingly, using electronic means). Upon completion of the observing run he or she would return to the home institute, happy if the run had been successful, less happy if technical problems or bad weather had hampered the observations. Back home, the plates would be scanned, transforming the analogue images into numerical digital information and the data processed to obtain physically meaningful information about the observed object. The end point of the process would be a paper in a refereed journal. This was, and had been, the normal procedure at observatories all over the world, but by the early 1990s, a revolution was in progress. Several factors came together: firstly, with the new large telescopes, observing time had become both an extremely costly and valuable commodity. Expensive, because of the considerable investment in capital; valuable, because the number of large telescopes was small and therefore, the requirement to use the observing time in the most efficient manner was an obligation towards science[1]. Secondly, silicon-based technologies had ushered in highly efficient detectors with digital output, and this, coupled with the information technology revolution that had opened up entirely new ways of communicating even across huge geographical distances, enabled a much more efficient use of the scarce commodity that was telescope time. Thirdly, this new generation of telescopes and instruments were exceedingly complex systems that could only be operated safely and efficiently by highly trained specialists and, fourthly, at least at ESO, there was a growing awareness of the importance of operational issues — the continued maintenance and upgrading of telescopes and their instruments. All of these considerations came together, driving a new operational concept for the VLT — a concept that would make it the most efficient astronomical observatory on the planet.

In 1988, an ESO internal working group, chaired by Peter Shaver, began to consider some of the operational aspects. In September 1989, the group presented its thoughts on the matter. *"In order to preserve the flexibility inherent in the VLT concept, it was considered imperative that no operational mode be 'designed out',"* wrote Shaver. *"In particular all the major observational modes — classical (astronomer at telescope), remote (astronomer in Europe) and service (by ESO staff in Chile or Europe) — [should] be fully accommodated in the design of telescope and infrastructure. Flexible scheduling, however, was seen as a major objective from the outset."* (Shaver, 1989). He ended the article rather cautiously: *"These are just the summary recommendations ... written comments from members of the community are most welcome."* Any suggestion that astronomers should not themselves be present at the telescope or not carry out their own observations was likely to invoke strong reactions. However, things

[1] This was nowhere more strongly pronounced than in the case of the NASA/ESA Hubble Space Telescope, which was not just a truly unique facility in itself, but also, because of its deployment in space, had to be operated in a remote mode. It was therefore no coincidence that the seeds of an operational paradigm change were laid at the Space Telescope Science Institute, led by Riccardo Giacconi.

had begun to change. We recall the successful trial runs in 1984 regarding remote control of the MPG/ESO 2.2-metre telescope. At that time the tests had been carried out locally in Chile, but first tests from Europe had been undertaken in March 1986. The results had been encouraging and so, in July 1987, regular observations from Garching began. A couple of months later, the 1.4-metre Coudé Auxiliary Telescope also became remotely controlled from Garching. In February 1991, ESO leased a 64 kbps satellite link for remote observations with its newest addition: the NTT, and in June 1992, a test was made with remote control of the NTT from the Osservatorio Astronomico di Trieste. The NTT was well suited for this kind of operation and it was also fitted with versatile multi-purpose instruments that could be controlled by computer. This kind of operation was called second-level remote control, because the link went via the ESO Headquarters in Garching. The test was successful: *"During the three nights of the final tests over 30 hours were devoted to astronomical observations and, as can be inferred by the users' comments, the system proved to be very easy and flexible to operate, considering also that most of the observers had no experience in the use of EMMI,"* concluded the team in an article in the *The ESO Messenger* (Balestra *et al.*, 1992). The astronomers carried out a wide range of observations, including studies of a T Tauri star, the abundance of lithium in dwarf stars, planetary nebulae candidates in galaxies of the Virgo Cluster, as well as more distant objects including Seyfert galaxies and clusters of galaxies[2].

The dream was to make remote observing available to many institutes in Europe and to maintain this option for the VLT as well. While ultimately this was not realised, it provided the first element of a comprehensive operational paradigm for the VLT. Astronomers did not *have* to travel to South America to observe. This might have made observations less romantic, but the expected savings in time and cost were seen as a blessing, especially for very short observing runs.

The NTT provided another illustration of the need to rethink existing observing practices. As we recall, the telescope had already demonstrated its true capabilities during the night of first light, delivering images with a resolution of 0.33 arcseconds. This not only showed how good the telescope was, but also how good the site was in terms of seeing. Such good seeing did not occur all the time. In fact, it happens only during a few days — or rather nights — each year. Looking at the many observing proposals, on the other hand, it is equally true that while some are critically dependent on optimal seeing conditions, others are much less so. But how could

[2] A preliminary account is provided in *The ESO Messenger* by Franchini *et al.* (1992).

those observations that needed optimal conditions be scheduled for nights when such conditions prevailed? In the classical mode, where astronomers were allocated telescope time on specific dates, often determined many months in advance (and they then travelled to Chile to be there for these nights), there was no way that observations could be scheduled to match the actual atmospheric conditions. On the other hand, the extensive campaign of site testing that Marc Sarazin had undertaken for a decade had an important byproduct: a much better understanding of the behaviour of the atmosphere, the frequency and time variations, which began to enable observers to predict the observational conditions, albeit on the time-scale of a day or so. All of these constituted strong arguments in favour of flexible scheduling. With these new opportunities, it became possible to plan a night's observations optimally, ensuring maximum utilisation of the telescopes and matching the observations to the prevailing weather conditions. The huge advantage for the scientists was the marked improvement in the chance of getting good data[3], but the inevitable downside was that under this new operational regime, astronomers would apply for telescope time, but not themselves carry out the observations. Instead, the observations would be made by a dedicated team of astronomers at Paranal, who would then send the observational data to the proposers. This was more than just a technical change: it was fundamental. Previously ESO's main "product" had been telescope time, now the main product became scientifically useful data. This had a number of consequences: observations had to be planned more comprehensively in advance, and in a standardised way. Observers lost a freedom they had so far cherished; to use the allocated telescope time as they thought fit. Now the programme had to be defined (using dedicated software provided by ESO) so that others (service observers) were able to execute the observations. The required observing conditions also had to be clearly specified. Whilst putting new demands on those applying for observing time, the new operational regime meant concentrating a lot of the effort at ESO[4] rather

[3] An early example of this was a project by a research group led by Yannick Mellier of the Institut d'Astrophysique de Paris to measure something called cosmic shear. This means investigating the weak distortion of the images of faraway galaxies induced by gravitational effects, in other words gravitational lensing. With a reasonable sample of observations, it is possible to assess the amount of matter in the entire Universe, which is why this method has sometimes been described as "weighing the Universe". But such observations require extremely high image quality, which translates into using the best telescope under the best atmospheric conditions. Since it is necessary to observe some of the most remote galaxies, a large telescope is necessary, and finally to study many galaxies, one needs enough observing time. In fact, the group needed about 100 hours of excellent seeing, allowing observations of some 50 individual fields, providing image data of more than 70 000 galaxies (Mellier *et al.*, 2000). Such a mammoth programme is only realistic with flexible scheduling of observing time.

[4] By 2006, the staff complement of the Data Management Division (DMD, see next page) had grown to 130 people. Subsequently the division was split into two, one for Data Management and Operations and one for Software Development, led by Fernando Comeron and Michèle Péron, respectively.

than with the individual researchers, but it also entailed significant advantages for science. For example, it would now be possible to maintain a central archive and with the new IT-based tools that could be expected, such an archive could be made widely accessible to the scientific community. Data mining — re-using observational data that had been collected already and possibly for other purposes — became feasible. There would be other advantages too, to which we shall shortly return.

≈

In 1991, ESO began experimenting with flexible scheduling on the NTT (Breysacher & Tarenghi, 1991). However, the first step towards a more compehensive system — into this promising new world, so to say — was taken with the establishment of the Data Management Division in January 1994. Initially — on an interim basis — it was led by Piero Benvenuti, who remained Head of the ST-ECF and who was obviously well acquainted with Hubble's observational procedures as well as the archive. At the suggestion of Dietrich Baade, Benvenuti set up an internal working group to consider the issue of an online data flow system for the VLT. The group, which was led by Preben Grosbøl, with Bruno Leibundgut, Jason Spyromilio, Joe Wampler, Bob Fosbury and Dietrich Baade as members, presented its report in the spring of 1995. The introduction justifiably warned: *"Some of the … concepts may be new and unfamiliar to astronomers… ."* It was in fact an outline of the operational changes to come. In the same year, the new head of the division was found: Peter Quinn, an Australian astronomer, who had previously worked at Caltech and afterwards at the Space Telescope Science Institute in Baltimore. Meanwhile he had returned to his native Australia to work on the MACHO project. This was a research project to test the suggestion that a part of the dark matter in the halo of the Milky Way could be made up of small, dark objects — such as brown dwarfs or planets — or as they were called Massive Compact Halo Objects, or MACHOs. The research tool was the world's then largest CCD camera mounted on a telescope at the Mount Stromlo observatory. This camera delivered huge amounts of data, so both from his time at the STScI and in the context of the MACHO project, Quinn was acquainted with the issue of managing scientific data on a large scale. In August 1995, he arrived in Garching.

At ESO, Quinn found fertile ground. As already indicated, several people had already worked on the ideas and solutions that would tie into the idea of a seamless data flow within an overall data management system. All in all the group comprised a dozen or so people, each of whom had dealt with individual aspects of data handling and processing. The new Data Management Division would develop its activities on this basis.

At the time when Quinn and his team set out to realise their thoughts about an integrated end-to-end system for science users, the VLT software had already been developed. The end-to-end operating system, for which only rudimentary ideas existed at the time, had to be put on top of this. How would this work?

In this situation, Joe Schwarz proposed that the NTT could be used to test VLT software, *de facto* turning the NTT into a UT0. In the course of time, the entire system could be tested and de-bugged and valuable experience gained in an operational environment[5].

Matching the telescope software with the end-to-end system was not an easy task. Whilst discussions about how to do this and how the various requirements should be balanced against each other were unavoidable, the two people in charge, Peter Quinn and Jason Spyromilio, agreed that both systems would have to fit together. One night, over pizzas and beer, they came up with what might appear as an unholy analogy: the Shark's Teeth. The meaning was that the two systems should interlock exactly like the teeth of a shark when the jaws close. Whatever people thought of this comparison it was adopted as a guiding idea, also perhaps because it implied that implementing the overall system would involve pain. The Shark's Teeth became part of the logo used internally by the Data Management Division in the years to come.

Another analogy was invoked in illustrating how the end-to-end system would actually work. The data would flow from the original observing proposal and, when approved, be packed into Observing Blocks — or OBs as they became known — and sent to the staff astronomers and telescope operators at Paranal. After observing the results would be sent back, now in the form of astronomical data, passing through an elaborate system of quality control, to end up on the user's desk in a science-ready format, as well as being stored in a rapidly growing science archive. Peter Quinn adopted a circular flow chart, which because of its shape, became known as the Egg. The Egg

[5] Of course, there were a few odd — and innocent — surprises. When the team started to test the software they immediately noted that something was wrong. They discovered that the software was based on the geographical coordinates of Paranal and not La Silla. After changing these parameters, it worked well.

A proud ESO team receiving the Computer World 21st Century Achievement Award for Science: (From left to right) Preben Grosbøl, Michèle Péron, Peter Quinn and David Silva.

of Columbus? In any case, the paradigm change was not just about efficiency[6]. It also enabled new science to be done. Furthermore, aside from the delivery of science-ready data, the associated technical data, such as instrument calibration data, made it possible for ESO to continuously monitor the performance of the various instruments and provide timely feedback to the maintenance crew at Paranal, as well as to the

[6] The development of this new system began in 1996, initially in cooperation with the Canadian Astronomy Data Centre that had also worked with the Gemini telescope project. For the science archive, the work involved the ST-ECF and the Centre de Données Astronomiques de Strasbourg (CDS), France.

scientific users. Provision of these data is crucial for the scientific users to be able to reduce the observational data and interpret them in a physically meaningful way. In 2005 ESO was awarded the prestigious (US) Computer World Prize in recognition of the contribution to science of this end-to-end system.

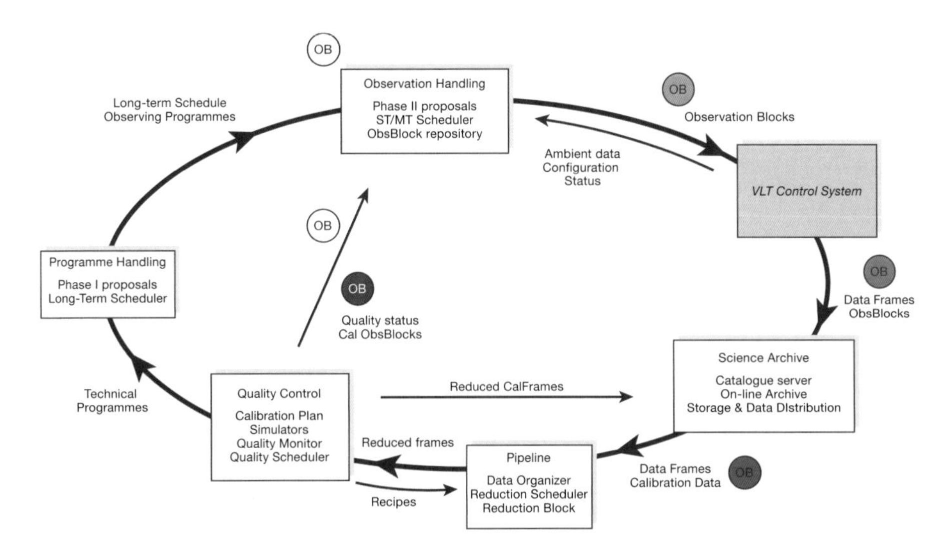

The data flow system, here depicted as "an egg" became a core element of the VLT operating paradigm.

≈

With the introduction of new and highly efficient observational modes a comprehensive digital archive of scientific data was becoming a reality. ESO's Science Archive Facility would eventually store all raw data obtained at ESO in Chile, as well as those obtained with the Hubble Space Telescope and, upon the entry of the United Kingdom, the raw data from the UKIRT Infrared Deep Sky Survey (UKIDSS). In the earliest phase, however, it basically covered observations with the NTT. During the first year, in 1991, it contained 1.6 gigabytes (GB) of data, at the time a respectable number. The archive at ESO grew quickly. By 1998, the data holdings had grown to 0.2 terabytes (TB). Three years later, it had grown to 9 TB. Hardly surprisingly, this was nothing compared to what would happen when Paranal reached its full observing capacity. In 2005, ESO decided to open the archive to users worldwide. By 2007, the archive held 70 TB of data. According to the ESO Annual Report, during this year, the archive received 11 000 unique requests from scientists involving 16 TB of data. Of those requests, almost 9000 related to observations with ESO telescopes, comprising *"12.3 TB in nearly three quarters of a million files"*. The addition of the two survey telescopes at Paranal would, however, completely dwarf these numbers, with VISTA alone adding some 150 TB per year to the data collection. The dramatic increase in data volume not only meant a constant race in terms of increasing the data storage capacity but also that the available space at the Garching Headquarters would no longer suffice. In June 2007, ESO therefore opened a new data centre[7] in rented rooms at the nearby Max-Planck-Institut für Plasmaphysik.

The next step would be to connect the scientific archives in different places to enable astronomers to obtain and compare multi-wavelength data, and to do so over the internet. Projects to facilitate this became known in Europe as the Astrophysical Virtual Observatory (AVO) and in the US as the National Virtual Observatory. Co-funded by the European Commission, the European project was led by Peter Quinn at ESO, but also involved ESA, the AstroGrid consortium (in the UK), the CDS at the Université Louis-Pasteur in Strasbourg, the TERAPIX astronomical data centre at the Institut d'Astrophysique de Paris and the Jodrell Bank Observatory in the UK. The AVO project was launched in 2001 and in June 2002, a conference jointly organised by ESO, ESA, NASA and the National Science Foundation (NSF) took place at the ESO Headquarters with the title "Towards the International Virtual Observatory". Tying together archives obviously entails a host of technical and scientific problems, for the objective is to be able to access instantly scientifically comparable data from different facilities. Together with their colleagues from particle physics, astrophysicists were among the first to embrace the idea of "e-science", with its new vocabulary of federated archives, grid and cloud computing, etc. using high-capacity data networks spanning continents and evolving at a breathtaking pace. At the time of writing, astronomy is on the verge of yet another revolution... .

[7] The current holdings comprise 265 TB (status May 2012).

270 days before VLT first light — members of the AIV Team gathered in front of the office container next to UT1 on Paranal. From left to right: Peter Gray, behind Gerhard Kretschmer, in front of him Krister Wirenstrand, Toomas Erm, Juan Osorio, Mauricio Pilleux, Marc Sbaihi, Erich Bugueno. (Photo: Gerd Hüdepohl)

Chapter III-13

The Countdown

"Everything went according to plan...."

Richard West,
ESO Press Release, 12 December 1997.

In the early afternoon of 6 December 1997 the cargo vessel *MV Tarpon Santiago* approached the port of Antofagasta. In its hold: the first VLT primary mirror. The mirror was loaded onto the ship on 11 November at the port of Le Havre. Under the command of Captain Dimitris Pangalos, the vessel had crossed the Atlantic Ocean on a route 220 nautical miles longer than normal, but one that provided the best chances of calm seas, passed through the Panama Canal and down the western shores of the South American continent. It had made a brief call in Iquique[1], the Chilean port close to the Peruvian border, before the final overnight voyage to Antofagasta. In the afternoon, preparations for the unloading began and, around 21:30, the container was lowered onto the transporter that was waiting on the pier. Unloading a 43-tonne container swinging back and forth on the steel cables from the ship's crane is not a simple task, especially if the set-down is supposed to be shock-free. It wasn't. For the last few centimetres it came down with a hard, nerve wrecking bump. Had the mirror, inside the specially designed container, survived? How many bumps had it already been exposed to on the high seas? And what lay in store for the 130-kilometre road trip to Paranal? The assembled people will have thought about that — the ESO staff from Garching and Chile, the members of the company in charge of the transport, Gondrand (of France), the representatives of the insurance company and the media people. But they could take comfort from the fact that even if this was a first — in a way it actually wasn't!

As mentioned in Chapter III-5, ESO had used a dummy mirror, made of concrete, for testing. On 31 October, just five weeks before, a test journey had been carried

[1] A small team from ESO's Education and Public Relations Department had boarded the ship there, while European television crews from Deutsche Welle and the Swiss SF Television were waiting in Antofagasta.

out with the dummy mirror[2]. The test had included the sea voyage and thus also the unloading sequence, just as had happened with the real mirror. And as now, the container had been fitted with accelerometers to detect and measure any shock to which it might be exposed during the entire voyage. While the precise data were continuously recorded, an interim check was possible with the help of small lamps, indicating the general status. On arrival at the port of Antofagasta, all the warning indicators were shining a bright, ominous red. The test trip, however, continued to its final destination at Paranal, after which the data from the accelerometers were downloaded and analysed. The first glance was sobering: at some point during the voyage, the dummy mirror appeared to have been exposed to a one-millisecond-long acceleration of a staggering 10g in three axes, a puzzling measurement. What on earth had happened? An initial inspection of the ship's logbook did not reveal any unusual occurrences. On the contrary: at the time of the incident, the vessel had been several hours out of the port of Havana in calm seas. Was the reading faulty? Eventually the cause of the problem was discovered: radio interference from walkie talkies used by the sailors on board the vessel. Both ESO and the insurance company were relieved! As in Europe, the test journey of the dummy mirror had provided proof that the procedure was safe, and thus could go ahead. But this time it was the real thing and everybody knew, there would be no room for mistakes!

On the Sunday morning, 7 December, the heavy-duty transporter was set in motion, escorted by local police and a retinue of cars. It moved at walking pace through the streets of Antofagasta, taking the coastal road and then, just south of the town, the steep uphill road towards the Pan-American Highway. The accelerometer readings provide precise indications: the lorry started to move at 08:25 and stopped at 18:30 after reaching what was then the dirt road B70. It was time for the overnight stop. It had been an exciting day. The drive through Antofagasta had been followed eagerly by its citizens, who had lined the streets as if had it been a state visit. Even when driving through the sparse countryside after La Negra, people — seemingly coming out of the desert — assembled along the roadside, waving and cheering, as the convoy passed by. There was an air of joy and anticipation, but ahead lay 60 kilometres of rough, unpaved desert road.

The next morning the convoy started again, moving forward at a speed of 5 km/hour. Three heavy-duty bulldozers were now in front of the lorry, levelling what was hardly more than a dirt track. Where the inclination of the road was too

[2] At the same time the first real mirror cell had been transported to Paranal.

The first VLT primary mirror arriving at Paranal.

great, the transporter would make use of its special lifting device, tilting the container (± 10 degrees) to keep the mirror as level as possible. A further night was spent en route, and on 9 December, at 17:00 hours, the convoy reached the gate at Paranal. It was then time for a first visual inspection of the mirror, again followed by a careful analysis of the accelerometer data by Matthias Hess, the ESO engineer in charge of transporting the mirror. Everything was fine. Shortly afterwards, the M2 unit and M2 beryllium secondary mirror arrived as well.

Clearly, for the citizens of Antofagasta the VLT meant the passage of heavy loads. Back in August 1997, the mirror coating vessel had also arrived in Chile. Manufactured by the German company Linde A.G., this huge vacuum vessel had a diametre of 9.4 metres and the load had a total weight of 100 tonnes.

Max Kraus, the "man moving things", sticks his head out through a hole in the protective plate at the Nasmyth focus of UT1, during the installation of the encoders. Kraus has been the engineer responsible for a wide variety of equipment for the VLT, e.g., the mirror transport and handling equipment (the M1 air cushion carriage, the M1 lifting platform and the M1 handling tool in the mirror maintenance building), the VLT Cassegrain instrument carriage, the mechanical part of the Auxiliary Telescopes and the transport system, the ALMA antenna transporters and many other big and small projects. (Photo: Peter Gray)

With the arrival of the mirrors, the countdown towards first light had begun. Tarenghi had put together a very able and experienced team to help him. The AIV team, led by Peter Gray and among others including Marc Sbaihi, Gerd Hüdepohl and Toomas Erm, had already arrived at Paranal a year earlier. AIV stands for assembly, integration and verification, meaning that the team would be responsible for the on-site telescope assembly and would also ensure that everything was done according to plan and fulfilling specifications. The group was installed in containers next to the enclosure of Unit Telescope 1, the first to be completed. Gray had joined ESO from the Steward Observatory in Arizona, where he had been involved in the MMT, LBT and the Magellan telescope projects. With a good-natured demeanour and a broad Australian accent, he seemed a perfect fit for the job of chief troubleshooter, working with many different kinds of people and transforming what was still a rough building site into a sophisticated science facility. Due to various delays in the project, the AIV phase had to be compressed and only really began after the

M1 cell and the primary mirror arrived (Gray, 2000). At that time, more people from Garching began to arrive. Like Tarenghi himself, several had been associated with the NTT project, including Jason Spyromilio, who had overseen the NTT Big Bang project and Lothar Noethe, who had worked closely with Ray Wilson in developing active optics. Naturally, the team included the key people who had brought the VLT forward. A detailed account is provided in the September 1998 issue of *The ESO Messenger* (Tarenghi *et al.*, 1998).

Adjusting the enclosure doors on UT1 — Cristian Juica and Erich Bugueno, both members of the VLT AIV team. (Photo: Peter Gray)

The first tests were conducted from an interim control room that had been set up in a container at the base camp. During the months of January and February, the testing of the telescope's mechanical structure took place, initially with the dummy mirror and a dummy mirror cell mounted to balance the telescope. The tests went well and the advantages of using the same control software as on NTT, and tested during the NTT Big Bang, became clear. As expected, the test phase also revealed residual problems that had to be tackled. These included oscillations in altitude tracking and stronger than anticipated wind effects on the mechanical structure. Wind effects were expected and were to be compensated by a field stabilisation system involving the secondary mirror, but originally the implementation had not been foreseen before later in the year. Now it was decided to try to implement it from the outset.

In March, the secondary mirror unit and the primary mirror cell were installed, though still with the dummy mirror. Telescope control was now carried out from a console inside a small hut inside the telescope enclosure, so that the telescope remained in sight of the operating team. By mid-May, the telescope control moved again, now to the proper control room, located in the separate building below the telescope enclosures on the southern side of the mountain.

While attention was concentrated on the telescope tests, a major problem occurred at the base camp, or rather in the mirror maintenance building. Here, the first primary mirror was waiting for the final process before it could be mounted on the telescope: the application of the thin aluminium layer that would transform it to a highly reflective mirror. The aluminium layer on VLT mirrors is applied by a process known as sputtering. The base material is a bar of aluminium with a very high degree of purity (99.995% pure). With the help of a magnetron, a tube able to create intense heat by means of electric and magnetic currents (most commonly used in microwave ovens), an argon plasma is created in the vacuum chamber. Positively charged atoms (ions) bombard the aluminium target causing atoms from the target to be ejected and reach the mirror surface. With the mirror rotating, it is possible to deposit a thin film of aluminium with a very high degree of uniformity across the entire surface. The thickness of the resulting aluminium layer is in the order of 80 nanometres[3].

Before the mirror coating began, a test firing of the system was conducted. The result was disastrous. Due to a magnetron failure the aluminium target was badly damaged and could no longer be used for a real coating run. Obtaining a new aluminium target from Europe in time appeared not to be possible. Instead, Michael Schneermann, the engineer responsible, located a company in Reno, Nevada, that might be able to step in at short notice. He flew to San Francisco and negotiated a quick delivery. It looked as if it might be delivered to Chile in time, but the clock was ticking[4].

Meanwhile, the secondary mirror was coated at La Silla. With a diameter of 1.1 metres, this could easily be handled by the mirror coating facility there. By mid-April, the mirror cell was dismounted from the telescope and brought down to the mirror maintenance building. The real primary mirror, still uncoated and protected by

---

[3] Readers interested in the technical details will find a full description is given in *The ESO Messenger* (Ettlinger *et al.*, 1999).

[4] An alternative solution that was considered was to spray a thin silver layer onto the surface of the mirror blank, but in the end, the idea was given up.

a plastic film, was mounted in the cell and brought up to the telescope. On 21 April everything was in place, but it appeared that first light would have to be done with an uncoated mirror.

≈

The VLT story is a fascinating mixture of meticulous planning, often years in advance, expressed in numerous charts, documents and even detailed operational manuals, reminiscent of the checklist used by airline pilots, and quick *ad hoc* decisions, taken on the spot if necessary. Even if the primary mirror was uncoated, and thus *de facto* offered a light-collecting capability similar only to that of a much smaller telescope, it could still be used for tests. Tarenghi, therefore, took the decision on the spot to try out the telescope. On the same night the optical engineers Paul Giordano and Francis Franza carefully removed the protective cover of the 53 square metre surface. When the first photons struck the polished surface of the mirror, technical first light was achieved. Now the telescope performance could be tested under real-life conditions: the pointing, the tracking, the active optics system and so forth. The first target was a globular cluster. It was quickly found, but the image was lousy. As so often, when dealing with exceedingly complex systems, we see that the human factor still plays a crucial role and that includes the propensity for errors. The globular cluster image was obtained with a custom-built VLT test camera, placed underneath the primary mirror at the Cassegrain focus. The test team, for the first time operating the VLT under real-life conditions, were, however, operating the telescope in the Nasmyth configuration! Once the mistake was recognised, the result looked very different. Buoyed by their success, the team moved the telescope to observe a single star. But they saw nothing. Initially the pointing model, which had been elaborated by Krister Wirenstrand, had been verified with an 8-inch Celestron telescope riding piggyback on the VLT telescope structure. So why was it not working now? Then the team realised that the test telescope had been fixed *on the side* of the top ring and was therefore not precisely centred, or as opticians would express it, on-axis. Furthermore, in this position, it did not correct for the field rotation associated with the alt-azimuth telescope mount. Again, once the error had been recognised, everything worked. To everyone's delight, the active optics system worked flawlessly and by the third night, images with a resolution of 0.4 arcseconds were obtained. As part of the tests, the mirror was deformed in several ways, producing odd images of the star and then corrected again. A proud Tarenghi called the Director General to report this success. But Giacconi, undoubtedly happy that things were going well, did not lose his focus: *"Stop playing with the telescope,"* he barked through the telephone at Tarenghi, *"And do something sensible with it!"*

≈

By mid-May, Tarenghi made what was hardly more than a pitstop at the ESO Headquarters in Garching for last-minute meetings. Meanwhile the new aluminium target had been flown in from the US, and Tarenghi took a gamble: we coat the mirror! On 18 May the mirror cell was taken down to the mirror maintenance building, the mirror itself was detached overnight and inserted into the vacuum chamber the following day. With the target date for the official first light just a few days away, this had become a real race against time. Conducting another test ahead of the real aluminisation was out of the question. Peter Gray was requesting authorisation to go ahead and fire the magnetron. He reached Tarenghi at Frankfurt Airport and got the go ahead. The trip from Frankfurt to São Paolo is a 13-hour flight. Tarenghi was used to it, but this particular one will have appeared unbearably long to him, yearning as he was to learn about the outcome of the first coating. Landing in the early morning hours at Guarulhos International Airport, he received the news he had hoped and longed for: the coating had been successful. Everything had worked as foreseen and on 21 May the mirror was back at the telescope.

The primary mirror with its cell is brought up to UT1 after the successful aluminisation on 21 May. In his logbook, Peter Gray noted: "Started operation at 5 am this morning. Moved outside on air cushions 20 min, loaded onto trailer and rigged 40 min. Departed MMB [Mirror Maintenance Building] at 6 am as planned. 1.5 hr transport, arrived UT1 at 7:30 am." (Photo: Peter Gray)

We have seen that by 25 May, the telescope had already been working for some weeks, albeit with an uncoated mirror. So when did first light really occur and how is it defined? At ESO (and in many other places), first light is defined as the moment when the scientifically meaningful observations *can* be made. Before a new telescope reaches that stage, many tests and much fine-tuning is usually required, but the amount of fine-tuning needed for the first of the VLT Unit Telescopes was sensationally small — even to the disbelief of many inside ESO. As we have seen already in the case of the Hubble Space Telescope, first light is the crucial moment when the transition from engineering to science begins. The images produced at first light obviously are testimonials of the potential of the new telescope and everyone will look at these images accordingly. For funders of a telescope, first light images are also important to justify the investment *vis-à-vis* the public. Perhaps surprising to lay people, first light is *not* about science — despite the criterion already mentioned. Why is that? When a new telescope comes into service, especially with such a capability as the VLT, lots of scientists will stand ready to use it and, in the highly competitive world of astrophysics, who would not like to be the one to publish the first scientific paper? First light images are therefore often selected in such a way that they fulfil the requirements for a technical assessment, demonstrate the potential for capturing stunning images, but not so that they can immediately be used for real scientific analysis. The first light event, however, is normally followed by a period known as commissioning. This is the phase where the telescope begins to produce science data, while still undergoing final adjustments and characterisation. The data obtained during this time are normally made available in an open, but orderly, way to the entire scientific community.

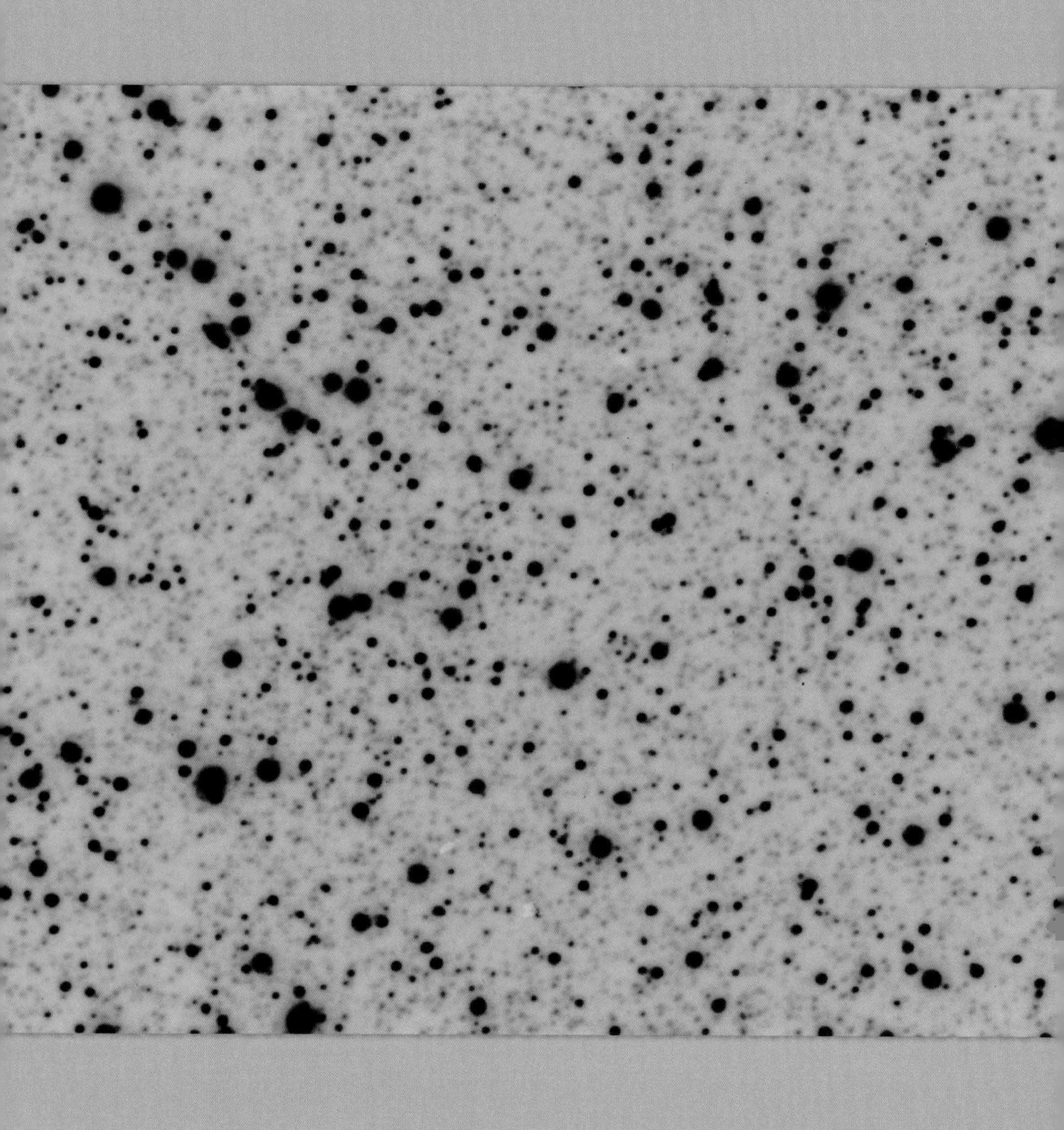

Obtained on 16 May — i.e. with an uncoated mirror — this 10-minute exposure of a small part of the globular cluster Omega Centauri demonstrated well the qualities of the VLT. This object was also chosen for the NTT first light (see p. 192) and seems to be a favourite among telescope builders for testing. As the other early images, this pictures was obtained with the VLT Test Camera developed by Martin Cullum.

Chapter III-14

Clear Skies, at Last

"A Great Moment for Astronomy!"

ESO Press Release, 27 May 1998.

It looked like it would be a lousy day. When Massimo Tarenghi glanced out of the window of his simple steel container abode, he saw a sky covered with clouds. Not a sight cherished by astronomers. Yet it was not necessarily the biggest of his problems that day.

As if he had needed any reminder, the calendar told him that this was the date that had been set, months in advance, for VLT first light. During the early phase of the project, serious delays had accumulated. But with a revised project plan, the management had tied itself to a strict construction schedule, with first light in the first half of 1998. Monday, 25 May, had been chosen. This was the day.

During the previous night the telescope had undergone its last tests, and things had looked good, until an earthquake detector activated the mirror protection system. There had been no earthquake, so the sensor had seemingly gone off at random. Returning to an observational mode, which required repositioning of the mirror, had taken several hours. Then, as the night neared its end and the night shift prepared to close down, the mirror cover, designed to protect M1 during the day, jammed. The operator, undoubtedly exhausted after an intense period of work, pressed the button again; and again. Then with a sudden jerk the cover came loose, jumping out of its tracks and pulling down with it a heavy metal beam. The beam dropped a couple of metres and came to a halt just millimetres above the primary mirror.

The cover was now hanging down, fixed only on one side and dangerously close to the primary mirror. Observing in this state was obviously out of the question. Should they postpone first light? After a brief telephone conversation with Giacconi, Tarenghi and a small team decided to try to keep to the deadline. Working from the morning hours and through the day the team rolled back the cover bit by bit, working a few centimetres above the delicate mirror. They succeeded. By 6 pm the cover had been pulled back. Miraculously as the Sun began to set, the cloud layer also began to lift. The telescope systems were started and the ventilation louvres opened. Soon the first stars appeared in the evening sky.

During the day of first light, a thick layer of clouds covered the sky at Paranal, but as night fell, the clouds disappeared. (Photo: Peter Gray)

In the control room, located in the separate building below the telescope platform, the atmosphere was intense. At the consoles were Roberto Gilmozzi, Krister Wirenstrand, Jason Spyromilio, Anders Wallander and of course Massimo Tarenghi. Looking over their shoulders were Peter Gray and Jörg Eschwey, the old ESO hand responsible for the site construction, two representatives from ESO's information service, Herbert Zodet and the author, and — like flies on the wall — the camera teams from two European television channels, Deutsche Welle and Swiss Television, SF. To outsiders, this hardly looked like a festive occasion. The members of the observing team looked tired, unshaven, and with black rings under their eyes. They had clearly not slept for a long, long time. But now was the moment of truth. The

VLT project had been underway for 20 years. The project had been given green light 11 years ago and since then, an estimated 5000 people had, in some way or other, been involved in this project. For few, this had been a job like any other. Most had been caught by the deep fascination of the potential of this telescope and had worked accordingly, whether scientists, engineers, opticians, technicians, shipping people, drivers or administrative clerks[1]. Site testers had endured incredible hardship over the years, construction workers had toiled amidst swathes of thick dust from the blasting under the merciless Sun at Paranal. In completely different circumstances, administrators and policy makers had fought for this project with governments and even in the courts. ESO's management had shepherded the project through harsh storms and, when in such storms, individual staff members gave up, had managed to replace them with new talent. And even for people who had not been directly involved, this was something very special. We will remember, how people had gathered along the desert stretch of Pan-American Highway, waving and clapping their hands as they watched the first primary mirror rolling past them in its container. But with the world as onlooker, this was the moment of truth. Would the VLT really pass the test, and stress, of its world premiere? Would this exceedingly complicated machine with its multitude of subsystems, with parts manufactured in so many different places in

Photo left: The first image appearing on the computer screen (left to right): Herbert Zodet, Krister Wirenstrand, Massimo Tarenghi and Jason Spyromilio. Photo right: An elated Peter Gray in the VLT Control Room. (Photos courtesy: Peter Gray)

---

[1] An interview with Dimitris Pangalos, captain of the ship that brought the first primary mirror to South America is surely representative of the feelings of many people outside the astronomical community, who had been involved in the project: *"As captain, many times I have used the stars to establish our position, and now I am glad to carry this mirror that brings the stars closer to us. I feel fantastic thinking that by carrying this mirror, I am part of this great undertaking."* The interview was conducted by the author on the day of the ship's arrival to Antofagasta.

Europe, be able to live up to the hopes and expectations of the astronomers, who were eagerly waiting for results back at Headquarters in Garching? Would ESO's Director General — who had witnessed the drama of the Hubble Space Telescope's entrance into the world of science first hand and who now carried the ultimate responsibility for the VLT — this time be satisfied? Would this be as good as the spectacularly successful NTT first light? Or would it be a sign, as many people outside ESO believed, that this was after all too big a challenge for this organisation? As darkness enveloped Paranal, the shutter on the VLT test camera, fixed at the Cassegrain focus, was opened for the first time. Shortly afterwards, the first image appeared on the monitor, to be scrutinised by some very critical eyes. They were not disappointed.

≈

In Europe, serious preparations for the first light event began several months before the magical date. The preparations included the production of comprehensive background materials for use by the media (photos, texts, broadcast-quality video footage), the organisation of a symposium for European science writers with some 40 professional science journalists, agreements with selected television channels about coverage of the event including provision of a feed to the EBU[2] satellite-based news exchange, and, of course, involving the astronomical communities in the member countries.

More technical, and more complete, background information was provided in the form of the 200-page *VLT White Book*, published on the web and in printed form just before the first light event.

As the day came near, at the Garching Headquarters and under the leadership of Richard West a small task force of young astronomers — the first light image processing team — was put together to handle the scientific data as they were transmitted from Chile and turn them into real images for use by the media.

First light formally occurred on Monday, 25 May, but in Europe it was past midnight, and so, Tuesday 26 May. In the late afternoon of that day, when the data had been carefully reviewed, it was decided which images should be included in the series of first light photos that was to be released the following day. Nine were selected and

[2] The European Broadcasting Union, the cooperation organisation of the European public service television channels.

One of the first light images was this picure of NGC 6302, a planetary nebula that also has become known as the Bug Nebula. It was a 10-minute exposure and had a resolution better than 0.6 arcseconds.

mass production of the prints began and continued during the night. Some of the pictures demonstrated the excellent optical and mechanical performance of the telescope, as is one of the purposes of first light. Others were simply pretty pictures, such as the image of the Bug Nebula that instantaneously became a worldwide hit. Once the pictures were ready, in the early morning hours of 27 May, astronomers travelled to same-day press conferences in each of ESO's eight Member States as well as Portugal[3] and Chile, organised with the help of ESO's Council delegates and in many places attended by ministers or high-level civil servants from the national ministries. In Germany, the press conference was held at the ESO Headquarters, with a

[3] Portugal had a sort of associated status at the time. See also Chapter IV-8.

proud Giacconi declaring: *"This is a great day for world astronomy. The new concept that is embodied in this telescope is that of active control of a thin, monolithic large mirror that has been developed at ESO, tested at the NTT and then embodied in the VLT, permits us to observe the sky ... in the optical infrared with unprecedented angular resolution and sensitivity."* In a videoconference from Paranal, Massimo Tarenghi cast aside his tempered reaction, exclaiming: *"You will have to come to Paranal and spend the night with us and I can tell you that you will have a beautiful night — one of the best nights of your life!"* It was as if, in one moment, he also shed the tremendous burden of stress that he had felt during the preceding months and years.

The conference room was full during the first light press conference at ESO in Garching. (Photos: Hans Hermann Heyer)

There was a resounding response from the press. *"The Universe is coming closer,"*[4] wrote Hansjörg Heinrich and Ulrich Schnabel in *Die Zeit*, the German weekly. Dieter Reimers was quoted in the *Darmstädter Echo* for saying *"[This is] the beginning of a new century"*[5]. The Danish newspaper *Jyllandsposten* carried the headline *"Dream Voyage into Space"*[6]. Many papers simply focussed on the achievement and saw it as a reason for celebration. *Limburg Dagblad*, for example, carried the title *"Pictures from Giant Telescope Received with Champagne"*[7], while others saw the VLT as a cause for national pride. Thus *Corriere della Sera* brought a headline saying *"World's Largest*

[4] *"Das All rückt näher".*

[5] *"Der Beginn eines neuen Jahrhunderts".*

[6] *"Drømmerejse ud i rummet".*

[7] *"Beelden reuzentelescoop met champagne ontvangen".*

Telescope speaks Italian"[8], referring to the admittedly strong Italian signature on the VLT project. Some papers also saw the triumph of the VLT in the context of American–European rivalry: *"Europe's answer to Hubble,"*[9] or spoke of *"The European Conquest"*[10]. The latter issue gained a somewhat sharper edge in some papers. The reason was that on 28 May, the day after the ESO press conferences, NASA announced what, from the media's point of view, amounted to nothing less than a scientific sensation — allegedly the first-ever direct photo of an extrasolar planet, obtained with the Hubble Space Telescope (NASA, 1998). The announcement was not based on a peer-reviewed paper, as is normally the practice for serious science news, but the result was going to be announced at the American Astronomical Society meeting that took place at the same time. Even if the announcement was received with some degree of scepticism, at least in the US media, it was obviously better news than a story about a European telescope. In Europe, conversely, widespread speculation suggested that this was a deliberate ploy to outdo the Europeans. Again, *Die Zeit* (in a free translation) wrote *"the suspicion is not far away that this exciting announcement was carefully timed."*[11] *Deutsche Welle,* the German television channel, which had followed the VLT project over quite some time, chipped in with similar comments. Even the *Financial Times* commented: *"Not to be upstaged, however, NASA's Origins Programme ... announced on the same day ESO released the news and images of its first light that NASA astronomers had observed what they believe may be a 'possible planet' outside the Solar System."* Alas, we may never know if this was indeed a case of what would later be colloquially known as press spin — or a pure coincidence. When the object was shown simply to be a star, NASA later withdrew the story quietly, and ESO took away a lesson or two of its own from the episode[12].

Just three days after first light, on 29 May, a major astronomy exhibition opened at the Paris landmark, Cité de la Science et d'Industrie. ESO's public relations department had outdone itself, mounting a large stand with the first light images. And during the exhibition, key VLT people gave talks to a large and excited audience mostly of amateur astronomers.

---

8 *"Parla italiano il più grande telescopio del mondo".*

9 *"Heet Europese antwoord op Hubble".*

10 *"La conquête européenne".*

11 *"... liegt der Verdacht nicht fern, daß die aufsehenerregende Veröffentlichung zeitlich abgestimmt war."*

12 Interested readers are referred to an article by Ray Villard, "Publicising a Science Discovery: It's All in the Timing — Two Case Studies", *Communicating Astronomy to the Public*, May 2008. When, in 2004, ESO went public with what it thought could be the first direct picture of an extrasolar planet, based on an accepted scientific paper, it was careful not to make any definitive statements.

≈

Sixteen days after first light, Council met for its regular summer meeting. Here too the satisfaction was very visible. Regarding the VLT, so often the subject of concern, this time, it was smiles all round! As is customary for Council, it expressed this formally by a resolution:

"The ESO Council,

recognising *the great dedication, resolution and competence with which ESO staff has worked to realise the VLT Project over a period of more than a decade,*

acknowledging *the excellent results that have been achieved already at the moment of first light, which show great promise for the years to come,*

expresses *its profound gratitude and admiration to all ESO staff at all levels in Garching, Paranal, La Silla and Santiago for having contributed to placing ESO at the forefront of astronomy and establishing this organisation and its mission as a model of successful European collaboration in the minds of the people in the member countries and elsewhere,*

conveys to the management and the staff its best wishes for the successful continuation of the VLT project and other endeavours to be undertaken by ESO in the future."

In spite of the NASA skirmish, the VLT first light had also been noticed by a wider audience in the US. *Popular Science* included it as one of winners of the Best of What's New Prize, awarded every year. The award ceremony took place on 13 November 1998 at the Tavern on the Green in New York's Central Park. Massimo Tarenghi received the prize on behalf of ESO, with the author of this book accompanying him. *Popular Science* had invited all winners to display their products at the event. In the case of ESO, this could of course only be in the shape of photos, and since any spectacular astronomical photo put on display in the US would almost certainly be taken for a Hubble picture, ESO printed a slogan on its pictures. It said: *"ESO. Astronomy made in Europe"*. This became ESO's institutional by-line for almost ten years, both signifying the importance of European cooperation if Europe wants to play in the top league and, of course, as a sign of the quality of work originating from ESO.

José Mariano Gago at the VLT Control Room. (Photo: Peter Gray)

A visit to Paranal by Portugal's science minister, José Mariano Gago, himself a particle physicist coincided with the VLT first light period. Gago's visit, on 19 July, was not simply inspired by his scientific interests, but also because Portugal was preparing to become a member of ESO. The success of the VLT would soon stimulate the interest of many other countries in joining, a topic that we shall deal with in Chapter IV-8.

≈

The Paranal Observatory was inaugurated eight months and ten days later, on 5 March 1999. During the previous night, the first Unit Telescope had produced images with a resolution of an impressive 0.25 arcseconds, while the second Unit Telescope had entered into its first light sequence, with technical first light achieved on 1 March. The inauguration ceremony was attended by the President of Chile, Eduardo Frei Ruiz-Tagle, and high ranking representatives of the ESO Member States and Chile as well as of their scientific communities. For President Frei, this must have been a special occasion. In 1969 his father, Eduardo Frei Montalva, at the time President of the Republic, had inaugurated the La Silla Observatory together with Olof Palme, representing the ESO Member States. Following speeches by Henrik Grage in his function as ESO Council President and of President Frei, the particle physicist and Nobel laureate Carlo Rubbia gave a challenging scientific lecture[13] before the President and the invited audience, and afterwards, ESO's Director General gave a prize to Jorssy Albanez Castilla, a 17-year old school girl from Chuquicamata near the town of Calama. The prize, consisting of a 6-inch telescope was awarded for the successful proposal of names for the four 8.2-metre Unit Telescopes that became Antu, Kueyen, Melipal and Yepun, meaning the Sun, the Moon, the Southern Cross, and Venus in the indigenous Mapoche language spoken in parts of Chile.

[13] In Riccardo Giacconi's candid memoirs *"though not understood was warmly applauded by the assembled and rather stunned dignitaries"* (Giacconi, 2006).

The days are long gone when astronomers peered through telescopes themselves. Nonetheless, for the benefits of the guests at the inauguration of the Paranal Observatory, a large screen had been mounted at the Nasmyth focus of UT1 to allow visual observing. (Photo: Gerd Hüdepohl)

The many visitors, most of whom had come from afar, were shown around the observatory and the central control room was filled with people, as the telescope controllers demonstrated the power of active optics by changing the figure of the primary mirror in such a way that the image of a star morphed into the letters V, L and finally T. Among the visitors in the crowded control room was Catherine Cesarsky, who had been designated to take over as ESO's sixth Director General only a few months later.

From the inauguration of the Paranal Observatory. Photo left: Massimo Tarenghi, Henrik Grage, Marta Larraechea Bolívar, wife of the Chilean President, Mirella Giacconi and Eduardo Frei Ruiz-Tagle. Photo right: Jason Spyromilio and Massimo Tarenghi demonstrate the capabilities of active optics to the inauguration guests in the VLT Control Room.

Like Riccardo Giacconi, Catherine Cesarsky had a wide-ranging background, rather than one focussed on optical astronomy. She had, among other things, headed the theoretical group of the Service d'Astrophysique of the Commissariat à l'Energie Atomique (CEA), and then been director of its Direction des Sciences de la Matière (DSM), covering physics, chemistry, astrophysics and earth sciences. As already mentioned, she had also been involved in infrared astronomy as the principal investigator of the ISOCAM camera onboard the Infrared Space Observatory of the European Space Agency. She had furthermore served as French delegate to the ESO Council[14] and, later as a member of the International Visiting Group that reviewed the organisation. Thanks to her upbringing in Argentina, she speaks fluent Spanish and naturally also has a good understanding of South American issues. At the same time, she is a keen follower of European affairs in general and of European science policies in particular, and, as it turned out, she proved to be a formidable political operator. In joining ESO as Director General, her focus was obviously on completing the

The inauguration of Paranal was followed by a visit to the ALMA site at Chajnantor. En route to Chajnantor, the guests made a stop at the Tropic of Capricorn. Seen here are Lodewijk Woltjer, Catherine Cesarsky, Michel Dennefeld, Pierre Léna and Gustav Tammann. (Courtesy: Pierre Léna)

[14] And had also, in 1994, been Vice President of the Council.

VLT and the VLTI. With ALMA on the horizon, a key challenge was to secure the funding, which almost inevitably meant convincing additional countries to join the organisation. Very much on her mind was the desire to strengthen the links between ESO and the scientific community, to drive home the idea (again, perhaps) that ESO *belonged* to the astronomers of Europe. Also on her agenda was the aim of building up relations with the European Union and improving ESO's standing in Chile, which in spite of the turnaround in 1995/96, still needed to be carefully nurtured.

≈

In connection with the VLT Inauguration, ESO organised a symposium entitled Science in the VLT Era and Beyond at Universidad Católica del Norte in Antofagasta with 200 participants. The photo shows Immo Appenzeller presenting the early science results with FORS.

Whilst we shall look at this period in the coming chapters, we should finish this chapter with a note about the people who put together the VLT. The prologue of this book records the author's personal impression of the VLT first light achievement. It is clear that what we saw was without exception an example of outstanding professionalism. Yet, those who achieved this incredible triumph were not just hard-working nerds. They were also "boys" who, despite the tremendous pressure on them found ways to express a freewheeling humour. In the VLT control room, a Walter Cronkite-like voice is occasionally heard with the solemn announcement *"There's no cause for alarm. But there probably will be!"* This is not an earthquake warning or anything similarly serious, but simply occurs when a particular set of computer commands are given[15]. And this was not the only example. When a VLTI delay line (see Chapter III-15) moved, a recording of the underground train arriving in Garching was played. And should a delay line run into a limit, there was a sound of breaking glass and on the visualisation on the screen the delay line turned upside down.

Antu began regular science operations on 1 April 1999. From then on, for the VLT staff at Paranal, it was about living up to the challenges of an on-going operation, with as little downtime as possible. Herman Nuñez, who made this cartoon, suggests that some may have felt struck with awe at this thought. But of course, they managed!

[15] The sound file was borrowed from *Pinky and the Brain*, an American television cartoon.

Aligning the first tracks of the VLTI delay lines in the underground tunnel at Paranal. From left to right: Pablo Vergara, Jorge Perez and Carlos Bolados. (Photo: Peter Gray)

Chapter III-15

First Fringes of the Phoenix

"I have little doubt that interferometric observations will in due time become as common and easy to perform as normal observations are now at the VLT... ."

Catherine Cesarsky, upon the announcement of the successful first observations with the VLTI, *The ESO Messenger*, 2001.

In Chapter III-4, we followed the early history of the VLTI, from the 1977 ESO conference at CERN through the launch of the formal programme led by Jacques Beckers to the decisions taken in 1993 by the ESO Council to postpone the implementation of the VLTI for financial reasons. ESO had, however, continued with preparing the infrastructure so that the VLTI could be put back on track as soon as it became financially feasible. In the first instance this meant constructing the delay-line tunnel and the other underground tunnels as well as the building intended for the coherent (and incoherent) focus. Keeping key personnel, all of whom were eager to realise their dream of a large-scale interferometer, had proved to be much more difficult. Some had left, others decided to stay, hoping for better times. As mentioned, this hope was in part linked to the prospects of Australian membership of ESO. In January 1996, however, it became clear that this would not materialise. In the meantime, the VLTI group had used its enforced break for additional studies to be better prepared for a relaunch. This included investigations of thermal effects, as they could be expected in the interferometric tunnels, of the effect of telescope vibrations at the level of nanometres as well as of microseismic activities of the site. It would prove to be time well spent. In addition, an Interferometry Science Advisory Committee with *"astronomers who have a background in high angular resolution astrophysics and interferometry, was established ... to review the development of, to define a key science programme for, and to recommend necessary conceptual changes to, the VLTI."* (von der Lühe *et al.*, 1997)

In late 1995, Giacconi had begun to realise that the Australian bid might not lead anywhere. Even so, at the time the financial pressure on ESO had eased somewhat, allowing him to "borrow" a sum of ten million deutschmarks from the VLT instrumentation budget, thereby enabling a modest restart of the VLTI programme, at first with two delay lines and two siderostats. The additional contributions by CNRS and the Max-Planck-Gesellschaft remained at ESO's disposal, but uncertainty existed as to whether the amount would suffice to cover a third delay line and a telescope.

≈

In broad terms, the VLTI consists of the four large Unit Telescopes, the Auxiliary Telescopes and instrumentation placed at the combined (coherent) focus. The light from the different telescopes is fed into the combined focus via an elaborate tunnel system that lies under the surface of the telescope platform. The delay lines constitute the prime optical element that allows light beams to be combined coherently. During an observation with the VLTI array, light from a celestial source will not arrive at *exactly* the same time at the various telescopes. The time difference is obviously minute, typically of the order of 0.2 microseconds (equivalent to light travelling a distance of 60 metres). Furthermore, with the diurnal rotation of the Earth, this time difference changes continuously. The purpose of the delay lines is, quite simply, to equalise this constantly changing time difference, so that the light arrives at the common focus at exactly the same time, irrespective of which telescope it has been relayed through. Rather than talking about a time difference, interferometry experts normally talk about an optical path difference: the difference in distance that light taking the various paths has to travel before reaching the detector. The delay lines are an opto-mechanical system comprised of a set of mirrors, mounted in carriages that move along a 60-metre-long track in the main tunnel. The light is bounced back and forth between the mirrors, until the optical path difference is equalised. As already alluded to, the path difference changes with the rotation of the Earth, the distance between these mirrors must also change. This requires an incredible precision in both the absolute position of the carriages and their motion along the track (at a speed of 5 millimetres per second) to stabilise the optical path. Thus the carriages must keep their position with an accuracy of 5 μm and the stability of the optical path must reach 14 nm[1] over a period of 15 milliseconds. The position of the carriages is controlled by means of a laser. Equally stringent requirements are put on the rails. In fact, the rails are laid in such a way that they compensate for

[1] 50 nm root mean square (RMS) allowed for infrared observations.

the curvature of the Earth over the track length of 60 metres. What ESO needed, immodestly, could quite reasonably be described as the world's most advanced (and highest precision) electrical train!

≈

In March 1998, the contract for the delay lines was awarded to Fokker Space, originally a subsidiary of the renowned Dutch airplane manufacturer, while the optics were done by the Netherlands Organisation for Applied Scientific Research (Nederlandse Organisatie voor toegepast-natuurwetenschappelijk onderzoek, TNO) Institute of Applied Physics (TPD). In a very literal way, the VLTI was now put on track. Then, in July, the head of the project team at ESO, Jean-Marie Mariotti, suffered an untimely death after a short illness at the age of only 43. There is little doubt that, perhaps more than anyone else, Mariotti would have wanted the VLTI project to succeed without further delays, and so, shortly after this sad moment, ESO awarded its second major contract for the VLTI for the delivery of two 1.8-metre movable Auxiliary Telescopes[2] to AMOS in Belgium[3]. In the course of the VLT project AMOS had become almost a "house supplier" to ESO, providing much highly specialised equipment. AMOS stands for Advanced Mechanical and Optical Systems, but it had evolved out of an old Liège company, Ateliers de la Meuse, in Belgium's coal and steel heartland. Under the direction of Bill Collin since 1983, the company has worked with a wide range of aerospace and astronomy projects and has built up considerable, and possibly unique, expertise. As in the case of Schott and REOSC, the working relationship with ESO was excellent. It was driven by a high degree of professionalism, openness — and, quite plainly — a great deal of enthusiasm, from Collin himself, key members of his staff, such as Jean-Pierre Chisogne and Carlo Flebus, and all the way to the factory floor. Even so, the 1.8-metre telescopes were undoubtedly a huge and challenging task for this relatively small company. Readers will remember the early days of La Silla, when the largest telescope was a 1.5-metre telescope. It was housed in a huge building, several storeys high, topped by an impressive white dome. The telescopes that AMOS were to deliver were not only quite a bit larger than this, they would be extremely compact and able to move on tracks criss-crossing the telescope platform atop Paranal. They would carry with them the entire control electronics, service modules for cooling, air conditioning, auxiliary power, compressed air,

[2] The contract included an option for a third telescope, which was subsequently delivered. Eventually, in 2006 a fourth telescope was delivered to ESO, thanks to a special contribution by Belgium, Italy and Switzerland.

[3] The original conceptual study had been carried out by IRAM in Grenoble.

etc. as well as the coudé optical systems; they were like snails, on the move with their houses on their backs. But much faster. Repositioning a telescope would take less than three hours. Thirty docking stations across the Paranal platform offered access to the underground tunnel system with its delay lines, allowing the telescope configuration to be changed almost at will. Delivery in Europe of the first telescope was foreseen for June 2001 with the first observations with two Auxiliary Telescopes (ATs) at Paranal to happen in early 2002 (Koehler, 1998). Optimism had returned to the VLTI project. In the end, perhaps there was a bit too much optimism. The final testing phase in Europe took longer than anticipated. The treacherous issue of spherical aberration that had plagued the Hubble Space Telescope was fresh in everyone's memory. Elaborate matching tests were therefore caried out, but with an 11-mirror optical system they obviously took time. Concerns regarding the optical path length stability had to be addressed, and finally, the contract included tests "on the sky". For that purpose AMOS built a track on the factory premises, allowing a telescope to be rolled out for observations. When that happened for the first time, the test team was in for a shock. The telescope appeared to be dramatically out of the specifications in terms of image quality. A visual inspection by Flebus and Koehler brought reassurance. After rolling out the telescope, the operator had simply forgotten to disconnect the safety system used during transport and to protect the primary mirror in case of earthquakes. Four pads were holding the mirror, albeit in this safety position, not in the observational mode. This was a benign reminder, so often necessary in complex, high-tech operations, of the importance of never letting the human factor get out of sight.

In March 2003, the preliminary acceptance of the first telescope happened in Europe with delivery to Paranal in October. By January 2004 the telescope was ready, but awaiting its twin, which arrived towards the end of that year. During the night of 2–3 February 2005, they carried out their first observation. However, by that time, the VLTI had already seen first light. First light? Without telescopes?

Aside from the work at AMOS and Fokker other activities had happened in Europe relating to the project. A test camera, the VLTI Commissioning Instrument (VINCI), had been built by the Observatoire de Paris-Meudon and software had been written at the Observatoire de Toulouse. At the beginning of 2001 the camera was in place at Paranal, ready for the first fringes.

Meanwhile, the delay lines and other equipment were installed in the underground tunnels and in the interferometric laboratory. Above ground, where the auxiliary telescopes would eventually be placed, two 40-centimetre siderostats placed 16 metres

One of the VLTI siderostats. (Photo: Herbert Zodet)

apart, were used to collect the light and so, on 17 March 2001, at 22:00 local time, the VLT Interferometer was used for the first time to carry out an astronomical observation. The target was Sirius (Alpha Canis Majoris), which is the brightest star in the sky, although it is in reality a double star, or binary as astronomers call them. Considering the ups and downs of the VLTI project, it would be tempting to invoke grandiose language, for it was — literally — the light at the end of the tunnel. Interferometry people, however, are more down to earth. In their world, it was merely "first fringes", which is not to suggest that they did not enjoy their triumph. Andreas Glindemann, who had succeeded Mariotti as head of the VLTI programme wrote: *"The tension was intense when starlight was guided for the first time from the primary mirror of the siderostats, through the light ducts, the tunnel and the beam combination laboratory to the detector of VINCI. And, after a few nights, the result was spectacular. The very first result, the fringe pattern of Sirius.... This was a joyful moment and the champagne corks were popping. But it was also a touching moment when we kept a minute of silence remembering Jean-Marie Mariotti who was one of the fathers of the VLTI and who died much too early three years ago."* (Glindemann *et al.*, 2001).

Celebrating the first fringes at the VLTI. From left to right: Pierre Kervella, Vincent Coudé du Foresto, Philippe Gitton, Andreas Glindemann, Massimo Tarenghi, Anders Wallander, Roberto Gilmozzi, Markus Schöller and Bill Cotton. (Photo: Bertrand Koehler)

In October, light from two of the Unit Telescopes (Antu and Melipal) was fed into the coherent focus in an observation of Achernar (Alpha Eridani), a first magnitude star. Critics may say that using two 8.2-metre telescopes to observe such a bright object is curious, but it was the first time that the real VLTI came into operation, and it was a success. Indeed, it seemed that now the time was ripe for the VLTI to rise as a shining phoenix. One and a half years later, interferometric measurements were conducted with different pairs of the four Unit Telescopes. This permitted the use of different baselines and thus expanded the capability of the system. Towards the end of 2002, MIDI, the MID-infrared Interferometric instrument (8–13 μm) was also installed, allowing the combination of two beams, i.e. telescopes, but either from the UTs — or from the Auxiliary Telescopes, although, as we have seen, at the time they were still in the factory in Liège. November 2004 saw the first article in *Nature* based on VLTI data. It was a study, performed with MIDI, of the innermost parts of protoplanetary discs around three very young stars, i.e. stars with ages of less than 10 million years. Mere babies by astronomers' standards. *"The light from two [VLT] 8.2-metre Unit Telescopes separated by 103 metres on the ground was combined, providing a spatial resolution of about 20 milliarcseconds. This corresponds to 1–2 astronomical units (AU) at the distance of the observed stars; an improvement of more than a factor of ten in spatial resolution compared to the largest modern-day telescopes, in this wavelength regime."* the authors wrote (R. van Boekel *et al.*, 2004).

≈

In the preceding years, working in Nice, Serge Menardi had developed a prototype fringe sensor unit, from which the FINITO fringe tracker emerged[4]. A fringe tracker is required for "long" integration times, i.e. beyond a few milliseconds. Observations of faint objects, where collecting sufficient photons takes some time can only be done with such a corrective system. The fringe tracker measures the optical path difference (OPD) to ensure the required stability of the system and thus reduces fringe blur (Bonnet *et al.*, 2006). It has been compared to a telescope autoguider or, perhaps more appropriately, to an adaptive optics system[5] that reduces image blur. When it was installed in 2003, FINITO marked an important step forward in observational capability. But it was a step that created major challenges. In fact, with FINITO came the crisis, though not because it was a poor instrument. The ESO Annual Report of 2005 simply states that *"FINITO was delivered in 2003, but fringe tracking could not be demonstrated at commissioning."* Behind this simple, factual statement lay many headaches. The problem was due to vibrations. For one and a half years, members of the VLTI team, notably Martin Dümmler, Bertrand Bouvier and Henri Bonnet worked on the issue. Furthermore, in 2005, ESO put together a tiger team to identify the sources of the vibrations. Step by step they located and removed the culprits: high frequency vibrations from hydraulic pumps at the telescopes and instrument ventilation systems. Other corrective actions included the elimination of minute alignment errors of the delay line rails[6].

While these problems were being dealt with, the VLTI instrumentation programme continued. AMBER, the Astronomical Multiple BEam Recombiner, was delivered in 2003 and saw its first fringes in March 2004 with two Unit Telescopes. And by June, a third Unit Telescope was added, following the installation of another delay line.

In February 2007, a section dedicated to AMBER science and featuring eleven articles, appeared in *Astronomy & Astrophysics*. The topics covered included observations of circumstellar discs around stars (in which planets may exist or be in formation) and studies of stars in their late evolutionary stages, amongst them Eta Carinae, a star in the southern Milky Way that has held astronomers spellbound for decades and which, astronomers believe, may one day develop into a supernova. Optical

[4] FINITO was originally supposed to become the first light instrument.

[5] Correcting for the first Zernike mode (piston), though not for higher-order distortion.

[6] FINITO was finally offered for visiting astronomers in April 2007.

inteferometry had clearly moved out of the niche of measuring the diameters of a few bright stars.

The last of the first generation VLTI instruments, PRIMA, the Phase Referenced Imaging and Microarcsecond Astrometry facility, was delivered to Paranal in July 2008 (van Belle *et al.*, 2008) after eight years of development in an ESO-led collaboration with the Max-Planck-Institut für Extraterrestrische Physik and Max-Planck-Institut für Astronomie, the Landessternwarte Heidelberg, the Observatoire de Genève, the École Polytechnique Fédérale de Lausanne, the Institute of Microtechnology of Neuchâtel, the Laboratoire d'Astrophysique de l'Observatoire de Grenoble, and Leiden University (Universiteit Leiden). Other participants included TNO and Thales Alenia Space. This list of partners illustrates not simply the complexity of this technology and ESO's role in harnessing expertise wherever it may be found, but also how know-how from such highly specialised R&D efforts can percolate through partnerships, thereby potentially finding new applications in other areas. PRIMA, allows observations of much fainter objects and the simultaneous observations of two objects separated in the sky, but perhaps most interesting, it would now become possible to reconstruct high-resolution images of objects in the sky by a technique that astronomers call phase-referenced imaging. Obvious targets for PRIMA are detailed studies of exoplanets that have been detected by radial velocity measurements, but active galactic nuclei constitute an equally interesting area, as indeed do all areas in need of ultra-high resolution imaging in the infrared domain.

We shall end our account of the VLTI here, but this does not mean that the VLTI had reached a steady state. On the contrary, in 2005 two second generation VLTI instruments were chosen at a workshop at ESO. As tradition has it, they will become known by their acronyms, MATISSE and GRAVITY (General Relativity Analysis via Vlt InTerferometrY), but there is every reason to believe that they will become famous for the scientific results they will deliver[7].

Before we close this chapter, let us take a step back from technical details of our story. Two statements mark the progression of the VLTI and both appear to be rather accurate. The first one is the remark by Woltjer, cited in Chapter III-11, about his low

[7] MATISSE, the AperTure mid-Infrared SpectroScopic Experiment, is a second-generation version of MIDI with extended wavelength coverage. Gravity could be seen as the successor to PRIMA, with a smaller field of view, but a much improved resolution. This will make it ideal for detailed observations, e.g., of the innermost regions surrounding the black hole at the centre of the Galaxy (Gillessen, 2010), where the gravitational field is exremely intense, hence the name.

"confidence in its early realisation" and the second one by Cesarsky at the beginning of this chapter. The realisation of the VLTI was undoubtedly a major challenge, but it also proved to be more than the icing on the cake or simply an elusive dream. First of all it comes close to the ultimate engineering challenge. But this has been solved with bravura, and the VLTI is delivering frontline science and the future plans show that it is supported by a vibrant and enthusiastic research community. It may still not be a routine mode of operation in the plug-and-play sense. But it has made great strides forward towards that goal, and it has done so because of the systematic approach that is enabled by the long-term funding and long-term planning that ESO can offer. Andrea Richichi and Alan Moorwood expressed it this way: *"It took the work of dozens of engineers and astronomers, a decade of planning, design and testing, hundreds of thousands of lines of code, the (occasionally unorthodox) use of telescopes from 40 centimetres to 8.2 metres in diameter, and a lot of sweat in the dry Paranal air.... It is undeniable that the VLTI constitutes today the term of reference for interferometry, in terms of sensitivity, angular resolution and accuracy."* (Richichi & Moorwood, 2006).

Aerial view of Paranal. The volcano in the background is Llullaillaco. (Photo: Gerd Hüdepohl)

Chapter III-16

Not just a Telescope, an Observatory; Not just an Observatory, a Home

"Although ideal for astronomical observations,
the site is hostile to human habitation
in that it is totally void of flora and fauna,
it is subjected to earthquakes and high wind velocities,
and the nearest water source is 130 km away."

Report by COWI Consult, 1992.

One of the more delightful aspects of flying from Santiago de Chile to Antofagasta is the possibility of grabbing a right-hand side window seat, allowing an unforgettable view of the Andes chain including the Aconcagua, the highest mountain in the Americas. Further north, just a few minutes before reaching Antofagasta, the plane passes over the Paranal Observatory. The four enclosures that protect the 8.2-metre Unit Telescopes stand out clearly against the desert background, their aluminium shielding sharply reflecting the rays from the Sun. Glittering jewels on the mountaintop indeed. A little to the southeast, the Paranal base camp can be seen. It is less spectacular than the telescope buildings, yet it remains clearly visible in the crisp air. Today, the telescopes are at the centre of everybody's attention. But for many years, the base camp was the place where things happened. As we have mentioned, Paranal lies in a deserted area. Before ESO's site search began, few people, if any, will have walked the slopes or rested at the top to enjoy the breathtaking view of the Pacific Ocean, or perhaps looked towards the east, where the towering volcano Llullaillaco resides, 200 kilometres away, on the Argentine border. Establishing an astronomical observatory in this location was a dream, but also a logistical nightmare. Everything, including water, had to be brought to the site — normally from Antofagasta — over a primitive desert track.

The civil engineering planning for the observatory began in 1990. For the purpose, ESO placed a contract with COWI Consulting Engineers in Denmark to develop the concept for the entire infrastructure. Clearly a functional observatory comprises much more than telescopes and advanced instruments such as those that we have previously described in this book. The facilities included mechanical and electronic workshops; facilities for the supply of electrical power, compressed air, cooling liquid, and communications; a mirror maintenance building; a large warehouse (in the Spanish-tinged ESO tradition described as the *Bodega*), accommodation for staff and, of course, the necessary roads[1]. Jens Kierstein Hansen led the work for COWI, while the ESO team comprised Lorenzo Zago, Marco Quattri, Jörg Eschwey and Peter de Jonge. Sustainable energy sources were originally considered for the power supply: after all, it is hard to find any place on the surface of this planet with more sunshine than Paranal. Wind is also abundant. However, the solution turned out to be too expensive for ESO and consequently the idea of "the Green Observatory in the Desert", as it was described, was abandoned. Early plans for a hotel also had to be shelved due to the budget problems that emerged around 1993.

As mentioned in Chapter III-I, on-site work began in September 1991. At the time, the base camp, 300 metres below the summit, comprised a mere nine containers including two bathrooms. With time the camp was first expanded and then supplemented by a separate camp for the contractor staff. At its peak the work force reached about 600 people.

By 1997 the financial situation at ESO, though still strained, allowed the issue of accommodation at Paranal to be revisited. Clearly, the conditions of the container village were not such that this could be permanent solution, even if it had been the only option for many years. The idea of a hotel — or Residencia as ESO preferred to call it[2] — which had been part of the original plan for Paranal was revived. In late 1997, following a pre-selection process involving more than 30 architectural companies, ESO invited nine companies to present their proposals for adequate accommodation facilities at Paranal. One of the companies that decided to respond to ESO's call for proposals was Auer+Weber in Munich. Like the other invited bidders, the company sent an architect, in this case Dominik Schenkirz, to Chile shortly before Christmas 1997.

[1] In addition, around 1996/97, a visitors' centre for the public was added to the observatory.

[2] ESO had traditionally spoken about its hotel at La Silla, only to see it entered into various popular tourist guides and, consequently, having to deal with the understandable disappointment of tourists trying to book rooms at La Silla.

The Paranal base camp, photographed in December 1997. The tall building to the left is the mirror maintenance building with the aluminising plant. To the right of the MMB is the warehouse. The low container complex in the front is the power plant. Note also the VLT M1 transport container. (Photo: Herbert Zodet)

Schenkirz was fascinated by the pristine landscape and felt that any conventional multi-storey building would look awkward and ill-placed in this meagre environment with its stark, almost graphical contours and intense colours. Returning to Munich, in just three weeks he and Phillip Auer came up with an astounding solution: bury the building in the ground. The result was an L-shaped building that effectively only has a single façade, overlooking a depression in the landscape, but also offering a view towards the Pacific Ocean. There were several advantages to this concept: the underground placement made the building energy-friendly, offered protection from the wind and addressed the problem of light pollution — the old enemy of optical astronomy — while at the same time creating a completely self-contained environment, agreeable to human beings. The raw concrete walls coloured with an iron-oxide pigmentation, created a gentle, yet desert-like sensation. In contrast, an internal tropical garden under a 35-metre glass dome overlooked a blue swimming pool, as an attractive optical centrepiece, but also an effective means of keeping the humidity at an acceptable level of 35%, as opposed to the extreme dryness of the outside. The shallow glass dome, the only feature of the Residencia visible on arrival at Paranal, led the thoughts in the direction of an observatory. The building both stood in contrast to the harsh desert environment (and therefore offered mental relief for its inhabitants), yet was in complete harmony with the surroundings. It was both a signature building and a low-cost project, providing attractive living quarters of 108 individual bedrooms, meeting facilities, a restaurant, recreational areas and, in general, a place for social interaction amidst the natural isolation dictated by the desert. It was, in short, a winning proposal. This was at least what the selection team, comprised of Massimo Tarenghi, Daniel Hofstadt and José Manuel Soares, a Portuguese architect and advisor, felt. Giacconi went along with the proposal and in June 1999, Council gave its blessing to the project. Construction began towards the end of the year and the building was completed in 2001. Fittingly, for a building that aims to be a self-contained oasis in the Atacama Desert, the unique interior furnishing and decoration was done by the Chilean architect Paula Gutierrez. The reception area also features two *mixographias* by Roberto Matta, the Chilean painter, philosopher and poet.

The 167-metre-long façade of the Residencia faces the Pacific Ocean with its impressive sunsets that makes the ferrous oxide pigmented concrete walls glow. The building covers a total of 12 000 square metres. It was constructed by the Chilean company Vial y Vives at a total cost of 12 million euros. (Photo: Gerd Hüdepohl)

As time would show, it fulfilled its purpose perfectly. More surprising, perhaps, was that the building, so isolated and available exclusively to ESO staff and official visitors, would catch the attention of the outside world. But it did. In 2004, the architects received the Premio Internazionale Dedalo Minosse alla Commitenza di Achittetura, and the Leaf award in two categories ("New Buildings" and "Overall"). In 2009, Jonathan Glancey, architecture and design writer for the UK broadsheet newspaper *The Guardian*, placed it on the list of top ten buildings of the decade. Numerous articles in professional journals and popular magazines — from *Wallpaper, Casabella, Bauwelt, Via Arquitectura, Opus C* to *BMW Magazin*, the Lufthansa in-flight magazine and many others — reported on the ESO Residencia and, as we shall return to in Chapter IV-10, the building featured in the 2008 James Bond movie under the name La Perla de las Dunas.

The cupola covering the tropical garden stands out against the sunset sky. The rectangular structure provided a huge design challenge in view of the requirements for earthquake resistance. (Photo: Peter Gray)

Part IV:
Towards New Horizons

Artist's impression, created in 2003, of ALMA. (Image by Herbert Zodet)

Chapter IV-1

ALMA

"It is the observation of [the] southern molecular clouds that have shaped much of the millimetre radio astronomy, yet we persist in observing them from the northern hemisphere through many air masses of atmospheric attenuation."

Roy Booth, 1994.

At the beginning of this book we briefly reviewed the status of astronomy in the mid-1950s. Since then — and now approaching the turn of the millennium — much had happened. One of the most important developments in the meantime was the "opening" of the electromagnetic spectrum, i.e. the possibility to observe celestial objects at practically all wavelengths. For example, observing an object in the short-wavelength high-energy domain yields different information from looking at it in the optical band of visible light or in the infrared. Each waveband is like one chapter of a book: to understand the book as a whole, one needs to read all the chapters. In this case, the information gathered would constitute what could be called the *Great Book of the Universe*. The opening up of the spectrum was greatly facilitated by the possibility of using space-borne observing platforms, as the Earth's atmosphere is opaque to much of the radiation that arrives from deep space. This is fortuitous, since much of this radiation — gamma rays, X-rays and ultraviolet radiation is dangerous for life. In this sense the terrestrial atmosphere acts as a protective wall for us. At the same time, this wall has windows that also allow us to peer into the depths of space from the ground. Conveniently, one such window is in the wavelength domain that we call visible light. Other windows allow longer wavelengths to reach us, notably in the radio domain. In between, there are further rather narrow windows, which nonetheless provide interesting possibilities for the study of celestial objects and help enrich our understanding of their nature and the physical processes that go on inside them. Astronomers call one such window submillimetre and millimetre astronomy (itself divided into a number of subwindows), after the wavelengths in question, i.e. from about 0.3 millimetres to 3.5 millimetres. The domain is also described by the frequency range 84–960 GHz[1]. Submillimetre and millimetre astronomy has often been described as the astronomy of the cold Universe.

Thermal emission from all cold objects is at far infrared or submillimetre wavelengths. But in a wider context, it is the study of origins. There are a number of particularly interesting research areas linked to this specific wavelength domain. At the near end of the cosmic distance scale, trying to understand the formation of stars and planets still raises many questions. Stars are formed inside dense clouds of gas and dust that literally shroud the birth process. Visible light cannot penetrate these dense clouds. Infrared radiation can do so to some extent, but even longer wavelengths allow us to see right into the heart of the clouds and thus to study what happens on these stellar maternity wards. In Chapter III-8, we have already alluded to this. Submillimetre and millimetre wavelengths are also suitable for mapping clouds of complex molecules, the presence of which has been verified in the interstellar medium in the Milky Way. At the far end of the distance scale, submillimetre and millimetre astronomy enable us to observe distant galaxies in their early stages. As these remote galaxies formed after the Big Bang strong ultraviolet radiation from young stars heated the interstellar dust. This energy was subsequently re-emitted in the infrared, but due to the redshift, the wavelengths have stretched and this ancient signal can now be observed in the millimetre domain.

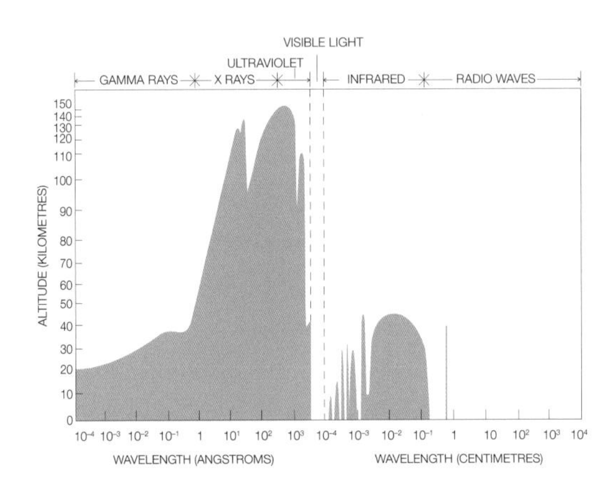

This drawing shows the transparency/opaqueness of the atmosphere to electromagnetic radiation from outer space. The border indicates the 50% transparency level. The atmospheric windows that can be explored from the ground are clearly visible in the optical, near infared, submillimetre and millimetre wavelengths as well as in the longer frequency radio domain (after Pasachoff, 1979).

≈

In the late 1960s and early 1970s, there were relatively few astronomers interested in submillimetre astronomy and this research area was most certainly a niche activity.

[1] The frequency coverage of ALMA.

Lacking dedicated telescopes for this wavelength domain, submillimetre astronomers in the southern hemisphere had therefore to resort to using optical telescopes. As early as 1978 Thijs de Graauw had used a 1 millimetre receiver on the 3.6-metre telescope. Later, in 1981–1983, de Graauw's group used CAT for millimetre-wave observations[2]. Tom Wilson had also measured carbon monoxide (CO) at a wavelength of 2.6 millimetres using the 3.6-metre telescope. The 1-metre telescope has also been used in this way. It was obvious that this hardly constituted an optimal way of using these telescopes as can be seen in a 1985 article in *The ESO Messenger,* appearing under the title "Of Pearls and Swine": *"Who would play a Stradivari violin at a country dance? Or who would mix very old Scotch with Coca-Cola? Not us. But we do things that look equally improper in the eyes of many astronomers... ."* (Krügel & Schulz, 1985) But relief was on the way. Back in 1986/1987, ESO had expanded its telescope park at La Silla to include a submillimetre telescope, the 15-metre Swedish–ESO Submillimetre Telescope (SEST), which we have already mentioned. Among other things, it had mapped CO emission in the Magellanic Clouds[3], observed carbon monoxide molecules in the expanding shells around carbon stars and scanned low-mass star-forming regions. Sitting at the far end of the southern flank of La Silla, this telescope had rendered good service to the scientific community but given the rapid expansion of the research field, it was slowly becoming dated. More importantly, the move towards larger optical telescopes with their improved imaging capabilities drove forward the development of submillimetre telescopes that would give a similar performance. But we once again remember that image resolution depends not only on the size of the telescope mirror but also on the wavelength. Achieving the same resolving power in the longer wavelength domain would therefore require significantly larger telescopes — or antennas as they are also called. This could not realistically be achieved with a single dish, but called for an array of several, interconnected telescopes.

In 1991 the SEST community began to think about an interferometer. The original idea focussed on placing an additional telescope on a hill adjacent to the SEST site at La Silla, an idea that was discussed at a meeting of the SEST User Group at ESO in May. According to the subsequent report in *The ESO Messenger, "a lively*

[2] These observations also allowed the sky transmission in this wavelength domain to be assessed and paved the way for the later deployment of the SEST. *"La Silla appears to be a good site for millimetre-wave observations, with 230 GHz transmissions very often between 80% and 90%,"* a report in *The ESO Messenger* said (Brand, 1982). As it turned out, La Silla was also a good site for higher frequency (shorter wavelength) observations down to about 1 millimetre (299 GHz).

[3] A key programme carried out by Marc Kutner, Monica Rubio, Roy Booth, François Boulanger, Thijs de Graauw, Guido Garay, Frank Israel, Lars Johansson, James Lequeux and Lars-Åke Nyman.

discussion ... followed on the idea of a millimetre-wave interferometer at Paranal, introduced and stimulated by a talk by R. Booth which outlined a possible concept consisting of 10 metre × 8 metre antennas.... He considered that such an array could be built for about 50 million deutschmarks and suggested that we in Europe should consider ways of raising money in order to achieve a useful millimetre array on Paranal as soon as possible." (Shaver & Booth, 1991). Booth and his colleagues contacted Harry van der Laan, who had taken over as Director General of ESO. Van der Laan reacted positively, but cautiously, to their ideas suggesting that a study group, with Booth as chairman, should be established to consider the issue. The first meeting of the group took place at IRAM in December of that year. During the meeting, news broke of the detection by American astronomers[4] of carbon monoxide with a redshift of 2.3. The news was a bombshell. Never before had CO been detected at this distance. The prospect was that it might be possible to study CO emission (and other kinds of atomic and molecular emission), not just from the Milky Way, but also from distant objects. This would give new insights into the conditions in the early Universe, including star formation and chemical enrichment of the interstellar medium in early epoch galaxies. It would, however, require much more collecting area and also better resolution than hitherto envisaged — inspiring the participants to think along grander lines, notably in terms of an array with a collecting area of 7000–10 000 square metres.

New facilities in the millimetre domain were not only on the wish list of European astronomers. US plans for a major new millimetre array (MMA) began in the early 1980s, but in the course of developing ideas for the project, it became clear that the funding body (NSF) wished to see international partners involved. In Japan astronomers also began to make plans for a similar facility.

In Europe, Booth's working group began to draw up plans for what they called the Large Southern Array (LSA). In May 1994 the ESO STC gave its blessing to ESO involvement, leading to a collaboration between ESO, IRAM, the Onsala Space Observatory and the Netherlands Foundation for Research in Astronomy (Shaver & Booth, 1998)[5]. The agreement led to the establishment of a formal LSA Study Group that presented its findings in 1997. The report suggested either an array of 40 8-metre antennas and 25 to 35 15-metre antennas or an homogeneous array of 50

[4] Bob Brown and Paul Vanden Bout.

[5] Later, CNRS, the Max-Planck-Gesellschaft, the Nederlandse Onderzoekschool Voor Astronomie (NOVA), the UK Particle Physics and Astronomy Research Council (PPARC), the Swedish Natural Science Research Council (Naturvetenskapliga forskningsrådet, NFR) and, finally, the Instituto Geográfico Nacional of Spain joined the European partnership.

to 60 antennas of about 12 metres in diameter (Grewing & Guilloteau, 1997). But why in Chile? Of course the involvement of ESO might seem to make this a natural choice, but, as for optical ground-based astronomy, the selection of the right site is critical to scientific success. The southern hemisphere provided observational access to the centre of the Milky Way — a key reason for ESO's presence in the first place — but the dryness of the Atacama Desert, especially the options of high sites with sufficient space, would prove to be even more decisive in choosing a location for millimetre astronomy than for the optical case.

And so, the search for a site began. La Silla was no longer under consideration, and neither was Paranal. The choice was also linked to progress in receiver technology, which made it realistic to study the sky at frequencies approaching 1 THz, albeit only from a high altitude site. The ambitions of the community, keen to observe in the high frequency bands, had been awakened. Because of the relationship between atmospheric transparency and elevation in this wavelength domain, there are rather stiff requirements for a ground-based observational site. A dry (in terms of column density water vapour above the telescopes), high altitude site with a flat area for extended array deployment was needed and, for obvious logistical reasons, so was reasonably easy access to the area. This pointed towards sites on the Altiplano, the area that straddles the borders between Chile, Peru, Argentina and Bolivia.

One person at ESO had a particularly well-developed knowledge of Chile's II Region, Angel Otárola. He had trained as an engineer and joined the SEST team in September 1990 as a SEST operator. Before that, however, as a student he had worked with German geodesists in establishing the very first GPS network in Chile to study the dynamics of the Andes range. Together with Lars Baath (of the Onsala Space Observatory), he undertook a first exploration to the area in 1991. He also assisted the early Japanese site search conducted by the Nobeyama Radio Observatory and led by Naomasa Nakai. These early efforts also involved Wolfgang Wild and Nick Whyborn. In May 1993, a small team comprising Kimiaki Kawara, Nagayoshi Ohashi from Japan, Whyborn and Otárola, reached the Chajnantor area, although snow prevented them from fully exploring the site.

US astronomers had already looked at sites in the southwest of the continental USA, Hawaii (Mauna Kea), and then in Chile[6]. In April 1994, a search party from the US

[6] They were not alone. Since 1992 Japanese astronomers had carried out tests at various sites in northern Chile for their Large Millimeter Array. By 1997, Chajnantor had also emerged as the site of choice for the Japanese project.

National Radio Astronomy Observatory visited the highest sites surveyed by the Harvard Smithsonian Center for Astrophysics (CfA) for the SMA[7]. The most remote of the sites visited by the team, and the only site close to comparable to Chajnantor, was near the border town of Ollagüe, *"where the train once stopped as it went from Calama, Chile, to La Paz, Bolivia"*, as Paul Vanden Bout, director of the NRAO, later recalled. Chajnantor was not on the list. The high ground near Chajnantor was visited at the end of the trip, *"partly by intent and partly as an excuse to see San Pedro de Atacama"*. Otárola, who was part of the team, had suggested a closer look at Chajnantor area. It appeared to be a good place, offering a huge plain for the array and an elevation that would allow observations in the high frequency (submillimetre) domain. Following further trips to the area NRAO therefore began site testing in 1995.

Site testing at Pampa de Pajonales. Standing in front of the container and the photovoltaic modules that provided power to the interferometer and 183 GHz water vapour radiometer (left to right): Lars-Åke Nyman, Daniel Hofstadt, Angel Otárola and Peter Shaver. (Courtesy: Angel Otárola)

ESO had considered a site at a somewhat lower elevation of 3800 metres above sea level (msl) at Pampa Pajonales, southeast of the Llullaillaco volcano. This site met

[7] The Submillimeter Array (SMA) is a radio interferometer with eight antennas, operated by the Smithsonian Astrophysical Observatory and the Academia Sinica Institute of Astronomy and Astrophysics. It was eventually placed on Mauna Kea in Hawaii.

the requirements for observations in the millimetre domain, but was problematic for higher frequencies. The site-testing team included Angel Otárola, Roberto Rivera Maita, Guillermo Delgado and Lars-Åke Nyman. By 1998, it had become clear that the American and European (as well as the Japanese) projects were likely to merge and that a site would be needed that allowed for higher frequency observations. The partners homed in on the plateau at Chajnantor.

For a while, La Casa de Don Tomás, a typical hotel in San Pedro de Atacama, became the central operating point for ALMA activities. This picture, taken in 1997, shows preparations for a visit to the Chajnantor Plateau. (Courtesy: Lars-Åke Nyman)

At the beginning of June 1998, the joint Onsala–ESO site-testing group deployed a weather station, two interferometers and water vapour radiometers on Chajnantor to test its suitability for submillimetre astronomy (Otárola *et al.*, 1998).

≈

The prospect of major ESO involvement in a non-optical ground-based astronomy project was not welcomed by everyone in the ESO community. Although the ESO Convention simply uses the term astronomy — thus enabling the organisation to undertake activities in all spectral domains that can be exploited from Earth — many had come to perceive ESO as an optical astronomers' club and saw this development as a paradigm shift. While many saw the new initiative as an important and very positive step, others considered it as a distraction from what they thought should be ESO's focus. For Giacconi, the case was clear: he had himself been involved with X-ray astronomy, gamma-ray astronomy and optical astronomy..., *"For me, it's all astronomy,"* he said, but there were other reasons as well. It would certainly take a while before the next optical project reached maturity and ESO wanted to retain the unique project expertise it had acquired in the course of VLT project. Furthermore, Giacconi realised that a joint US/Japanese project might become a reality and he did not want Europe to be excluded. At the summer meeting of the ESO Council on 4–5 June 1997 the topic of ESO participation in the LSA and options for international cooperation were discussed. The meeting concluded that ESO should continue with *"exploratory talks ... on an international level"*.

Giacconi did not sit on his hands. On 25–26 June 1997, with a European delegation[8] — he and Paul Vanden Bout, also with a delegation, met at the NRAO headquarters in Charlottesville, Virginia, to discuss the possibility of merging the two projects. The advantages seemed clear and by the coffee break the major items were resolved, paving the way for a joint document. The basic elements of the paper were to work towards a joint project of 64 12-metre antennas, located at a high (submillimetre-friendly) Chilean site, funded on a 50/50 basis by Europe and the US. There was also a pledge to begin negotiations with Japan and an understanding that ESO and NRAO would do the job, hence no new organisation was to be created. Given the somewhat *ad hoc* nature of the meeting, however, it wasn't clear whether the document could be signed on the spot and both realised that the difficulty. Giacconi said: *"Paul, I don't know if you can sign this paper, but I am almost certain that I can't. So why don't we do it!"* And so they did. ALMA had been conceived[9, 10]. Surely not the first child to be conceived without the explicit prior consent of its grandparents!

The ESO Council acquiesced. The LSA and MMA projects established joint working groups to set out the basic parameters, including the issue of wavelength (or frequency) coverage, the question of a homogeneous or an inhomogeneous antenna array, how the project could be organised, and finally, a costing. Their proposals were endorsed by NRAO and the LSA consortium in November 1997. In June 1998, at the meeting that also celebrated the VLT first light, the ESO Council endorsed the outcome of the LSA/MMA studies and *"request[ed] the executive to continue*

[8] The delegation comprised Torben Andersen, Roy Booth, Riccardo Giacconi, Michael Grewing, Stephane Guilloteau, Karl Menten, Dietmar Plathner, Peter Shaver, and François Viallefond.

[9] After the meeting, one of the American participants, Ed Churchwell, noted: *"Although there are outstanding unresolved questions, I believe that this was a watershed meeting... ."*

[10] The name, however, was only decided in March 1999 at a meeting at ESO between NSF, ESO and a UK delegation. It concluded an elaborate selection process, with more than 30 options, organised by Bob Brown. He also presented the choices to the participants. *"The committee, in silence, stared at the list on the projector screen for what seemed to me to be several minutes. Finally, Ian Corbett said 'ALMA is the best. I like acronyms I can pronounce',"* Brown later recalled. The committee agreed.

LSA/MMA studies and develop a proposal for the design and implementation 'of a large mm/sub-mm array' in Chile to be submitted to Council in June 1999." It also appointed a negotiating team *"to enter into discussions with NRAO and NSF to study the conditions of a union of the LSA and MMA into a single common project... ."*

≈

The European LSA project, now expected to merge with the MMA, was supported by several organisations in Europe, rather than just by ESO. The European scientific community of millimetre/submillimetre astronomers were somewhat fragmented at this time. France, Germany and Spain had come together in support for IRAM and naturally saw this organisation as their main facility. British and Dutch astronomers used the James Clerk Maxwell Telescope (JCMT) in Hawaii and had strong links with the US, which had facilities such as the Owens Valley Radio Observatory in California, at the time with an array of six ten-metre antennas. It is therefore not surprising that discussions took place at several levels and in several groupings, sometimes even in parallel. With the UK still not a member of ESO, Ian Corbett of PPARC coordinated a series of meetings among funding agencies at INSU in Paris. The meetings began in 1998, but ceased with the formation of the European ALMA Coordination Committee and the ALMA Committees with the US and Canada (Corbett, 2011). This notwithstanding, at the extraordinary ESO Council meeting in September 1998, it became clear that several countries[11] had made their support for the new project conditional on it being implemented through ESO. At the meeting, the ESO Member States made a commitment to channel the funds for their participation into a possible joint project through ESO. At a follow-up meeting in September, a European Co-ordination Committee (ECC) *"to direct the European effort"* as well as European Negotiating Team (representing the prospective funding partners) were established. Furthermore a European Science Advisory Committee, chaired by Karl Menten, was set up, and the first key personnel were recruited (Shaver & Kurz, 1999). Stephane Guilloteau became the European project scientist and Richard Kurz, who had joined ESO in March 1998 from a position as project manager of the Gemini project, became the European project manager. On his watch work began in preparation for the initial phase of the joint project together with his US counterpart Bob Brown. At the same time, the European effort had to be streamlined. Many institutes were keen to participate in the project. At one point, the number approached 25, but the individual contributions were almost exclusively in-kind and

[11] Belgium, Denmark, the Netherlands, Sweden and Switzerland.

varied considerably in scope. Patchwork solutions, it seems, are often the outcome in European scientific collaborations and they can be very successful. Nonetheless for a joint project of this scale and scope, this was hardly a sustainable solution. In the end, the agreement for the three-year design and development phase (Phase One) was signed by only seven European organisations, whilst NSF signed for the American side. The agreement was sealed at a ceremony at the NSF Headquarters in Arlington, Virginia, on 10 June 1999[12].

≈

Before we move on, we need to address the issue of terminology. So far this book has used the language of optical astronomers. Optical astronomers use telescopes, radio astronomers usually talk about antennas, optical astronomers observe in wavelength domains or bands, radio astronomers prefer to talk about frequencies. Unsurprisingly, when dealing with ALMA, we encounter new technical terms. The main components of ALMA are the antennas, the "front ends", the "back ends" and the correlator. The front ends include the receivers and the associated cryogenic casings — in other words what optical astronomers would call the detectors. The back ends are the systems that digitise (and otherwise process) the radio signals before they are fed into the correlator, which is effectively a supercomputer. In simple terms the correlator has the same function as the coherent focus of the VLTI — it is where the signals from all the telescopes (or antennas) are brought together in phase.

≈

One of the key elements of the development phase was the procurement of two prototype antennas, with each partner ordering one. Even uninitiated readers will be aware that radio dishes are hardly a new and untested developments. But, just as we have seen in the case of optical telescopes, because ALMA was about doing new science, the requirements for the ALMA antennas were much stricter than for existing telescopes. This pertained to the accuracy of the surface of the antenna dish, to the precise pointing accuracy of the antenna, and to the ability to carry out fast switching

---

[12] Astronomers in Japan had also been thinking about a major project, called the Large Millimetre and Submillimetre Array (LMSA). In 1998, they decided to work towards a merger with the LSA/MMA projects. In parallel to the ALMA Phase One Agreement, a joint Europe-US-Japan resolution was therefore signed at the meeting. At the time, it was thought that this would lead to an array of no less than 96 antennas. In a separate development Canada joined the partnership in 2002 under the umbrella of the NSF.

between the source being observed and a calibration field[13], among other things. Furthermore — since the antennas were supposed to operate in the open air without a protecting dome — it was also clear that the extreme conditions to be expected at the Chajnantor Plateau with intense solar irradiation, strong winds, swift temperature changes and so forth placed demanding technical requirements on the telescope. For their prototype, the Americans selected Vertex RSI as the contractor. Established in 1963 in Duisburg (Germany) as Krupp Antennentechnik[14], it had been taken over by Vertex RSI, an American company, in 1993. It was undoubtedly a prudent choice, since Vertex RSI was considered not just a very experienced company, but certainly one of the world-leaders in the antenna market.

Like the Americans, ESO issued a call for tender. At ESO, Torben Andersen[15], who had also been involved in the 32-metre EISCAT antenna at Spitsbergen, had developed a first design back in 1997. His suggestions were complemented by contributions from IRAM, notably by Dietmar Plathner (who had worked for ESO in the 1970s), but also by David Woody from Caltech. All this work resulted in a pre-design that served to guide the bidders. Among these was the Italian company EIE[16] in Mestre. EIE had not built antennas before, but their proposal was innovative, making extensive use of carbon fibre and also proposing direct drive systems for the antennas, similar to those that had been successfully used for the NTT and VLT projects. EIE had been involved in both projects and ESO had confidence in the company and its engineering skills. The estimated weight of the antenna was 43 tonnes[17], considerably less than half of the Vertex antenna and possibly with better dynamical characteristics. Selecting them would provide a real technological alternative to the Vertex antenna. It is worth pointing out that the thinking behind the choice of two prototypes was not simply that one might prove superior to the other, but rather that it might be possible to combine the best of both solutions for the full array. Both contracts were awarded in February 2000.

---

[13] We have already seen this mode of observation used in the infrared, with a chopping secondary. In the case of the ALMA antennas the requirement was that the antenna dish should be able to move two degrees and settle down again, ready for observations within just 1.5 seconds.

[14] The location of this company is hardly a coincidence, given the presence of the Max-Planck-Institut für Radioastronomie in the region. Over the years, Vertex had supplied a number of antennas to radio astronomy observatories (Effelsberg, Pico Valeta, etc.) (Menten, 2009).

[15] Together with Franz Koch and Niels Jessen (from the Risø National Laboratory).

[16] In partnership with the company Costamasnaga s.r.l..

[17] The final mass of the antenna after design modifications grew to about 78 tonnes, still significantly less than the competing design.

Another key aspect of Phase One was to put together the European contribution to the project beyond the development of a prototype antenna. Important elements were software development — the primary software development was done by ESO but important contributions were made by the UK Astronomy Technology Centre and IRAM — and the assignment of receiver bands. As we have mentioned the wavelength domain offers a number of narrow windows — or bands — in which observations can be conducted from the ground. The ALMA project aimed at covering no less than ten bands, although not all would be available to astronomers from the start. The individual receiver bands were given numbers. Accordingly, band 1 covers 31–45 GHz, band 2 67–90 GHz, band 3 covers the frequency range 84–119 GHz (3.5–2.5 mm wavelength)[18], band 4 125–163 GHz, band 5 163–211 GHz, band 6 211–275 GHz, band 7 275–370 GHz, band 8 385–500 GHz, band 9 602–720 GHz and band 10 787–950 GHz. The European participants agreed to cover bands 7 (1.1–0.8 mm wavelength) and 9 (0.5–0.4 mm wavelength), reflecting the scientific interests of the European scientific community. The Rutherford Appleton Laboratory in Didcot, UK (RAL) and Chalmers later agreed to also cover band 5[19]. IRAM was tasked with developing the receiver for band 7; NOVA/SRON took on band 9, while RAL developed the cryostat. Important work under Phase One also included correlator development activities carried out by Alain Baudry and his team at the Université de Bordeaux together with the Dutch ASTRON.

≈

In parallel, preparations for the construction phase, Phase Two, were going forward. At the June 2002 meeting in London, the ESO Council formally approved *"European participation through ESO in the baseline Bilateral ALMA Phase II at the 50% level"*. It was not common practice at ESO to move from one project phase to the next before the first finished. But speed was of the essence because of the character of the ALMA partnership.

---

[18] The lowest frequency band to be implemented initially. The other early bands are 5, 6, 7 and 9.

[19] This frequency band is rather difficult because of atmospheric water vapour, but ALMA observations in this band offer interesting synergies with the ESA cornerstone mission Herschel, flying at the same time as the early operational phase of ALMA. Because of its experimental nature, but also because it was not part of the rebaselined ALMA, only six receivers were foreseen in the so-called ALMA enhancement programme, financed through the European Commission's 6th Framework Programme. Later it was decided to equip all antennas with band 5 receivers.

On 25 February 2003, Catherine Cesarsky and Rita Colwell[20], signed the agreement on behalf of ESO and Spain, and the US National Science Foundation (NSF) and the National Research Council of Canada, respectively[21]. The third partner, Japan, was not able to join the agreement at the time and so within the ALMA community this historic document became known simply as the bilateral agreement. The joint ESO/NSF press release was naturally more upbeat: *"Green Light for World's Most Powerful Radio Observatory"* it announced. The cost of *"650 million euros (or US dollars),"* as the text said (reflecting the euro/dollar exchange rate at the time), *"is shared equally between the two partners"*. It is worth reflecting on this. There had been many scientific collaborations between Europe and the US, but this may have been the first major project — at least in astronomy — in which Europe was not the junior partner.

As already foreseen in the original concept, the antennas could be moved between 225 stations[22]. They could be placed in different configurations, either spreading out over a distance of 16 kilometres on the Chajnantor Plateau or in more compact mode[23], with the signals being sent by fibre optic links to the correlator. We recognise the idea of movable telescopes from the VLTI Auxiliary Telescopes, though in the case of ALMA, the concept is taken to a much larger scale[24]. The ALMA Observatory would be operated from the ALMA Operations Support Facility (OSF) at an elevation of 2900 metres. ESO committed itself to providing half of the antennas, the cryostats for all the front ends, the receivers for two bands, substantial work on the back end as well as software. Further ESO deliverables included computers, parts of the correlator, the antenna transporters, the preliminary base camp, the OSF and the power station as well as sufficiently wide access roads from the public road between San Pedro de Atacama and Toconao and between the OSF and the high site, all in all a distance of 47 kilometres.

The testing of the first prototype started in April 2003. The site in Socorro, New Mexico, is well known to radio astronomers. It is the home of the Very Large Array radio telescope (VLA). To the ALMA project it became known as the ALMA test

---

[20] By proxy.

[21] The agreement foresaw that ESO would act as the sole European partner, although it would work with the institutes in Europe that had been involved so far. On the North American side, the practical implementation would be entrusted to the NRAO, which itself was managed by Associated Universities Inc. (AUI). In 2006 Japan (with Taiwan) became full members of the partnership with the signing of a tri-partite agreement.

[22] The number was reduced to 196 as part of the descoping that also led to a reduction in the number of antennas.

[23] In the compact configuration, the antennas are located with separations of only 15 metres.

[24] Of course movable antennas were nothing new in radio astronomy.

facility (ATF). The first antenna to be mounted at the ATF was the Vertex antenna. What had by now become known as the "European antenna"[25] was running late; some necessary design changes had become apparent and the original EIE partner company Costamasnaga had been forced to file for bankruptcy. Finding a new partner and finalising the necessary agreements had taken the better part of a year, but a new consortium was eventually established between EIE and Alcatel Space Industries of France. In May 2003, the European antenna was finally ready for shipment to the US. In order to save time it was decided to ship key parts of it by plane into New Mexico. There aren't many freight planes that can cope with such a large piece of equipment, but a solution was found: the converted Airbus A300 transporter (with the fitting nickname Beluga), originally built to ferry large aircraft components between the Airbus manufacturing sites in Germany, France, Spain and the UK, was able to accommodate a structure of roughly half the size of an ALMA antenna dish; and so, the receiver cabin and reflector backup structure (the so-called BUS) were assembled in two halves and loaded onto the plane. On 16 May 2003 it took off from Milan's Malpensa airport, to arrive at Albuquerque four days later. After assembly on site, the antenna was ready for testing in January 2004.

Loading the prototype antenna into the Airbus Beluga transporter. (Photo: Massimo Tarenghi)

The prototype test phase turned out to be more difficult than anticipated. Since the test phase had to be completed within a predefined period of time for cost reasons, the late delivery of the antennas was clearly problematic. Furthermore, testing antennas with the accuracy required for ALMA had not been done before. To some extent the test techniques had to be further developed to verify the performance of the

[25] Perhaps unavoidably, the two different antennas became known as the "American" and the "European"antennas. This was unfortunate, firstly because it had political connotations, but even more so because it was not entirely correct. Thus, the Vertex ("American") antenna had been developed in Germany and major components such as carbon fibre structures, the antenna control systems, computer systems and drive systems, of the production antennas were also manufactured in Europe.

antennas at the level of the ALMA specifications. And finally, the ATF site choice had been a compromise: the great advantage of the VLA site was that it offered the full technical infrastructure necessary. The disadvantage, however, was that the site was far from optimal when it came to testing the antennas at higher frequencies. The first set of tests was concluded in April 2004, but they were inconclusive. In its report, the Antenna Evaluation Group executive stated that *"it was not able to definitely confirm that the designs would satisfy the ALMA technical specifications"*. (van der Kruit, 2012). More tests would be needed. This was a major problem. The North American side feared that a delay could jeopardise the project since the US Congress might raise objections. A cost increase could also be expected, although as we shall see shortly, this had probably already become inevitable at that time. They therefore pushed for a quick decision.

Despite these difficulties, the ALMA partners took the next step by the end of 2003, moving towards the procurement of the full set of antennas. According to the bilateral agreement, each partner would deliver half of the total number of antennas. There would, in other words, be two separate contracts to industry. Though not aimed for, it entailed the possibility also of two different types of antennas, albeit with the same performance. In the following chapter, we shall briefly address the governance issues that led to this decision. Here, it suffices to note that on both sides of the Atlantic bids were received for consideration by the respective partners. The tenders opened in May 2004, however, were alarming if perhaps not totally unexpected. Over the preceding years, commodity prices had increased considerably and by much more than expected[26]. Even more so, because of the boom, the labour market in Chile, a major exporter of commodities, was under pressure, leading to substantially higher site development costs. Implementing the full programme, as foreseen in the bilateral agreement, appeared to be no longer financially feasible. Savings could be made either by reducing the number of antennas or the number of receiver bands. From a scientific point of view, the latter option appeared decidedly unattractive. Fewer antennas, on the other hand, would in the first instance mean a loss in sensitivity that in principle could — at least partly — be compensated for by longer integration times and fine tuning of operational modes.

---

[26] Although discussions about ALMA had begun earlier the official cost estimate was based on US dollars at their 2000 value. However, in the early years of the new millennium, the world economy was booming. The rapid growth in demand in the emerging economies had sent commodity prices skyrocketing. For example steel prices rose from about 350 US dollars/tonne in the year 2000 to a peak of 760 US dollars/tonne in late 2004 (Source: *Purchasing Magazine*).

The result was a re-baselining exercise that examined how a reduction in the number of antennas might impact the science, and established the minimum number of antennas thought to be needed to achieve the primary science goals. At a meeting in October 2005 in Garmisch-Partenkirchen, the idyllic Alpine ski resort, the project was redefined. It would now comprise 50 12-metre antennas, 25 to be delivered by each party, at an estimated cost of 750 million US dollars[27]. However, Japan had joined the project in September 2004, bringing in its own plans for an array of four 12-metre antennas and 12 8-metre antennas. The bilateral ALMA project thus became Enhanced ALMA, with a total of 66 antennas, about the same number as the original European/North American project had foreseen.

In parallel, in April 2005, a joint antenna evaluation group concluded that both the proposals by Vertex RSI and EIE/Alcatel Space/MT Mechatronics, submitted in response to the call for tender, were technically acceptable. The companies that had built the prototype antennas had of course also followed the testing at Socorro themselves. The results of the performance tests were obviously relevant for the industrial bidders, but the tests were, as we have seen, only concluded during the tendering phase. The slowing of the procurement due to the prolonged testing at the ATF in Socorro beyond 2004 brought the need for renewed negotiations with the bidders, whose offers were only valid for a certain period of time.

For the North American side, the situation was becoming critical, but it was no easier for the Europeans. Among the issues at stake were different opinions on industrial policy questions, such as how to ensure the required performance of the antennas. The ESO procurement system demands industrial contracts with strict compliance criteria and associated penalties if specifications are not met. The US procurement rules were somewhat different, based on the principle of best effort. The two partners also had differing views on how the expected maintenance costs should be factored into the selection, with American rules seeing a fairly sharp division between construction and operation costs. On the other hand, the geographical distribution of subcontractors to the main supplier was perhaps more of an issue for the European countries. And finally, company mergers during this phase complicated matters. Vertex RSI was taken over by General Dynamics, while Alcatel Space and Alenia had merged[28]. Taken together, this made for a complex procurement process and because

[27] This also meant an increase in cost of about 90 million euros (albeit with a considerable contingency) to each of the principal partners, ESO and NSF. Though it was painful, both parties accepted this.

[28] Only to be absorbed by Thales, the French industrial giant, a little later.

of the fluid situation, the bidders to the ESO contract were allowed to amend their offers before the final decision was taken.

Whereas the North American ALMA partner had already selected its supplier, Vertex RSI, placing its contract in June 2005, ESO needed more time. Several ESO Member States had become nervous about the cost and requested a cost review to ensure that the project was still deemed affordable. By the late summer, the issue had been resolved and ESO, too, could now move towards an industrial contract for its share of the antennas. The contract would be the largest ever signed in ground-based astronomy in Europe. The ESO management, the Finance Committee and Council had their focus on this crucial decision. At the same time, the two main bidders to the ESO call put up a dogfight with few holds barred — as is often the case with large industrial contracts. In the end, the partnership that had developed the European antenna emerged as the winner, with the lowest bid. On 7 December 2005, ESO signed the 147 million euro deal with the consortium led by Alcatel Alenia Space and including EIE and MT Aerospace (Germany). The contract foresaw the delivery of 25 antennas, but contained an option for an additional seven antennas, as did the North American contract with Vertex RSI.

Signing of the antenna contract between ESO and the AEM Consortium. It was the largest individual contract with industry ever signed by ESO. From left to right: Thomas Zimmerer (MT-Mechatronics), Hans Steininger (MT-Aerospace), Patrick Mauté (Thales Alenia Space), Catherine Cesarsky, Gianpietro Marchiori (EIE), Vincenzo Giorgio (Thales Alenia Space) and Ian Corbett. (Photo: Mafalda Martins)

≈

Just two weeks later, ESO signed another important industrial contract, this time with the German specialist vehicle maker Scheuerle Fahrzeugfabrik GmbH, *"a world-leader in the design and production of custom-built heavy-duty transporters."*[29] According to this contract, Scheuerle would build two giant transporters to move the ALMA antennas around on the Chajnantor Plateau as well as between the observing site and the operations facility at 2900 metres altitude. Like the antennas they carry, the transporters are technical masterpieces. Each transporter is 10 metres wide, 20 metres long and 6 metres high and weighs 130 tonnes. It is powered by two 500 kW diesel engines, as powerful as Formula One racing engines. Unlike Formula One cars, however, the ALMA transporters move at no more than 12 km/hour when carrying their payloads. The altitude takes it toll: because of the lower oxygen level at 5000 metres, the engines lose about half of their power, but what is more critical is controlling the vehicle during its descent from the high site. To ensure safety, several brake systems are installed. The suspension system, designed partly by ESO, is also unique. Since the transporters must be able to move with extreme precision,

The first ALMA transporter during a dance with a Smart car at the official presentation at the factory on 5 October 2007. (Photo: Hans Hermann Heyer)

[29] ESO Press Release, 22 December 2005.

both between the antennas and in deploying individual antennas on their respective pads, they have advanced sideways steering, positioning lasers, an anti-collision system and cameras to aid the driver. Finally, the specially trained drivers have their own oxygen supply, when they operate the vehicles. The first of the twin transporters was presented to the media at an event at the factory on 5 October 2007. It gave a very memorable demonstration of its impressive maneuverability, performing a remarkable dance to music!

Whilst the antenna contract clearly constituted a major and highly visible milestone in the ALMA project, work had also begun in many other areas. On 6 November 2003, a groundbreaking event for the ALMA observatory took place at the future location of the Operational Support Facility, following the acquisition of the one square kilometre of land that would be jointly owned by ESO and AUI.

Breaking the ground for ALMA: in November 2003 a formal ceremony marked the start of the construction at the Chajnantor site. Seen are Piet van der Kruit (ESO Council President), Wayne VanCitters (Director of the Astronomy Division at NSF) and Massimo Tarenghi, the new ALMA Director.

Much discussion had also focussed on the development of the necessary electronics, including the front ends with the receivers and cryogenic systems. It was decided that each partner would establish a regional Front-End Integration Centre (FEIC), i.e. one in the US, one in Europe and one in Taiwan, which had become a partner of

Japan. The Rutherford Appleton Laboratory was awarded the contract as the European FEIC.

Since a lot of the technological development took place in different regions of the world, it was vital to ensure that ALMA would remain a single, coherent programme. Cohesion was achieved by setting up integrated product teams with members from each partner. Teams were established in all the major activities: management, site, antennas, front ends, back ends, correlator, computing, system engineering and integration, and science, as well as outreach. Furthermore, in early 2003 a Joint ALMA Office was established in Chile, with staff from all partners. Although ESO was willing to host the office, the initial decision was to place the office in a separate location and thus office space was found on the 18th floor of the prestigious El Golf building[30] in Santiago. In 2010, however, the office moved to a newly constructed building, covering 7000 square metres including car parking, on the ESO premises in Vitacura.

Another element to be delivered by ESO was the OSF, which comprises 6000 square metres of buildings, equivalent to a soccer field. The OSF is the place where antennas are assembled and tested, and maintenance is carried out, and from where the telescopes on the high site will be operated. It also includes an 8 MW multi-fuel[31] power station, adequate living quarters for the ALMA staff as well as a visitors' centre. The access roads were completed in 2006.

≈

While construction was proceeding, planning had begun for the operational phase. To aid the worldwide user community, it was decided to establish ALMA regional centres (ARCs) in each of the regions participating in the project (Asia, Europe and North America). In a manner reminiscent of the tasks of the ST-ECF (see Chapter II-4) these centres would form the interface between the astronomers and the ALMA observatory and support the entire process from proposal submission to the delivery of observation data. In Europe further decentralisation was foreseen so the European ARC is comprised of geographically distributed nodes coordinated by an ARC manager at ESO in Garching. The coordination centre at ESO started operations in 2006, led by Paola Andreani.

---

30 As more space was needed, ALMA subsequently rented additional rooms in the nearby Alsacia Building.

31 Using either diesel or liquefied petroleum gas.

In March 2007, the first observations were carried out with two ALMA antennas, not from Chajnantor, but using two prototypes at the ALMA test site in New Mexico. The first of the transporters arrived in Chile in February 2008 and the first ALMA antenna, from Japan, was moved up to the high site in September 2009. By March 2009 all components for the first AEM (European) antenna had arrived in Chile and integration could start. The antenna was accepted in early 2011. ALMA was becoming a reality.

≈

The very first antennas on the high site, though, were not ALMA antennas. In 2000, the US Cosmic Microwave Background Imager, an interferometer that measured fluctuations in the cosmic microwave background was installed, followed in 2002 by a single 10-metre Japanese dish, ASTE[32, 33]. The third was European, although also not forming part of the ALMA array. As mentioned, by 2003, the SEST was facing closure. However, as the ALMA prototypes were built, one of which by Vertex Antennentechnik, this created a possibility for acquiring a copy of the prototype at a very reasonable price. The option to place this as a successor telescope to SEST at Chajnantor, at twice the elevation of La Silla, was clearly interesting. This project was instigated by Karl Menten and the outcome was a joint project between the Max-Planck-Institut für Radioastronomie, the Astronomisches Institut der Ruhr-Universität Bochum in Germany, the Onsala Space Observatory and ESO to place a 12-metre single-dish antenna at Chajnantor[34]. In June 2001, the partners signed a Memorandum of Understanding, followed shortly afterwards by an order to Vertex for the antenna. Since ESO was becoming involved in a giant array of submillimetre/millimetre-wavelength telescopes, what could justify the installation of a single antenna? Firstly, the new antenna was foreseen to combine a relatively wide field of view with a new and powerful bolometer. This would allow for early studies of the submillimetre/millimetre sky and help to prepare the scientific community for ALMA. In a sense, it could be compared to the survey telescopes often used in conjunction with large optical telescopes. Secondly, having an ALMA prototype antenna at Chajnantor early on would enable ESO to gain operational experience with the antenna under the prevailing conditions. Thirdly, the new antenna

[32] Atacama Submillimeter Telescope Experiment.

[33] Albeit at the adjacent Pampa La Bola, at a slightly lower altitude.

[34] The antenna was purchased by MPIfR, with OSO and MPIfR providing instrumentation and ESO being in charge of operations.

would use the control software developed for ALMA. This provided an opportunity to test the software over a period of years, before the ALMA array became active. And finally, perhaps as a byproduct, the telescope team would build up a significant body of knowledge about the observing conditions, including all relevant meteorological data, which would be of great value to ALMA. Service-mode operation, in the manner of the VLT, was foreseen for ALMA to allow for flexibility in the execution of observation programmes. But this is greatly facilitated by a good understanding of the weather patterns and their fluctuations. Given all these reasons, the new project could hardly have been given a better name: APEX, the Atacama Pathfinder Experiment.

Since APEX is a stand-alone telescope, i.e. not part of any array, some design modifications from the original ("American") ALMA antenna were necessary. As a result APEX features Nasmyth foci (in addition to its Cassegrain focus) and a chopping secondary mirror. These modifications were required for single-dish operation for the array receivers and bolometers. Thus the antenna has a total of three instrument stations, one at the Cassegrain focus and two at the Nasmyth foci. The main instruments foreseen were basically heterodyne receivers and bolometers. Because of technical limitations in heterodyne instruments, the original (high frequency) signals are converted to lower frequencies by combining them with a local oscillator, before they are amplified. Heterodyne instruments can detect individual spectral lines of interest to astronomers. Bolometers, on the other hand, have been described as little more than extremely sensitive thermometers and are therefore used for observations of the electromagnetic continuum.

As the originally foreseen instrumentation was delayed, instruments were transferred from SEST to be used for the early performance tests. The commissioning instrument, however, was the new FLASH two-channel heterodyne instrument developed in Bonn that covered the frequency ranges 420–500 and 790–830 GHz. Furthermore, during the early years, APEX 2b, a 350 GHz heterodyne receiver built by Chalmers Tekniska Högskolan in Sweden was used. CHAMP$^+$ was another heterodyne instrument, built by the Max-Planck-Institut für Radioastronomie together with the Dutch NOVA/SRON and JPL in the US. But the workhorse instrument of APEX is the Large APEX BOlometer CAmera (LABOCA), a 300-element bolometer array operating at 870 µm and covering a field of 11 arcminutes — especially suited to detecting high-redshift dust emission, i.e. the study of distant galaxies. LABOCA was developed jointly by the APEX partners in Bonn and Bochum but also with involvement from the Friedrich-Schiller-Universität Jena.

The APEX antenna at Chajnantor. (Photo: Hans Hermann Heyer)

The first project manager was Rolf Güsten from the Max-Planck-Institut für Radioastronomie, with Jorge Melnick acting as the ESO interface. Carlos de Breuck became ESO project scientist and, in 2003, Lars-Åke Nyman, who had previously been station manager for SEST, became station manager and thus responsible for the operations of APEX.

Telescope installation began in 2003. A hard winter, not to be disregarded at 5000 metres above sea level, caused some delay, but by October the telescope had passed the preliminary tests[35]. First science observations were carried out in 2005 and the formal inauguration took place on 25 September 2005. As early as July 2006, the first 26 scientific papers based on APEX observations appeared in a special Letter issue of the European journal *Astronomy and Astrophysics*. Among the early results was the first detection of CF^+ (fluoromethylidynium) (Neufeld *et al.*, 2006); adding this to the growing list of complex molecules that have been found in space in the past, using telescopes including SEST. Other papers presented results from studies of dense clouds in the Milky Way, some of which were known to be star-forming regions, providing new information both on interstellar chemistry and star formation processes.

[35] ESO Annual Report 2003.

LABOCA saw first light in August 2007. One of the most spectacular early science results was a joint 300-hour ESO–MPG survey of a 30 × 30 arcminute region that was named the LABOCA Survey of the Extended Chandra Deep Field South, with the somewhat curious, and perhaps slightly misleading, acronym LESS. Certainly, the old saying of "less is more" seems entirely appropriate here. The survey became the reference work in this area for quite some time. We shall encounter the Chandra Deep Field again later in the book.

Jörg Eschwey and Juan de Mella at the ALMA OSF construction site in December 2004. (Photo: Hans Hermann Heyer)

Along with the construction work on Chajnantor, work began on a base camp at Sequitor on the southern outskirts of San Pedro de Atacama. The base camp is a small compound with a control room, offices and laboratory facilities, and 16 dormitories and leisure facilities. The buildings are generally low adobe-style structures, in the traditional style of the village, with the exception of the control-room building, which is a small tower on top of which is mounted a microwave antenna that is used as a link with the APEX antenna on the high site.

Chapter IV-2

Into New Territory

"Caminante, no hay camino;
se hace camino al andar."

Antonio Machado in *Proverbios y Cantares*, 1912.

It is difficult to overestimate the impact of the ALMA project on ESO and on the European scientific community in the field. Far from being just an addition to the normal ESO programme it changed the character of ESO as an organisation, transforming it into a multi-programme organisation and integrating what for ESO was a new area of science and technology with its own specific culture. At the same time, at the European level, the structuring effect on the scientific community was considerable, bringing together the entire community into a system similar to the one for optical astronomy, i.e. with ESO and the national institutes[1] forming partnerships and working towards a common goal. Last but certainly not least, ALMA exposed ESO to the challenges of being a partner in a global project that was far more complex than anything it had done in a European context. The issues included the bringing together of vastly different cultures and traditions as well as administrative and legislative frameworks — well established, but rigid — that were difficult to reconcile. These factors affected all aspects of the project from funding to construction and operation.

One of the fundamental reasons behind the success of ESO is that the organisation functions as a framework for the implementation of activities that the Member States wish to carry out together. Because the organisation exists and possesses a legal personality, Member States can transfer funds to it and the organisation can use these funds both for procuring equipment and for its operation. This may sound trivial, but it is important to understand, because there are other models for international cooperation as well. When it comes to global, or quasi-global (as in the case of ALMA), projects, it can be extremely difficult, if not impossible, to create a common pot, from

[1] And, of course, IRAM.

which a joint project can be financed. But other ways of working together offer themselves as possibilities. One solution is that each partner retains the power over its own purse strings, i.e. a system in which no money is normally transferred between the partners. This in turn almost inevitably leads to a scenario where the partners deliver equipment and services to the central project: in-kind contributions, as they are called. The basis of such a system is a set of items to be delivered, based on their commonly agreed value to the project. Since value and cost are not necessarily congruent, it remains for each partner to control its costs. The advantage of this approach is that it is easier to handle politically by the individual partners, but — aside from the aspect of cost control — it puts stringent requirements on coordination so that in the end the project functions as a single entity. It also poses management questions that the partners must address, such as: how is coordination organised and who is responsible? How will the operations be implemented? And how does the project interact with third parties? There is nothing unusual about these questions, but at the same time, there have been few scientific collaborations on this scale and there is certainly no commonly established rulebook for such undertakings. It is therefore not surprising that the original idea for the ALMA partnership had to develop into a more sophisticated model. The first step in that direction occurred in 2002 when the agreement about Phase Two was being prepared. To ensure coherence, it was decided to establish a dedicated ALMA Board[2] comprising representatives of the partners, along with some independent members, as well as a Science Advisory Committee. At a meeting in Venice, the partners also decided to set up a Joint ALMA Office (JAO) to coordinate the project. Paul Vanden Bout was appointed interim ALMA Director with Massimo Tarenghi as project manager. With the signing of the bilateral agreement, in 2003 the JAO was installed in Santiago and Massimo Tarenghi assumed the post of ALMA Director, Tony Beasley became project manager and Richard Murinowski the project engineer. Further coordination measures followed with the establishment of the Integrated Product Teams. With time, this governance structure would be tested: the balance between the partners, who were responsible for the deliverables and the joint structures, put in place to ensure coordination and coherence, was a fine one and it was not always easy. For example, all major procurements, although paid for by the individual partners, had to be approved by the full ALMA Board, potentially raising issues over the sovereignty of the executives. Despite unavoidable difficulties, however, the system delivered what it was supposed

[2] From the European perspective, since Spain at the time was a partner of ALMA but not a member of ESO, a European ALMA Board was also established to coordinate Europe's position in the project. Naturally this forum would also serve to discuss ALMA matters in much greater detail than it would be possible at the level of the ESO Council.

to, underpinned as it was by a shared desire to achieve the goals of the project and in many cases strong bonds of friendship between the scientists involved.

Yet for a while, important questions remained open: which operational model should be chosen for the observatory? What would be the legal conditions for operating the joint facility in Chile? And who should act as employer for the staff?

The first meeting of the ALMA Science Advisory Committee was held in Chile, on 10 September 2001. Seen from left to right: Rafael Bachiller, Al Wootten, Yasuo Fukui, Richard Crutcher, Neal Evans, Robert Brown, Leonardo Bronfman, Mark Gurwell, Pierre Cox, Christine Wilson, John Richer, Geoffrey Blake, Peter Shaver, Jack Welch, Ewine van Dishoeck, Richard Kurz, Arnold Benz, Yoshihiro Chikada and Naomasa Nakai.

≈

For ESO, it would seem obvious that the existing *Convenio* with the Government of the Republic of Chile constituted a suitable solution. It had withstood the test of time, and with the 1995 amendments was put on a new footing. AUI, for its part, had signed an agreement with the Universidad de Chile[3] and thus operated within a different, if largely similar, legal framework. It is hard to avoid a sense of *déjà vu*. In Chapter I-3 we saw how different operating conditions played a role in 1963, when

[3] The AUI agreement with the Universidad de Chile was similar to those of the Carnegie Foundation (operator of the Las Campanas Observatory) and AURA for the CTIO at Cerro Tololo. Furthermore, the In 1963 foreign observatories had become subject to a national law, regulating the conditions for their activities in Chile.

discussions took place between ESO and AURA about sharing a site in Chile. In 1963, the differences led to a decision to go separate ways. Now, with a new international partnership, with its specific characteristics and requirements, an innovative solution was needed. In any event, Insulza, who had been so helpful in past, had left the Foreign Ministry and the new team had to be convinced of the importance of the new project and its benefits for the Antofagasta region as well as the scientific community in Chile. On ESO's side, the key person was Daniel Hofstadt, the experienced ESO representative in Chile, whereas AUI was represented by Eduardo Hardy. They were backed up by their respective superiors, Catherine Cesarsky as ESO Director General and Riccardo Giacconi, who, after his term as ESO Director General, had become head of AUI. Thus, in June 1999, ESO and AUI approached the Chilean government regarding support for a joint venture to establish ALMA and obtain access to the land needed on the Chajnantor Plateau. The Foreign Ministry, it appeared, was of the opinion that a new and specific agreement would be required for ALMA.

In April 2001, things appeared to begin to move, not least thanks to the then Assistant Foreign Minister of Chile, Christian Barros. The Chilean government established an inter-ministerial team comprised of representatives of the Ministries of Science, National Properties, Foreign Affairs, the regional authorities, the Environmental Agency, and others, to negotiate with the ALMA partners. For the land acquisition, the outcome of the talks was that the partners would be granted a renewable concession for 50 years for the use of the land. The recipient of the concession, however, would have to be a Chilean company[4]. It was important to note that the Chilean Government accepted the principle (albeit implicitly) that both ESO and AUI would be able to construct and operate ALMA under their current status and no further mention was made of the need to establish a separate legal entity for ALMA in Chile. From the Chilean side, the project was sanctioned by the President of the Republic, Ricardo Lagos, who welcomed ALMA. Even so, progress at the lower levels took time. Encouragement came in the form of a visit by the German Foreign Minister, Joschka Fischer, who paid a visit to the Paranal Observatory on 6–7 March 2002. In bilateral discussions with the Chilean authorities, he stressed the importance that Germany attributed to the new project, also in the context of Chile's broader relations with Europe. However, whilst undoubtedly still in favour of the project, in May 2002, the Chilean government had come to the conclusion

[4] The company was established and is the owner of the one square kilometre of land at the OSF and the concessionaire for the Array Operations Site (AOS) and road corridor between the sites.

that the ALMA project required an amendment to the respective accords with ESO and the AUI.

In July 2002, the discussions began and covered all the pertinent issues including collaboration programmes with the II Region and the science agency, CONICYT, the access of astronomers in Chilean Universities as well as the environmental impact procedure. Later that month the amendment to the *Convenio* was drafted. It was signed by the ESO Director General and the Chilean Foreign Minister María Soledad Alvear on 21 October 2002. It enabled ESO to *"open a new centre of observation"* within the framework of its existing legal status and specified 10% of observation time for Chilean scientists in as well as the cooperation commitments with the Antofagasta Region and CONICYT. The amendment would, however, need parliamentary ratification.

≈

In 2006, ESO established a small site museum next to the road leading up to the ALMA high site on Chajnantor. Two shepherds used to live at this site with their stock of animals. Their dwellings, long since abandoned, were restored by ESO and descriptive panels were installed that portray the life of the people in this high altitude environment. Seen here are (from left to right): Pedro Cruz, representative of the community of Cucuter, Marcela Hernando Pérez, at the time Intendente of the II Region, the Mayor of San Pedro de Atacama, Doña Sandra Berna, and Felix Mirabel, then ESO's representative in Chile.

In parallel, both the ESO[5] and the AUI representatives were extensively involved in building smooth links with the scientific, the regional and the indigenous communities who were concerned about the ALMA project. Working with these communities was new for ESO since, in the desert sites of La Silla and Paranal, the local population had at most comprised a few families engaged in small-scale mining. The ALMA site and its nearby oasis village of San Pedro were different. The area is home to the Lican Antai people, who like many other indigenous peoples display a

[5] Daniel Hofstadt was ably assisted in this task by a small, but immensely dedicated team comprising Esteban Illanes, Laura Novoa and Mary Bauerle, all ESO staff members in Santiago.

growing readiness to assert themselves and their culture *vis-à-vis* the surrounding society. A number of anthropological and archaeological artefacts were found on the site along the access roads to the Chajnantor Plateau. This was, and remains, a sensitive issue. ESO contracted specialised assistance for it, and made a number of mitigation commitments for the preservation of these archaeological items.

Furthermore, the Altiplano, with its stunning natural beauty, is a delicate ecosystem with a unique, yet threatened, population of fauna and flora. Environmental protection was clearly very important. A number of endemic and unique plant species such as the giant cactus, *echinopsis chilensis*, grow in the slightly less dry region below Chajnantor, at between 2700 and 4500 metres elevation. Viscacha rabbits and other mammals, such as llamas, vicũnas and guanacos, the native South American camelids, roam the area. Furthermore, the road leading up to the ALMA site crosses a national flamingo reserve.

For this reason, ESO and AUI agreed to carry out an environmental impact evaluation as part of the negotiations. ESO took the lead in the process, which proved to be rather complex, requiring a number of specific authorisations and the building of political goodwill through public diplomacy. On the Chilean side 13 different administrative entities were involved in the procedure, including the municipality of San Pedro de Atacama, the local indigenous communities, the health, public works, national parks and road authorities and the tourism authority. ESO solicited expert help, for example to monitor the high altitude wildlife and to assess the impact of road activities on their behaviour, a task that was carried out by an archaeologist, Ana Maria Baron from San Pedro and a German biologist, Michaela Heisig[6].

By March 2003 the environmental validation resolution was granted within the framework of Chilean environmental law. A major obstacle for the access to the Chajnantor land was removed. In March the land concession was also granted through a Public Property Ministerial decree and ESO/AUI were now able to sign the land concession contract.

As mentioned, the amendment of the ESO agreement for ALMA had to be ratified at the Chilean parliament and a new round of lobbying had to be carried out

[6] ESO went to extraordinary lengths to meet the stringent environmental requirements, including carefully moving and replanting cacti and refraining from blasting activities in connection with the road in order to protect the wildlife, such as the viscachas, already mentioned. Later studies confirmed that these efforts had borne fruit: if anything, the population of these animals has increased.

at the parliament. At the recommendation of its Foreign Relations Committee, the lower chamber (the House of Deputies) approved the amendment on 8 May 2003, followed on 10 June by the Senate. With the subsequent signing of the cooperation contract with the Antofagasta Region and CONICYT — after four years of patient negotiations and preparations — the road was now open to formally initiate ALMA activities in Chile.

On the final question of the ALMA staff, after much discussion, the partners in 2006 agreed that local ALMA staff in Chile would be hired by AUI. At the same time, it was decided to move the ALMA offices to the ESO compound in Santiago, where today they form an integral part of a vibrant scientific community.

The Chajnantor site, photographed in 2005. (Photo: Hans Hermann Heyer)

The VISTA telescope with Paranal in the background with the VLT and the VST. (Photo: Gerd Hüdepohl)

Chapter IV-3

Buds at Paranal

"... the advent of a new generation of wide-field digital detectors and dedicated telescopes has started what can be considered the 'renaissance' of optical surveys."

Massimo Capaccioli in
Questions of Modern Cosmology – Galileo's legacy
(D'Onofrio & Burigana, 2009).

If you stand on top of a nearby mountain and let your eyes glide over the La Silla mountain ridge, you will see a chain of white, eggshell buildings — the telescope domes. The view is interrupted only by the box-like structure of the NTT, bearing witness to the traditional evolution of an observatory: after an initial surge of telescopes, as time went by, new ideas came up and as money became available, more telescopes were added to the observatory. The growth of the telescope park is the result of years, sometimes decades, of development in astronomy and is typical of most 20th century astronomical observatories. But there are exceptions, and Paranal is one. One of the unique features of Paranal, i.e. of the VLT, is that everything was conceived as one complete, fully integrated and self-contained facility. That is, almost.

Chapter I-7 discussed the classic combination of a large telescope with a survey telescope, seen at Palomar, Calar Alto, Siding Spring and La Silla, and so on. The survey telescopes would, among other tasks, act as pathfinders, enabling astronomers to identify particularly interesting objects for study with the larger telescopes. However, the 8–10-metre-class telescopes that were entering into service could easily see objects that were 100 or more times fainter than the previous 4-metre-class telescopes. No similar increase in the capability of survey telescopes had occurred. In fact, for all the accomplishments of the VLT, and the pride that Europe's astronomers could justifiably feel, when it came to the systematic mapping of large parts of the sky, ESO was doing less well. In 1996, as Chairman of the working group looking at the future of La Silla, Johannes Andersen, wrote in *The ESO Messenger*, *"It is ... one of the findings of the [Working Group] that ESO does not possess a truly competitive wide-field imaging telescope."* (Andersen, 1996). The 1-metre Schmidt telescope at La Silla, which still used classical photographic materials, closed in 1998 and the only other telescope with some wide-field capacity was the MPG/ESO 2.2-metre telescope fitted with WFI, the wide-field camera mentioned in Chapter II-1. At ESO the WFI had been strongly pushed by Alvio Renzini, the VLT project scientist. For all its qualities, however, the WFI would not be sufficient either.

Luckily, plans were under way to address the problem. The solution came in the shape of two new telescopes, one covering the visible part of the spectrum and one the infrared: the VLT Survey Telescope (VST) and the Visible and Infrared Survey Telescope for Astronomy (VISTA). Neither initiative came from ESO, which had its hands full with the VLT. The VST was an Italian initiative, instigated by Massimo Capaccioli of the Osservatorio Astronomico di Capodimonte, in Naples, and strongly supported by Renzini. And VISTA came from a country that at the time was not even a member of the organisation, the United Kingdom.

Capaccioli took up the directorship of the Naples Observatory in 1993. His scientific interest in galaxies had brought him into contact with the world of wide-field imaging and he had been associated with Schmidt telescopes, especially the Italian Schmidt telescope at the Osservatorio Astrofisico di Asiago in the Dolomites. As director of the Naples Observatory, he was supposed to drive forward a project for a 1.5-metre telescope, to be placed at the Toppo di Castelgrande Observatory. However, Capaccioli had larger plans.

Despite its charm and its rich history going back almost 3000 years — not least as an outstanding centre of culture in the ancient world — the Naples of today is situated in a region of chronic economic depression. As such, it receives financial support from the Italian government and the European Union to foster new growth in the region. The support by the Italian government included *"existing centres of excellence in science and technology"* (Arnaboldi *et al.*, 1998). In this context a special grant of nine million euros was made available in 1995/96, half of which was originally earmarked for a new telescope. With this grant, Capaccioli saw an opportunity to build a major survey telescope, but he also realised that the grant was a one-off allocation, and it was therefore important to integrate this new telescope into an operational environment that would ensure its long-term efficient scientific use and the necessary maintenance. Along with a team of engineers, led by Dario Mancini, he set out to prepare a proposal to ESO — the VST. The VST was planned to be ten times better overall than the MPG/ESO 2.2-metre with WFI, because of its better resolution, the slightly larger aperture, the much wider field and the better site.

Constructing such a telescope would seem to be a huge task both for the small Naples team and for a region that is not renowned for its high-tech industry, but the Naples Observatory had already been a partner in the development of the WFI for the MPG/ESO 2.2-metre telescope[1]. In any event, the proposal was carefully reviewed by ESO, which led to a number of changes to the original project including the requirement for active optics, and in June 1998, an agreement was reached and endorsed by the ESO Council. According to the agreement Naples would deliver the telescope, whilst ESO would be in charge of the enclosure.

As had now become customary, it would be an alt-azimuth telescope, housed in a compact building. It was even possible to find a place for the telescope at Paranal itself — on the northwestern edge of the platform. At the time of planning of the VLT, much thought had gone into the positioning of the individual Unit Telescopes. This was partly driven by the needs of the interferometer, but was also to avoid the adverse effects on the wind flow caused by the buildings themselves. Luckily, it would be possible to place the somewhat lower and smaller VST building in the chosen location without causing undesirable wind effects. The idea was to build a 2.6-metre telescope with a field of view of at least 1 degree × 1 degree, and fitted with just one instrument, a 32-CCD mosaic camera. The camera would be developed under separate funding following a call for expressions of interest. The camera featured a staggering 256 megapixels (compared to 67 megapixels for the WFI on the 2.2-metre), and with a larger mirror and active optics, the telescope would both collect more light and focus it more precisely. According to the specifications, the telescope should be able to detect objects of magnitude 25.5 in 30 minutes of integration time and achieve an image resolution of 0.7 arcseconds. If this looked exciting, the timescale was even better: According to the original plan, the VST should be ready by 2001. But problems lay ahead.

As the project started, a call for tender was issued for the optics. Carl Zeiss Jena won the contract. However, the Zeiss bid foresaw that the primary and the secondary mirrors would actually be produced in Russia, by the Lytkarino Optical Glass Factory,

[1] Also, at least one company in the Naples region, the Tecnologie Metallurgiche Avanzate s.r.l., had produced parts for the azimuth tracks for the VLT Unit Telescopes.

also known as LZOS[2]. This was at a time when Russia was opening up commercially and beginning to offer her impressive technological know-how on world markets[3]. LZOS did a sterling job and in the spring of 2002, the mirrors were ready for shipment. To speed up the delivery they were shipped by air cargo from Moscow into Frankfurt airport, then transported by lorry to Hamburg and loaded onto a ship bound for South America. In early May, the ship docked at the port of Antofagasta and the boxes were put on a truck for the final leg of the journey to Paranal. Then, on 6 May, the boxes were opened. The mirror had shattered[4] — the worst nightmare of any telescope builder. Obviously, telephone lines were buzzing between Paranal, the ESO Headquarters, the Naples Observatory — and Moscow. Capaccioli showed a photograph of the broken mirror to Anatoly Samouilov, the president of LZOS. Although hardly a soft-hearted soul, the sight brought tears to his eyes. This reaction should not come as a surprise to anyone. As we saw it in the case of the VLT, developing instrumentation for astronomy is an enormous professional challenge. But the fascination of astronomy itself is an equally strong driver and true professionals, as cool-headed as they may appear, often develop an almost emotional relationship with the object of their task. In any event, in just 40 days, a contract for the delivery of a replacement mirror was signed with LZOS[5]. Even so, it was now clear that a long delay would be unavoidable[6].

Meanwhile, the VST camera, OmegaCAM, was being built by a consortium of Dutch[7], German[8] and Italian[9] institutes with funds from the Dutch Organization for Research in Astronomy, the German Federal Ministry of Education, Science, Research and Technology and the Italian Consorzio Nazionale per l'Astronomia e l'Astrofisica (CNAA) and Istituto Nazionale di Astrofisica (INAF) (Kuijken *et al.*, 2002). ESO's optical detector team provided the detector system. The work progressed according to plan, and by 2008, the new camera was delivered to Paranal.

2 The lenses for the wide-field correctors and the prismatic lenses for the atmospheric dispersion corrector were made by Schott.

3 The primary mirror was made of Astrosital, a glass-ceramic similar to Zerodur.

4 It appeared that the box had been affected by a rescue operation in connection with a fatal accident in a port of call between Hamburg and Antofagasta.

5 The secondary mirror had survived the trip, but it had slight defects. Luckily, they could be repaired by LZOS.

6 The new mirror was ready in the spring of 2006 and was airlifted directly to Chile.

7 The Kapteyn Institute (Kapteyn Instituut), Groningen and Leiden Observatory.

8 University Observatories of Munich, Göttingen and Bonn.

9 Padua and Naples observatories.

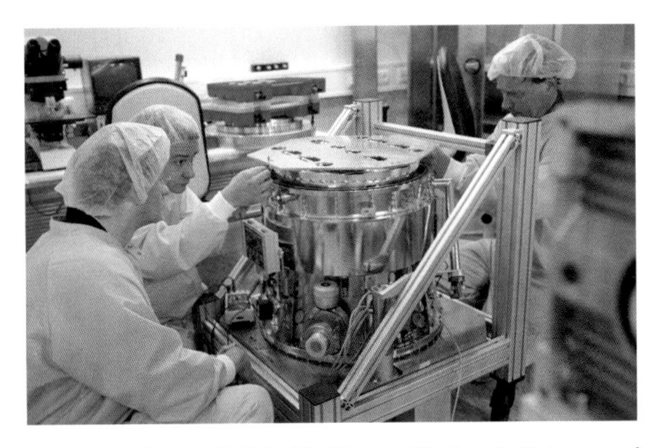

Preparing OmegaCAM: Olaf Iwert, Christoph Geimer and Sebastian Deiries. (Photo: Hans Hermann Heyer)

During the wait for the VST to be ready, it was assembled and stored under vacuum conditions until it could be mounted on the telescope.

The main mechanical structure and the mirror cell, in Naples in 2005. (Photo: Massimo Capaccioli)

The delay also led to a loss of momentum in the project. It became difficult for Capaccioli to retain the project team and to keep spirits high amongst his staff. The project was floundering. After a troubled time, help was given, partly by other INAF facilities (notably the observatory in Padua, Capaccioli's old institute) and partly by ESO, especially Jason Spyromilio, Roberto Tamai and Stefano Stanghellini. Additional industrial partners were also brought in and the way forward slowly emerged. A number of problems, especially with the active optics, led first to a redesign, and later a decision was taken to separate the construction of the M1 cell and M2 unit from that of the main mechanical structure. As a result, following an audit by ESO in October 2006, those large mechanical parts that were already finished left Salerno in June 2007 to be assembled at Paranal. However, the cell for the primary mirror and the M2 units were still pending. Finally, just before Easter 2009 they were shipped to Chile. Unlike the other mechanical parts, the mirror cell was assembled in Italy, complete with actuators and all other necessary subsystems and shipped in one piece. En route, the vessel suffered a major breakdown at sea and anchored off the port of Toulon for several weeks for repairs. When the precious cargo finally arrived in Chile, there was yet another shock: the assembled mirror cell had been flooded with salt water and was red with rust. There was little choice but to return the structure to Europe for a total overhaul. On completion of this work, everything was sent to Chile again, this time without mishap. In June 2011, the impressive first images from the telescope started to appear, giving some consolation to those who had struggled for so long to enhance the capabilities of the VLT with a wide-field imager.

Building telescopes, the reader will have realised, is no simple task; astronomers are always trying to catch more photons and exploit them better so they work at the very cutting edge of technology. Dealing with sophisticated equipment and making it work under tough conditions is a huge challenge. So delays are not rare occurrences, whether the reasons are rooted in tricky high-tech issues or in more mundane problems. The second telescope that we shall describe here would be no exception, either.

≈

When it came to wide-field astronomy, the UK has a proud tradition, notably with the UK Schmidt Telescope at Siding Spring in Australia as its strongest manifestation. But towards the end of 1990s, like the ESO Schmidt telescope, the UKST was becoming dated and UK astronomers interested in this area had begun to consider a next generation survey telescope. Initial ideas focused on a 2.5-metre telescope, possibly to be operated in a partnership with NOAO. A consortium of a large number of universities had coalesced around the project, but perhaps the two key people at the time were Jim Emerson (as member of PPARC's wide-field survey panel) and Mark Casali, in charge of new projects at the UK Astronomy Technology Centre (ATC) in Edinburgh. In 1998, the UK government established a one-off fund for research infrastructures. We shall return to the significance of the fund in Chapter IV-8. Here is suffices to note that a senior PPARC manager, Ian Corbett, not only encouraged Emerson and Casali to apply to the fund for money for the new survey telescope, but also to develop a much more ambitious project. The core team quickly expanded to include Andrew Lawrence (for the science case) and Adrian Russell (UK ATC Director). To meet the deadline for applications, the team only had a couple of weeks. Developing a new telescope project on such a timescale is obviously impossible, so they decided to submit a proposal based on a design recycled from the 4.2-metre William Herschel telescope in the Canary Islands. The application was successful, but it became clear that simply building the telescope as proposed was out of the question for several reasons, including the amount of available funds. So, a major redesign was carried out that ultimately led to a completely different telescope.

Before we continue the story of the telescope, we shall briefly address the issue of siting. As mentioned, some discussions had taken place regarding cooperation between the UK and AURA in the context of the Gemini telescope project. With Gemini South destined for Cerro Pachon, close to the Cerro Tololo Interamerican Observatory in Chile, this would seem to be a logical choice of site for the new survey telescope. However, in parallel, the relations between the UK and ESO had developed, opening up the possibility of choosing the Paranal area instead. Paranal itself would not be able to accommodate yet another telescope, but 1.5 kilometres north of the mountain, what in ESO language was colloquially known as the NTT Peak[10] was empty and seemed to be an almost ideal home for the new telescope. Talks between ESO and PPARC led to a decision in February 2000 to place the telescope "at Paranal" (meaning, of course, the adjacent hill), but with the understanding that operationally it would be considered as a part of the Paranal Observatory.

Initially, Massimo Tarenghi provided the key interface between ESO and the VISTA project, to be succeeded first by Jason Spyromilio and later by Martin Cullum. The link with ESO also meant that not only ESO was willing to provide considerable help to the project, but it also took a keen interest in ensuring that it was designed according to ESO standards or was at least integrable into the operational environment of Paranal. After years of painful budget cuts to UK astronomy the shortage of qualified system engineers had become acute. From ESO Stefano Stanghellini, who had just finished his main tasks regarding the last VLT Unit Telescope, agreed to move to Edinburgh and work with the ATC project team. His arrival led to major changes in the project. The telescope had originally been conceived as a classical telescope with passive optics. Stanghellini, however, brought with him his experience from the VLT: the concept of a modern lightweight telescope with a thin meniscus mirror. This meant choosing an active optics system rather than a traditional thick primary mirror. In early 2001 another major change in the design was the decision to go to an extremely fast primary mirror with a focal ratio of *f*/1. This led to a much more compact telescope with the advantages that go with it in terms of savings on the mechanical structure and the enclosure. However, producing a large *f*/1 mirror is far from trivial because the mirror is highly aspheric and polishing it is extremely difficult and time-consuming. In retrospect, there is no doubt that it was the right decision, but there was definitely a price to pay. The new design also created space problems for the infrared camera, but these problems were overcome by an ingenious design solution.

[10] See p. 180.

As part of the design process, Stanghellini proposed involving European industry by contracting out feasibility studies (to AMOS and to EIE s.r.l.) and undertaking a cost estimate. He also helped by drafting the first set of technical specifications for the telescope. In April 2001 Alistair McPherson joined as project manager and helped to bring the project from Phase A to the construction stage.

At the end of Phase A, VISTA appeared to be a state-of-the-art telescope with a potentially very attractive performance. But it was also one with a cost overrun. As we recall, the name VISTA stands for Visible and Infrared Survey Telescope for Astronomy, but one cost-cutting solution was to scrap the camera for the visible part of the spectrum. ESO did not find it difficult to accept this idea. After all the VST was expected to cover this and in any case, there certainly was an increased focus on the infrared, which was profiting from the rapid advances in the field of detectors. Alan Moorwood, admitably an infrared astronomy diehard, put it rather bluntly: *"If you're going to produce the best infrared camera in the world, why waste your time on a visible camera?"*

≈

Towards the end of 2001 the project moved into construction, and procurement became the issue. After calls for tender, the contract for the optics went to LZOS in Russia, and for the main mechanical structure to Vertex in the US. The latter choice raised a few eyebrows in the ESO Member States, who felt that European companies should have benefitted from this order. By that time, the UK accession to ESO had been agreed, but the VISTA telescope had been defined as an in-kind contribution and therefore remained the responsibility of the UK. The UK, however, was bound by European Union (EU) procurement rules and since Vertex was the lowest bidder, it received the contract. The enclosure was contracted to EIE, the company that had been involved in many ESO projects in the past (as well as the VST). The other major component, the infrared camera VIRCAM, was built by a consortium in the UK comprised of the Rutherford Appleton Laboratory, the ATC and Durham University. The telescope now being built promised to be an awesome science machine.

It is important to note that VISTA, like the VST, is conceived as a single-purpose telescope. Just like the classical Schmidt telescopes, it is fundamentally a huge camera, though with the three-tonne camera unit mounted at the Cassegrain focus. The heart of the camera — with the biggest infrared focal plane ever used in astronomy — is an array of 16 2048 × 2048 Raytheon CCDs (67 million pixels), cooled to a temperature of 70 K. The entrance window has a diameter of almost one metre and it was a major challenge to prevent the window misting up due to the huge temperature difference between the cool camera and the ambient air. Success was achieved by fitting a baffle with a coating that is non-reflective at the science wavelengths, but highly reflective in the thermal infrared and thus helping to keep the window clear. In terms of its science capability VISTA covers a field with a diameter of 1.65 degrees. The pixel size of the detectors equals 0.34 arcseconds, meaning that the telescope will utilise the best seeing conditions found at Paranal. With VIRCAM, the telecope will undertake six surveys, including a full-hemisphere survey some 25 times more sensitive than the Two Micron All Sky Survey (2MASS) undertaken towards the end of 1990s, and a survey of 0.75 square degrees of sky to exceptional depths[11]. The potential performance of VIRCAM is truly mind-blowing. In a typical night, it will collect a staggering 315 gigabytes (GB) of data. But all of this comes at a price: VIRCAM constitutes about one third of the entire cost of the VISTA telescope.

The VISTA building during construction in December 2004. Paranal is seen in the distance. (Photo: Hans Hermann Heyer)

[11] To magnitude 24 in the *K*-band.

There is no doubt that Europe's astronomers were looking forward to sifting through the VISTA data, but this project was also hit by severe delays. Both the main mechanical structure and the enclosure saw delays, but the primary mirror in particular took much longer to produce than expected. In the end, it was only delivered from the Moscow factory to Chile in April 2008, onboard an Antonov freighter. Final integration and commissioning then went rather smoothly, with technical first light achieved by the middle of 2008[12].

With the addition of the two survey telescopes, Paranal has become a mature observatory and closed the gap with respect to up-to-date survey telescopes. Night after night the telescopes are now producing a real flood of high quality science data, posing new challenges in terms of data storage and curation, not to mention the capacity of the scientific community to exploit the data. Yet with the ESO dataflow system and its ever-growing archive, open to scientists all over the world, ESO was well prepared for this development — more so than many other research institutes. But it is a continuing challenge and one that will keep growing in magnitude.

[12] ESO Annual Report 2008.

Chapter IV-4

Of Eponymous Birds and Euros

"Permit me to ask a 'provocative' question?
The VLT is now almost a reality and it
would be interesting to look into the more distant future....
Have you had any ideas about what might come after the VLT?"

Richard West at the Venice VLT conference,
October 1986.

Among the main strands in the history of astronomy is the interdependence between technology and science. As Martin Harwit expressed it in his book *Cosmic Discovery*: *"The most important observational discoveries result from substantial technological innovation in observational astronomy"*. As far as telescope development is concerned, over the last century we have seen something like an astronomical equivalent of Moore's law, i.e. a doubling of the telescope size every few decades, from the 2.5-metre (100-inch) Hooker Telescope in 1917, to the 5-metre (200-inch) Hale Telescope in 1948, to the 10-metre Keck telescopes in 1993 (Gilmozzi, *ibid.*). Of course the VLT plays its own part in this evolution. In the discussion of the NTT and the VLT, we touched on the "5-metre barrier", the fact that — for a number of reasons — telescopes with mirrors much beyond 5-metre in diameter were not feasible with conventional technology. Luckily, a number of new ideas had emerged that opened up the possibility of much larger telescopes. The main elements were the fact that alt-azimuth mounts became feasible (due to the rapid advances in computer technology), the introduction of actively controlled mirrors (pioneered at ESO with the NTT, and also heavily reliant on computer technology) and the idea of using mirrors composed of many individual segments instead of a monolith. The latter is normally credited to a suggestion in the 1930s by the Italian astronomer Guido Horn d'Arturo[1,2], but it was not implemented for large optical telescopes until 1993 with the Keck telescope on Hawaii. But if it were now technically possible to build much

[1] After World War II, Horn d'Arturo actually built a 1.8-metre telescope with a segmented mirror.

[2] Seeds may already have been planted by William Parsons, the 3rd Earl of Rosse, in the 19th century.

larger telescopes, and if — in principle — there was no apparent limit on how large, what would that mean for astronomy? Or, putting the question the other way round: how big a telescope would astronomers wish to have in order to tackle decisively some of the nagging questions that they could now ask, but not answer? One of the astronomers who began to think more seriously about this was Roberto Gilmozzi[3], an Italian scientist who joined ESO in 1994. Before that he had been responsible for science operations at the International Ultraviolet Explorer, the astronomical satellite operated between 1978 and 1996 by the European Space Agency and the UK Science Research Council together with NASA. After that, he had worked at the Space Telescope Science Institute in Baltimore. At ESO, as department head, he became involved in VLT instrumentation projects such as FORS and UVES, enabling him to complement his scientific and operational expertise with experience in managing complex technological projects.

At the time, discussions took place about what was then called the Next Generation Space Telescope (NGST)[4], an 8-metre optical telescope to be deployed in space[5]. Gilmozzi's thoughts increasingly focused on what complementary science could, or should, be done from the ground, once the NGST was launched. One might expect to discover a host of faint galaxies at the edge of the Universe, but to investigate them spectroscopically, a large ground-based telescope would be necessary. Another prospect would be to get a better understanding of whether exoplanets were ubiquitous or not. Once again, to probe a sufficiently large number of stars in search of faint signs of possible planets around them, would also require a large telescope. Other scientific questions, such as how we might improve our understanding of the formation history of galaxies by studying individual stars in a representative sample of galaxies and the study of so-called "turn-off" stars — the so-called "Virgo or bust!" science case — came to mind, not to mention the possibility of directly measuring the expansion of the Universe.

[3] At a 1996 conference in Landskrona, Sweden, Matt Mountain presented a paper on the scientific requirements for the next generation of telescopes. As a starting point he looked at what it would take to obtain spectra of the faintest objects in the Hubble Deep Field. Gilmozzi, went one step further — thinking of what would be needed to do spectroscopy on the faintest objects that would become observable with what was then called the Next Generation Space Telescope.

[4] The proposal was the outcome of a meeting at STScI, organised by Riccardo Giacconi in 1989.

[5] The NGST ultimately became the James Webb Space Telescope, a 6.5-metre telescope, which, at the time of writing is scheduled to be launched in 2018.

Gilmozzi's thoughts led him to conclude that a next generation ground-based telescope should have a size of around 100 metres. He began to ask the engineers at ESO about this. He later recalled what happened: *"They sent me away, telling me that I was crazy!"* But rather than accepting that his proposition was impossible, he returned with questions so that he could understand just why this would not be possible. *"What really changed was that the engineers started to talk among themselves and one day in the autumn of 1997, Philippe Dierickx, the optical engineer in charge of the VLT mirrors, came to me and said: I think this can be done! ... And then things really started. It was a magic moment like the collapse of the wave function..."* A small group of interested people — boffins one might say — started to meet after working hours, typically in the congenial surroundings of a Bavarian beer-garden (appropriately, these informal get-togethers became known as the 100-beers meetings). By the end of the year, the group had developed a general concept and a rough science case, although they did not yet feel ready to go public with their thoughts. The project even had a name. The first was WTT, meaning the Wide Terrestrial Telescope — although the joke about the Wishful Thinking Telescope could hardly be avoided. Philippe Dierickx then came up with the name of OWL, for OverWhelmingly Large Telescope, but also to associate it with the *"the eponymous bird's keen night vision"*, as he expressed it.

The excitement began to spread. It is therefore not surprising that news of this soon reached the ears of the Director General. Giacconi was not amused. To be sure, he had fought hard to keep everyone at ESO focussed on the task at hand: the completion of the VLT. In doing so — and with the single-mindedness that was characteristic for him — he had exhorted "his people" to work day and night. And he had himself put up with more than a fair share of serious problems — internally and externally — as we have already seen. Even so, he was quietly preparing ESO's engagement in what could become a global astronomy project — ALMA. And now some of his own staff seemed to have begun to stray off course and into new territory. He summoned Gilmozzi to a hearing of a kind that any normal staff member would have hoped never to experience. But as Gilmozzi started to explain the idea, it kindled Giacconi's scientific curiosity. The meeting ended indecisively. Somewhat later, while preparing for a conference in Hawaii on Astronomical Telescopes and Instrumentation organised in March 1998 by SPIE, Gilmozzi asked if he could present his ideas there. Giacconi agreed, provided that the ideas were to be presented exclusively as Gilmozzi's private ideas. *"But then a day or two later Giacconi called me back and said: I've thought about it. You should present this as an ESO project."* From then on, ESO would work towards the development of an Extremely Large

Telescope programme. Soon, a small project office was established with Gilmozzi as principal investigator and Dierickx as project engineer. Other members were Guy Monnet, Enzo Brunetto, Marco Quattri, Bernard Delabre, Norbert Hubin, Franz Koch and François Rigaut.

Gilmozzi's driving idea could be described under the headline "Size Matters". Perhaps more fittingly "Size is Everything". He was prepared to sacrifice elegance in the design, even some performance, in favour of light grasp, reflecting his scientific dreams. This approach was based on the idea that for a diffraction-limited telescope, the science gain is proportional to fourth power of the growth in telescope size. This, in turn, meant that the telescope would have to rely completely on the efficacy of adaptive optics.

≈

Gilmozzi was not the only one who had given thought to this, but his approach was more daring than anybody else's. We shall look at the details on p. 420 in this manuscript. Among the other scientists who had long thought about the next generation telescopes was Arne Ardeberg. Ardeberg had served as Director of the La Silla Observatory until 1984 and been strongly involved in the search for a site for the VLT. Meanwhile, he was back at Lund University (Lunds Universitet) in Sweden, where, since 1991, he had begun to think about a new large telescope, originally of 20-metre aperture, then 25-metre, then with 40-metre and ultimately, with a mirror 50 metres across. In 1993, he had set up a "telescope group"[6] to work on these ideas and, in 1996, he helped to organise an international SPIE conference on large telescopes in Landskrona, Sweden, where some initial ideas about the 25-metre project were presented. His ideas included what he called a "live optics" system, a kind of intermediate stage between active and adaptive optics, correcting not only for gravitationally induced deformations of the primary mirror, but also for higher frequency wind effects on the telescope (Ardeberg, 1996). By 1998, the ideas had not simply matured, but changed. For example, the original idea of a spherical primary mirror had been abandoned. The project had also grown. It was now for a 50-metre telescope, first called the Swedish Extremely Large Telescope[7], before — a couple of years later — adopting

[6] One of his collaborators was Torben Andersen, the engineer who had worked at the ESO Telescope Project division in Geneva and, after a stint at the Nordic Optical Telescope, served as systems engineer on the VLT project. We met Andersen in this book in connection with the Coudé Auxiliary Telescope at La Silla and, later, ALMA.

[7] Presented at a joint ESO/Lund University workshop at Bäckaskog Castle in June 1999.

the name Euro-50. Apart from Lund University, Euro-50 involved scientists from the Tuorla Observatory (Tuorlan Observatorio, Finland), the National University of Ireland, University College London and the National Physical Laboratory, UK[8]. It also had strong support from the Instituto de Astrofísica de Canarias (IAC). In this form, i.e., as Euro-50, the telescope was foreseen as a Gregorian telescope with a segmented *f*/0.85 primary mirror and a large deformable secondary mirror. A Gregorian telescope, named after its inventor James Gregory (1639–1675), is a classical two-mirror telescope with an intermediate focus between the primary and secondary mirrors. With the introduction of Cassegrain-type telescopes, the Gregorian was largely abandoned. That it now seemed to re-emerge, shows how telescope designers had begun to cast off established preconceptions and not only embrace new ideas, but were also revisiting ideas of the past[9].

But ESO harboured serious reservations about the Euro-50 design. The requirements for the precise centring of the optical elements are considerable. The Gregorian telescope design also needs a long tube, 85 metres in the case of Euro-50, imposing strong requirements regarding stiffness. There were also concerns about having a deformable secondary of the size proposed for the telescope. The potential costs were also not as clear as for OWL, for which the estimated costs had been verified in close collaboration with industry[10] throughout the planning phase. This may have been caused by the fact that the Euro-50 project group was much smaller than OWL's and working on a much more restricted budget. Finally, the early and seemingly unconditional selection of a particular site raised a number of questions, at least as seen from ESO's perspective[11].

But Euro-50 had its followers. To some, OWL was an engineers' project; to some it was too ambitious. The mere thought of a 100-metre telescope may have made some more than a little hesitant. Because of the particular circumstances under which the OWL project had emerged — within ESO and somewhat overshadowed by the VLT and the ALMA project — it had not been discussed widely in the scientific community, and the broader community had not bought into the project the way they had done in the past for the VLT. But in any case, European astronomy now

[8] We note that, apart from Sweden, the institutes all came from countries that were not (yet) members of ESO.

[9] Other modern telescopes that have reintroduced the Gregorian design include the Magellan Telescopes and the Large Binocular Telescope at Mount Graham in Arizona, USA.

[10] Some companies had actually carried out studies for free.

[11] ESO carried out detailed tests over several years in a number of carefully selected locations, including La Palma, before choosing a site for the E-ELT. The choice was only made in 2010.

had two alternative ideas on the table — would that also mean two telescopes? Not even the most enthusiastic astronomy diehard could consider this a realistic option. For a while, the astronomical community seemed split and some feared that this might seriously jeopardise any attempt to obtain public funds for a next generation telescope worthy of the name. Hoping for a solution to the impasse, a meeting was organised in early May 2002 in Turku, Finland.

What was then the European Community had, since the early 1980s, become engaged in the funding of R&D. The political instrument, which became known as the Framework Programme, had started very modestly, both in terms of budget and scope, focussing on very specific areas of technology development. Subsequent editions of the Framework Programme expanded their reach, and, the sixth version in 2002 also contained funds for developing new research infrastructures, even with some limited funds for construction[12]. The Euro-50 project team planned to apply, while ESO, at least for while, remained undecided. ESO is of course funded directly by its Member States and obtaining EC funding, with possible strings attached, was not necessarily part of ESO's operational model. Since the fifth Framework Programme funds had also provided for "networks" of science institutes in Europe[13], including two for mainstream astronomy — OPTICON (for optical astronomy) and RadioNet (for radio astronomy). The Turku conference, was organised by OPTICON. The meeting brought together astronomers from all over Europe including, of course, representatives of both ESO and the Euro-50 project. No decisions were taken, but it became clear that additional studies of many of the technological aspects of both projects were needed. This seemed to provide a way forward.

A follow-up meeting was therefore organised in Bologna on 12 July 2002. The participants were ESO representatives (Catherine Cesarsky, Roberto Gilmozzi and Guy Monnet), from Euro-50 (Arne Ardeberg, José-Miguel Rodriguez Espinosa and Torben Andersen) as well as Giancarlo Setti, Roland Bacon, Rafael Rebolo, Tim de Zeeuw and Gerry Gilmore (chair of OPTICON). Firstly it was decided that both projects should continue for a period of 18 months, followed by a review. Secondly, both parties would come together in a joint proposal to the European Commission,

[12] The programme enabled European Commission (EC) funding covering up to 10% of the project cost, however capped at 15 million euros, and therefore clearly a minor contribution to a project that would cost somewhere between 500 and 1000 million euros.

[13] In EC parlance, OPTICON was a so-called "an integrated infrastructure initiative".

seeking funds for studies of the generic technologies that would be needed by both projects.

The joint proposal — for "an ELT Design Study" — was submitted to the EC in 2004. It had no less than 39 partners and a budget of 42 million euros, of which 22 million were requested from the EC. Most partners therefore added their own contributions. Not surprisingly, ESO's contribution was the largest at 8.5 million euros. The main areas of activity were adaptive optics, wavefront control, optical fabrication, mechanics, system layout, site characterisation and instrumentation — clearly all of major importance for the next generation optical telescopes. As it happened, the proposal was chosen for funding, albeit with a much reduced contribution from the EU (8.4 million euros), causing a number of partners to drop out. In the end, about 25 institutes took part, with ESO hosting the project office. The study was completed in 2009[14]. By that time, the situation had become much clearer regarding Europe's aspirations in the field of ground-based optical astronomy. The steps on the way involved important decisions by the ESO Council, the international review of the OWL project, the actions undertaken in the wake of this review and finally, a major conference in Marseille in December 2006, leading to the agreement by the ESO Council to give the green light for a Phase B study for Europe's forthcoming Extremely Large Telescope (ELT).

≈

Both Council and the Scientific Technical Committee had of course followed the evolution of the OWL project. There was no doubt that from a scientific point of view an Extremely Large Telescope would be the next logical step. Others saw this in the same way, especially in the US. The project was presented at an OECD (Organisation for Economic Co-operation and Development) Global Science meeting in Munich during the first days of December 2003, together with other large astronomy projects worldwide. Yet ESO had its hands full with the completion of the VLT/VLTI and was gearing up to full participation in the construction phase of the ALMA project (see also Chapter IV-1), and these required massive resources. In the December 2003 Council meeting, a Council working group on Scientific Strategy, chaired by Ralf Bender, presented its thoughts, recommending that ESO should *"reduce La Silla to a minimum scientifically viable level to free resources, continue the planned upgrading*

[14] Although followed by an additional, if smaller, EC-funded (ELT PREP) study under the 7th Framework Programme. The budget was six million euros with five million from the EU.

of VLT, complete the VLTI instrumentation programme to allow imaging with UTs, construct and bring ALMA into operation, carry out a detailed design study for OWL and ultimately build an ELT, alone or in partnership." Gradually, a consensus was building in Council that *"ESO should seek to lead the development of an ELT on the shortest possible timescale"*[15]. At the same time, Council discussed the possible implications of the establishment of the European Strategy Forum for Research Infrastructures (ESFRI), a new body set up at the request of the European Council, in partnership between the European Commission and the Member States and associated states of the EU, to identify the next generation of large research facilities of pan-European interest. ESO's Member States slowly edged towards a decision, and in December 2004, the ESO Council passed a milestone resolution. It said:

"[The] ESO Council ... agrees that
...
– over the last decade, the continued investment of ESO and its community into the improvement of ground-based astronomical facilities has finally allowed Europe to reach international competitiveness and leadership in ground-based astronomical research,
– the prime goal of ESO is to secure this status by developing powerful facilities in order to enable important scientific discoveries in the future, ...
...
adopts the following principles for its scientific strategy:
...
– ESO's highest priority strategic goal must be the European retention of astronomical leadership and excellence into the era of Extremely Large Telescopes by carefully balancing its investment in its most important programmes and projects,
– the construction of an Extremely Large Telescope on a competitive time scale will be addressed by radical strategic planning, especially with respect to the development of enabling technologies and the exploration of all options, including seeking additional funds, for fast implementation... ."

In March 2005, ESFRI published an embryonic list of European projects with the Extremely Large Telescope included, thanks to the intervention of several Council delegates. Even if the European list did not release major funds, this was an important political step, since the list developed into a roadmap for European research infrastructure projects with the EU member states being encouraged to produce national

[15] Council minutes, December 2003.

roadmaps, aligned with the European one[16]. In the same month, after a six-year stint as Director of the Paranal Observatory, Gilmozzi returned to Europe to become head of the ESO Telescope Division, and lead the new project. Together with his colleagues he prepared a 700-page project proposal, the OWL Blue Book, analogous to the 1987 VLT Blue Book. For the telescope itself, the proposal was largely based on in-house work, with the involvement of European industry. However, the scientific community had been heavily involved in both the development of the science case and the first ideas about instrumentation. The OWL proposal represented both a radically new approach to telescope design and a continuation of the experience gained so far with the current 10-metre-class telescopes, e.g., as regards active and adaptive optics, the use of segmented mirrors and low-cost structures.

Artist's impression of the OWL telescope. (Illustration by Herbert Zodet)

[16] Large telescopes were not only the dream of European astronomers, but also of their American colleagues. In the US two projects emerged, the Thirty Meter Telescope and the Giant Magellan Telescope (GMT) of the Carnegie Foundation. To discuss issues of common interest, informal meetings of senior executives, typically at international airports such as Heathrow or Amsterdam, were organised in that period. Among the ideas considered were the sharing of technologies and a possible division of scientific tasks, implying developing complementary instrumentation (perhaps even mutually compatible) and the sharing or exchange of telescope time. The first of these meetings took place in 2004. The meetings, however, also discussed non-optical astronomy projects such as the Square Kilometre Array (SKA).

The 100-metre OWL primary mirror (i.e. a light-collecting surface of an impressive 6000 square metres) was to be a spherical surface, made up of more than 3000 identical hexagonal segments, each 1.6 metres in size. The secondary mirror, no less than 25.6 metres across, would have been made up of more than 200 flat 1.6-metre segments. Low-expansion glass-ceramics or silicon carbide was foreseen for the mirror elements. To compensate for the unavoidable severe spherical aberration, a four-element corrector would be located mid-way in the telescope structure, with two flexible 8-metre-class active mirrors (similar to the VLT primary mirror), a 4.2-metre focusing mirror and a 2.3-metre flat, fast steering adaptive mirror for first stage adaptive correction. In line with the idea of already identifying upgrade paths at the earliest stage, it was thought that the 4.2-metre mirror might be exchanged for an adaptive one at some point. The telescope tube assembly would be a steel structure of 15 000 tonnes on an alt-azimuth mount. Active optics, the particular optical design and the use of standardised elements, both for the primary mirror and for the mechanical structure, would all contribute to reducing cost and thus, it was thought, it would become possible to build this giant for the price of one billion euros.

The OWL project generated considerable interest in European industry. This picture shows Jean-Pierre Swings, Roberto Gilmozzi and Philippe Dierickx presenting the project at a meeting of the Walloon Space Cluster (of industries). (Photo: Hans Hermann Heyer)

In November 2005, an international review panel[17] worked its way through all the aspects of the project. The panel concluded *"that the team has demonstrated a plausible case that OWL is feasible and that a 100-metre telescope can be built and operated.... However, the committee has identified several areas where the risks involved in pursuing a 100-metre diameter telescope are sufficiently serious that the likelihood of significant schedule slip is high and there is a significant probability that the*

[17] Chaired by Roger Davies (Oxford) and with Jean Gabriel Cuby (L'Observatoire Astronomique Marseille-Provence [LAM]), Brent Ellerbroek (TMT), Daniel Enard, Reinhard Genzel (MPE), Jim Oschmann (Ball Aerospace), Larry Ramsey (Pennsylvania State University), Roberto Ragazzoni (INAF), Stephen Shectman (Carnegie Observatories) and Larry Stepp (TMT) as members.

required scientific performance will not be achieved. The estimate of the cost to successfully achieve the 100 metre goal remains substantially uncertain…."

OWL was, essentially, found to be a step too far. Yet both the science case for an extremely large telescope and the conviction that, within Europe, ESO would be the right organisation to realise it, remained. The committee therefore also recommended *"that the project [should] proceed to Phase B, which should start with a thorough re-evaluation as described above."* This meant moving towards a smaller telescope that was less complex, and technologically less risky, and, in particular, discarding the idea of a segmented secondary mirror in favour of a monolith, as well as aiming for a solution that would be more instrument-friendly. The reality of this, however, was a complete redesign of the telescope. The OWL review was clearly a milestone event, but it would have been easy to stumble over this stone. The challenge was now to retain the momentum, to keep up the enthusiasm and dedication of the project team, and to secure the full participation of the scientific community. As Director General, Catherine Cesarsky had long been concerned about what she considered to be insufficient community involvement. And she understood fully the need for speedy and decisive action. She therefore immediately rolled out a set of blitz actions on a breathtaking schedule, involving five joint ESO/community working groups on the key themes of science, telescope design, instrumentation, adaptive optics and site selection. The working groups were asked *"to synthesise and collate ELT capabilities in the specified topic area, noting existing community studies and ongoing efforts and to propose a basis for prioritising capabilities in the specified topic area, a list of key tradeoffs and an initial prioritisation of an ESO ELT capabilities."* The reference to "existing community studies and ongoing efforts" was of course a clear signal to consider good ideas regardless of their origin, but also that the project should now move on. The first meetings were supposed to occur as soon as mid-January 2006 with the groups delivering their reports by April. A consolidated report by May would form the basis for the next phase, to last until the end of the year, during which period a Reference Baseline Design would be elaborated by an E-ELT Project Office to be established at ESO and led by Jason Spyromilio, then Director of the Paranal Observatory. The process towards a final proposal would be accompanied by a core ELT Science and Engineering working group (ESE), comprised of the chairs of the five subgroups, an ELT External Advisory Board (EEAB) and, of course, the STC. Following the regular December Council meeting, in which the Science Strategy Working Group[18] (SSWG) was requested to support the effort, the SSWG chair, Tim de Zeeuw, set up

[18] We shall return to this body in Chapters IV-7 and IV-9.

a meeting on 21 December. By the end of that day, the invitations to the prospective chairs of the working groups went out[19]. Following their acceptance, already on the next day, invitations went out to those who had been identified as possible members of the respective working groups. Cesarsky's initiative was received with overwhelming enthusiasm by the scientific community. Although the invitations to participate in the working groups had only been issued just before Christmas and the meetings were to commence in mid-January, virtually everybody accepted the invitation — and the challenge. Keeping up the momentum, in January 2006 Cesarsky presented the plan to the ESO staff, who were just returning after the Christmas recess. She ended her presentation with a bold statement: *"We have the brains, we have the knowhow, we have the means — let's do it! (with the Community)"*. Just two days later, on 13 January, the first working group met. As planned, the draft reports were ready by the end of February 2006. The first meeting of the core ESE took place on 21 March and a second meeting on 21 April. The core group was co-chaired by Daniel Enard and Guy Monnet. This was followed by a workshop with all the working group participants on 27–28 April. By that time, major changes to the project had already been defined: the size of the primary mirror was set at 42 metres and a novel optical design by Bernard Delabre, to be known as "the five-mirror telescope" was proposed[20]. Within the same timeframe, ESO set up an ELT Standing Review Committee (ESRC), chaired by Roger Davies and with Bob Williams, Bengt Gustafsson, Matt Mountain, Jean-Loup Puget, Monica Tosi and Reinhard Genzel as members. The first meeting took place on 19 May. The recovery action had electrified the scientific community, and thus at a conference in Marseille from 27 November–1 December 2006, ESO was ready to present and discuss its idea with the wider community in the shape of some 250 participants. In a sense, this conference was reminiscent of the 1986 Venice Conference that preceded the VLT decision.

[19] Daniel Enard (Telescope Design), Marijn Franx (Science), Colin Cunningham, Gérard Rousset (Adaptive Optics) and Roland Gredel (Site Selection). The groups were co-chaired by ESO staff.

[20] Given the differences of opinion about the telescope design, the choice of the chair of the telescope working group was clearly critical. It fell on Daniel Enard. Enard was highly respected in the entire community. Furthermore, as he had left ESO in 1996 to work on the VIRGO gravitational wave detector project in Italy, he was rightly perceived as a neutral arbiter, and expected to forge a consensus in the group. He did. His seminal work at ESO on instrumentation and his contribution to the VLT notwithstanding, achieving consensus in this ELT working group while at the same time helping to launch a truly innovative design was perhaps the greatest service he rendered to ESO. It would also be his last, as less than two years after the decision to move to the ELT Phase B study, in August 2008, he died, aged 68.

The participants of the Marseille meeting on the E-ELT.

Anyone present in Marseille who might also have attended the 1986 Venice conference would have realised how the VLT project evolved after that meeting. Wisely, therefore, the aim of the Marseille meeting was not to settle on one single design solution. Rather the aim was to achieve consensus on some basic design requirements, embodied in two different design solutions, and to enable the ELT project office to elaborate on both solutions with the understanding that in the end, a definite choice had to be made. In their report from the meeting, Roberto Gilmozzi and Jason Spyromilio, described the starting point for the next phase: *"The telescope to be built should have a primary mirror of order 40 metres in diameter (42 metres was thought to be a good compromise between ambition and timeliness), should not be based on spherical mirrors, should have adaptive optics built into it and deliver a science field of view of at least five arcminutes diameter with a strong preference for larger fields. Furthermore the telescope was to provide multiple stable observing platforms while maintaining a focal ratio that would be favourable to instrumentation."* (Gilmozzi & Spyromilio, 2007). At the time, the question of the optical layout was left open — either a Gregorian telescope or the five-mirror design, which had evolved in the course of the preceding months and soon turned out to be the winning option[21]. So,

[21] The ESO 2006 Annual Report provides the reasoning: *"While the Gregorian design had advantages in a better theoretical performance in adaptive optics, a smaller mirror count and smaller central obstruction, the five-mirror design was considered to have advantages in the image quality across the focal plane, the control of the wavefront using laser guide stars, the flexibility of the focal stations, the deployment of atmospheric dispersion compensators, the smaller dome and the relative ease with which it could be upgraded to follow the development of technology in the future. Most critically, however, the five-mirror design was considered to have much lower risk in the area of adaptive optics by virtue of the separation of the field-stabilisation and adaptive-optics functions, a key recommendation of the OWL review."*

by early 2007 the preferred solution was for a 42-metre telescope with a segmented primary mirror[22] comprised of 906 elements, each 1.45 metres wide, and with a 6-metre secondary mirror. With the five-mirror design the M3 was a 4.2-metre mirror relaying the light to a double system of mirrors providing for the adaptive optics. The quaternary (M4) 2.5-metre mirror was fully flexible, with at least 5000 actuators operating at a frequency of 1000 Hz and finally the M5 was a 2.7-metre mirror that would correct for lower-order disturbances.

A major scientific driver for the ELT generation of telescopes has been exoplanet research. In this cartoon from 1997, Philippe Dierickx adds his own comment on the prospects in that area.

Given the importance of the E-ELT as the observational facility for optical astronomers for a long time to come, it is not surprising that Europe's astronomers engaged strongly in the discussions about the definition of this telescope. After all, what was at stake was their very professional future as scientists. With no lack of competing ideas and wishes that were passionately fought over, these discussions were not always straightforward. The outcome of the Marseille conference was therefore a major diplomatic victory for Cesarsky — bringing together a unified scientific community behind the revised E-ELT proposal. With this outcome, at its December meeting, the ESO Council approved the proposal to move into the detailed design study phase with the aim of completing it by December 2009. Much work and many difficult decisions lay ahead, including the critical choice of site, but the way forward had been cleared. In August 2007, when her term of office expired, Cesarsky could look back on an impressive set of achievements: the last and technically most challenging part of the VLT project, the VLT Interferometer, had been put on track, literally and also scientifically, the ALMA project had been approved and the major industrial contracts awarded, the OWL conceptual study had been converted into a concrete

[22] Unlike the US, Europe had little experience with segmented mirror technologies. However, at the time the 10-metre Gran Telescopio Canarias (GTC) was under construction in Spain. Once Spain joined the organisation, ESO's engineers gained access to this facility (see also Chapter IV-8).

Roberto Gilmozzi presents the E-ELT reference design at the Council meeting in December 2006. (Photo: Hans Hermann Heyer)

project for an extremely large telescope with the backing of the scientific community, and ESO had seen the largest influx of new Member States since its foundation in 1962. We shall look at the latter shortly, but also note that for a dynamic organisation like ESO, plenty of work of course awaited her successor, Tim de Zeeuw — the completion of the ALMA project, the challenge of transforming the E-ELT design study into a real, fully financed, project while also maintaining Paranal as Europe's foremost astronomy base. Yet it will be for another chronicler, in due time, to describe this part of ESO's continuing history.

It belongs to the traditions at Paranal that, after every M1 recoating, a photo of the team is taken. This particular one, in July 2007, also marked the farewell to Catherine Cesarsky, here surrounded by the team members, as Director General of ESO. (Photo: Gerd Hüdepohl)

An ISAAC observation from 2001 of Messier 16 (M16), a star-forming region in the constellation of Serpens. Also called the Eagle Nebula, the object is probably even better known as the Pillars of Creation thanks to the spectacular Hubble image. This VLT image, however, shows a much larger area in the sky, including the cluster of hot stars (NGC 6611, at the top of the picture), which gives the pillars their shape. (Observation by Mark McCaughrean and Morten Andersen)

Chapter IV-5

The VLT in Retrospect

"Astronomy is once again leading physics in posing the most profound questions on the nature of the universe we live in."

Riccardo Giacconi in his book *Secrets of the Hoary Deep.*

At the time of writing, the VLT is approaching its 15th anniversary. Over the many years of the project it has been the centre of attention and it remains the main facility of ESO today. In this chapter, we will try to summarise the impact of the VLT. Before we look at what the VLT has given us, however, we will look at the cost and in particular how it increased over the years. We saw that the original proposal cost of 382 million deutschmarks grew to something close to 1000 million deutschmarks (or roughly 500 million euros) and how both the Council and ESO as an organisation struggled under this burden. But what were the realities behind the increase? A closer look reveals a much less dramatic situation than the numbers above seem to suggest. Firstly, different accounting methods were used, and secondly inflation, over a period of more than ten years, took its toll. Additions to the programme did affect the final price, but as we have discussed, these additions were crucially important in ensuring the scientific output of the VLT. According to a calculation made in 1999, the year when the Paranal Observatory was inaugurated and the first Unit Telescope handed over to the scientific community, the total sum of external contracts amounted to 577 million deutschmarks. These contracts were awarded between 1989 and 2003. If the contract value of all of them were to be converted to 1999 values, the final number would be 612 million deutschmarks. This included the four 8.2-metre Unit Telescopes, the first generation of instruments, the first VLTI phase including two of the Auxiliary Telescopes, the site development and construction. This number must be compared to the cost estimate in the original proposal, which, if also updated to its 1998/99 value, would be 524 million deutschmarks, meaning that the real, net increase was in the 15–17% range, caused by scientifically justifiable additions.

But it is important to note that the estimated cost contained in the Blue Book proposal included neither manpower nor operation costs[1]. By 1999, 142 full-time equivalent personnel were occupied with the VLT. Including manpower, operations (until 2003) and general overheads, the cumulative cost at 1999 prices became 990 million deutschmarks. What may have been underestimated in the original proposal, therefore, was the personnel effort required. But readers are reminded of the dramatic changes that occurred in that period in the way astronomical research was conducted — enabled by new technologies but also with manpower implications. Although the cost rose, the scientific gains were, simply speaking, enormous. Building a telescope like the VLT without ensuring its optimum use for science would have been indefensible.

≈

The VLT cemented the position of both Schott and REOSC as world leaders in the field of telescopes. Schott has supplied its products to customers across the world, including NASA (the Chandra X-ray satellite and Constellation X), the airborne Stratospheric Observatory for Infrared Astronomy (SOFIA), the 10.4-metre GTC telescope on La Palma in the Canary Islands, the Chinese Large Sky Area Multi-Object Fibre Spectroscopic Telescope (LAMOST) and the 6.5-metre Magellan Telescope. REOSC polished the mirrors of the twin 8-metre Gemini telescopes and the mirror segments for the GTC, among others. AMOS, the Belgian company, grew into a major supplier of astronomical telescopes and instruments. And of course many of the technologies that the VLT required found their way into other areas as well. For example, the Shack–Hartmann wavefront sensors, with their lenslet arrays, found applications far beyond astronomy, in areas such as eye surgery and laser beam profile measurements.

≈

But the VLT was of course not built to develop new technologies, however interesting they may be. It was built to do science. So how has the VLT contributed to astronomy? It has become a truism that astronomy, probably the oldest of the sciences, is currently experiencing another golden age. To a large degree, the reason is that technology opens up new avenues of enquiry. Thanks to the incredible advances in technology that have occurred during the last 100 years or so, astronomy has entered a phase

[1] These costs were estimated seperately.

The first 8.2-metre blank at the Schott Glassworks being prepared for shipment to France in June 1993. Understandably the proud Schott team had placed a large banner behind the blank, saying "Größter Spiegelträger der Welt". (Photo: Hans Hermann Heyer)

where it is not just asking questions, but can actually aspire to provide real answers to some of these most fundamental questions. How old is the Universe? How did it evolve? How did the stars and planets form? Might this immense space that we call the Universe harbour life elsewhere than on our tiny planet? Such questions are unlikely to be answered by individual scientists or by using one particular telescope. The research enterprise is instead more like a symphony orchestra with every instrument contributing to the whole. The VLT has occupied a significant place in the current concert of astronomy and it has contributed to most areas within this science. If we look more closely at the numbers we find that, since the start of science operations and up until early 2012, observations with the VLT have formed the basis of more than 4200 research papers. It is therefore impossible to provide a synopsis of the research activities that will do any justice to the scientists and to the VLT itself in just a few pages. Nonetheless, the following paragraphs may convey both a sense of the importance of the work done with the VLT and the breadth of the research topics.

Fundamental constants constitute the numerical values of basic physical parameters such as the gravitational constant, the elementary charge, the mass of the electron, or the speed of light (in vacuum). We don't know why they have these particular values, but we realise that, were they any different, the world that we know would not exist. In a famous quote from Einstein, he resorted to almost poetic language, to discuss the mystery surrounding the fundamental constants: *"[They] are genuine numbers which God had to choose arbitrarily, as it were, when He deigned to create this world."*[2]. Constants, unsurprisingly, are not supposed to change, but how firm is this rule? From an astrophysical point of view, the Universe constitutes not simply

[2] Letter of 13 May 1945 from Einstein to Mrs Use Rosenthal-Schneider, cited in Mehra (1973).

a giant physics laboratory, but one that allows us to look back in time to the earliest epochs of the Universe. It follows that astronomy lends itself naturally to studying some of the fundamental constants, including the question of possible time variations in the value of the fine structure constant, which defines the strength of the electromagnetic interaction. Might it have changed over this huge timespan, as some scientists have suggested?

In astrophysics, spectroscopy is often the almost magical tool with which astronomers probe the depths of space to investigate the nature of the distant objects of their study. In this case, however, the object was the huge expanse *between* some of most distant and brightest objects known — the quasars — and us. Because of their brightness, quasars can be used as searchlights, literally shining their powerful light through intergalactic space and onto the matter found there. By revealing the wavelengths of atomic transitions in the high-redshift (or high-z), i.e. distant, Universe, spectra of quasars also show the physical conditions of matter at different epochs. Using the UVES spectrograph mounted on Kueyen over a total of 34 nights, Raghunathan Srianand, Patrick Petitjean and their collaborators explored 50 absorption systems (interstellar clouds of gas) at distances of between six and eleven billion light-years to test the suggestion that the fine structure constant might have changed over time. Their results provided a strong constraint on any variation of this constant over the last ten billion years (Srianand, *et al.*, 2004; and ESO Press Release 31 March 2004). But it is unlikely to be the last word on this subject[3]. As observational techniques improve, the question will undoubtedly be revisited, for science with its Mertonian norms is an evolutionary process with continuous checks on results as we try to construct an evermore comprehensive edifice of knowledge.

As readers will know, our current thinking regarding the evolution of the Universe is based on the idea of a Big Bang, which goes back to theoretical work carried out in Europe and observational results first obtained in the late 1920s in the US. The observational evidence rested on the fact that when spectra of stars and galaxies are obtained, spectral lines, which can be associated with specific elements, are displaced relative to where they are located in a laboratory spectrum. Mostly, these spectral lines are shifted towards longer wavelengths, a phenomenon that astronomers call redshift. This phenomenon is traditionally explained by the Doppler effect, known from everyday life — for example — by the change in pitch from an approaching or

[3] Based on data obtained with the Keck I telescope and the VLT, in 2010 John Webb and colleagues suggested that there could be spatial variations of the fine structure constant. Their findings are as yet unconfirmed, however.

receding vehicle, say a train or a car. In astronomy the Doppler effect is observable, for example, in studies of binary stars or extrasolar planets. However, in cosmology the underlying mechanism is different; cosmological redshift is a manifestation of a stretching of wavelengths because the very fabric of space has expanded over the travel time of the light. The higher the redshift (and thus the recession velocity), the larger the distance to the source, and, because the speed of light is finite, the greater the so-called lookback time. We see objects in the sky not as they are now, but as they were at a certain time in the past. But how far do we look back in time and how old is the Universe? The link between what is now called the redshift velocity and the distance (and therefore, the lookback time) is the Hubble constant. Determining the Hubble constant has arguably been one of the greatest challenges faced by astronomers in the 20th century. In Chapter I-1, we mentioned that just a few decades ago estimates of the age of the Universe ranged from 10 to 20 billion years, depending on the value assigned to the Hubble constant.

The main tools to determine distances in astronomy have been triangulation (for the nearest objects only) and the identification of stars that could serve as "standard candles" (see Chapter III-7). An alternative and independent way to constrain the age of the Universe — as astronomers would express it — is by measuring the abundances of rare elements in stars. By considering the natural decay of radioactive isotopes with sufficiently long half lives, it is possible to estimate the ages of the stars. This technique is similar to the long-established practice of carbon-14 dating, used in archaeology. Given the timescales involved, this would suggest searching for either the isotope thorium-232, with a half life of about 14 billion years, or — even better — uranium-238, which has a half life close to 4.5 billion years. In one such study, astronomers observed a series of extremely metal-poor stars, assuming that they must have formed during the earliest phases of the Milky Way. The observations were carried out in 2000–2001 with the Kueyen Unit Telescope and UVES, the high-dispersion spectrograph. The astronomers clearly hoped that the resolving power and efficiency of UVES, together with the light-gathering power of the telescope would enable them to detect the extremely weak fingerprint, or spectral line, of U-238 in the stars. They were not disappointed. This first observation led to an age estimate of 12.5 billion years for a particular star and provided an important clue, as well, regarding the age of the Universe[4] (Cayrel *et al.*, 2001; ESO Press release 7 February 2001). The result also demonstrated the power of the VLT. Yet astronomers knew that there was more to come. Much more.

[4] The current estimate of the age of the Universe is 13.7 billion years.

The observations had focussed on the oldest stars in the Milky Way. Yet, in spite of their impressive age, these stars could not have been the first generation of stars. This was because they contained elements that must have been created in preceding supernovae. So when did the first stars in the Milky Way form? In 2003, VLT observations led to a new estimate. Once more, astronomers used the presence of chemical abundances in stars to answer this question. But instead of looking at the decay of heavy, unstable elements, they now looked at the gradual build up of a light element, beryllium-9.

It is thought that beryllium-9 is produced by the fragmentation of heavier atomic nuclei from supernova explosions when they collide with hydrogen and helium nuclei. The concentration of beryllium in the interstellar medium will therefore increase with time. As new stars are born in interstellar clouds, they will also contain beryllium and the later these stars begin to form the more beryllium they will contain. Studying the oldest stars that we can find and looking at their beryllium content will therefore provide information about the time elapsed between the first supernova explosions and the formation of the first of the second-generation stars. And since stars that end their lives as supernovae are short-lived, we will begin to understand when the very first stars in the Milky Way were born. The idea was simple but, needless to say, these observations were difficult to carry out. As was the case with the uranium measurements, the overall amount of beryllium is small, and the spectral lines are very weak. Furthermore, they fall in the part of the electromagnetic spectrum close to the ultraviolet cut-off of the atmosphere. Even so, a group of astronomers using UVES managed to carry out the first-ever such observations of two stars in NGC 6397, a globular star cluster located 7200 light-years away in the southern constellation of Ara. The results suggested an age for the Milky Way of about 13.6 billion years, which, considering the uncertainties, is in agreement with the current estimate of the age of the Universe (Pasquini *et al.*, 2004 and ESO Press Release 17 August 2004).

The Big Bang theory received perhaps its most impressive confirmation in 1992 with the mapping of the afterglow of the Big Bang — the cosmic microwave background radiation (CMB) — by NASA's Cosmic Background Explorer (COBE). The observations showed minute fluctuations in the background temperature of 2.7 K, precisely as predicted by theory. Scientists speak about the CMB anisotropy, meaning that it is not exactly the same in all directions. Later measurements have been carried out with much improved resolution first by NASA's Wilkinson Microwave Anisotropy Probe (WMAP) and slightly later by ESA's Planck satellites, with results appearing in 2010.

But science is about constantly questioning established assumptions and testing them with new experiments or observations. In 2008, a team of scientists used UVES to perform a unique measurement of the cosmic background temperature at a more recent epoch. To do so, they looked for molecular hydrogen in a galaxy seen 11 billion years in the past. The galaxy itself can hardly be seen, but behind it is a quasar. With this quasar conveniently acting as a searchlight, as we have already seen in the study regarding the fundamental fine-structure constant, it is possible to investigate the galaxy and its chemical composition through the imprint it leaves on the spectrum of the quasar. As well as hydrogen and its heavy form deuterium, the group also noted the presence of molecular carbon monoxide (Srianand *et al.*, 2008). This allowed them to determine the temperature of the background radiation. According to the theory, the CMB temperature should have been around 9.3 K at this particular early point in the Universe's history. The measurements suggested a temperature of $9.15\ K \pm 0.7\ K$ — a good match. The observations also showed that the physical conditions of the interstellar gas in this remote galaxy are similar to those seen in our own galaxy, the Milky Way.

As mentioned, modern astrophysics sees the Universe as a giant physics laboratory. Here we find the most extreme conditions that we can imagine — the emptiest of vacuums, the most densely compressed matter, the lowest and the highest temperatures. Undeniably interesting, but it is hard to describe the Universe as a friendly place. Occasionally parts of it are doused with high-energy radiation, such as gamma rays. The phenomenon was first discovered during the Cold War, as the US military launched satellites to monitor Soviet compliance with the ban on nuclear tests in the atmosphere[5]. While no sign of a Soviet violation of the agreement was found, the satellites instead detected bursts of gamma rays coming from space. This led to attempts to identify the objects from which the radiation emanated and thus a search for the cause. Once a burst had been detected, one approach was to train ground-based telescopes on the particular part of the sky from which the radiation came and search for the possible source. But gamma-ray bursts are unpredictable and short-lived events and catching them needs quick reactions. Dedicated small telescopes across the globe are used for this purpose, including telescopes at La Silla. But larger telescopes, such as the VLT also have an important role to play. By now, we know that most bursts — or GRBs as astronomers call them — are located at very large distances from us,

[5] This is reminiscent of the chance discovery of radio waves from space by Karl Jansky, and the advent of radio astronomy, following investigations of static that caused disturbances to short-wave transatlantic voice transmissions.

billions of light-years away. In fact, observations of a GRB detected on 23 April 2009 using the VLT and the MPG/ESO 2.2-metre telescope at La Silla revealed a source seen as it was 13 billion years ago — in other words seen as when the Universe was no more than 700 million years old. This is the object with the greatest measured distance (Tanvir *et al.*, 2009). Given the amount of energy that we measure, the energy released in a few seconds during such an event must exceed that of the Sun during its entire ten-billion-year lifetime. During the late 1990s, astronomers began to believe that these dramatic bursts were somehow associated with supernova explosions, but GRBs remained shrouded in mystery.

On 29 March 2003 — at 11:37:14.67 UT — a NASA satellite detected a very bright gamma-ray burst. Optical identification was achieved with a 1-metre telescope at the Siding Spring Observatory in Australia, enabling VLT astronomers at Paranal to point the Kueyen telescope fitted with the UVES spectrograph at the source[6]. They determined the distance to the object, now given the unspectacular designation GRB 030329, to be an estimated 2650 million light-years — the closest GRB ever detected. Even for astronomers, used to studying the most distant objects in the Universe, proximity counts, as we will remember from SN1987A. This particular GRB therefore provided an unprecedented opportunity to gain further insights into the origin of the phenomenon. The first detection of an optical afterglow was that made by the Dutch astronomer Jan van Paradijs and his collaborators following the 80-second gamma-ray burst (GRB 970228) in February 1997. But GRB 030329 was nonetheless the event that established the connection between long-duration GRBs and the death of a massive star. Over a period of one month spectra were obtained with the FORS1 and FORS2 instruments — on Antu and Kueyen, respectively — while the object rapidly faded. The result showed convincingly that GRBs occur in connection with the kind of supernova explosions that astronomers now call hypernovae, the deaths of stars with 25 or more times the mass of the Sun. The XXL version of a supernova, one might say.

Gamma-ray bursts, however, come in two kinds — long bursts (lasting more than two seconds) and short ones. Short bursts also emit higher energy photons than the long ones. It therefore seemed clear that the physical processes leading to these two types of events might differ. It had also not been possible to detect an afterglow

---

[6] Gamma-ray bursts are short-lived and it is therefore important to react immediately once a flash is detected by a space-borne observatory. For this reason, in 2004 ESO implemented a rapid response mode for the VLT, by which the regular observations are interrupted and the telescope automatically carries out observations of the GRB source, shortly after detection (Vreeswijk *et al.*, 2010).

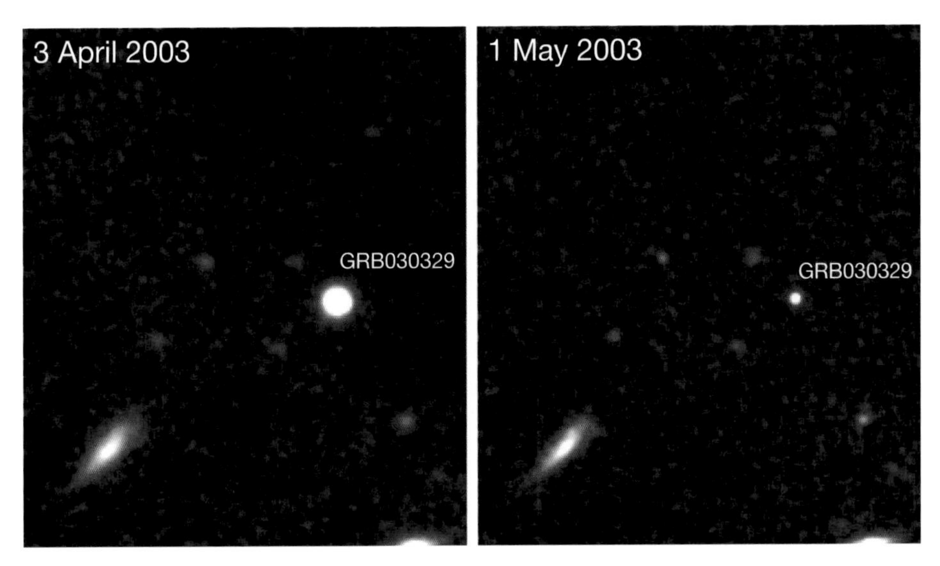

FORS observations showing the optical afterglow of GRB 030329.

after any short bursts, and thus to determine the exact position of objects, which was necessary for follow-up studies with ground-based telescopes. However, two short gamma-ray bursts, on 9 May 2005 and 9 July 2005 respectively, changed this. In the first case the observers used the VLT's Antu and Kueyen Unit Telescopes — once again with the twin FORS instruments. On 9 May, the Swift satellite[7] registered a burst of 40 milliseconds duration. This time it also detected an afterglow, which enabled the observers to obtain images of the area in question with FORS2. The object, known as GRB 050509B, was believed to be associated with an elliptical galaxy at a distance of 2700 million light-years (Gehrels *et al.*, 2005). Star formation does not take place in such an environment, and hypernovae, which are associated with high-mass precursors with short stellar lifetimes, are therefore also not expected to occur. Not surprisingly, there were no traces of a supernova explosion during the weeks after the burst. This gave support to the assumption of different causes for the long- and short-duration GRBs. The second event was a 70-millisecond duration burst. This time, the scientists — a team led by the Danish astronomer Jens Hjorth — were successful in identifying the fading source using the Danish 1.54-metre telescope at La Silla (Hjorth *et al.*, 2005). So what might be the origin of the short-duration bursts? Current speculation focuses on the merger of two of the most exotic objects in the sky — neutron stars. A neutron star is the remnant of a supernova explosion.

[7] Operated by NASA/ASI/PPARC.

It is an extraordinarily compact and dense object with a diameter of just some 10–20 kilometres, but a mass similar to that of the Sun or even larger.

Astronomers often speak about the cosmic zoo because of the many "species" of astronomical objects, some of which are truly strange. Neutron stars belong in this category. And so do black holes. But black holes are not just found in distant galaxies — as it turns out, we have one in our own stellar system, as well. So to study black holes, we are lucky enough to have a prime example in our own backyard. To be more precise, there is one at the centre of the Milky Way that is amenable to high-resolution studies with telescopes such as the VLT. As the scientists wrote in their research paper: *"The [Galactic Centre] is a uniquely accessible laboratory for exploring the interactions between a massive black hole ... and its stellar environment."* (Gillesen *et al.*, 2009). In that connection it should be remembered that one of the reasons why Europe's astronomers placed their observatory in the southern hemisphere was precisely to study the centre of the Milky Way in the most efficient way.

Black holes cannot be seen directly. However, their presence can be inferred by studying the surrounding region, including the behaviour of objects that are affected by the presence of this cosmic monster. Perhaps one of the most impressive studies that has ever been carried out in astronomy is that of the stars located at the centre of our galaxy. The study was undertaken by a team of astronomers led by Reinhard Genzel from the Max-Planck-Institut für Extraterrestrische Physik in Garching. Pinpointing and following stars at the densely populated centre of the Milky Way places extraordinary requirements on the telescope. Furthermore, since large swathes of dust that obscure the stars are found in this region, observations had to be made in the infrared, which allowed the scientists to peer through the dust. The team went to great lengths to carry out their study. We recall that they developed the SHARP instrument and, realising the potential of the newly commissioned NTT, convinced ESO to mount it on this telescope, although it was really not meant to accommodate such visitor instruments. The observations that were initiated with SHARP marked the beginning of a 16-year-long study, using some 50 nights of observations to meticulously map the motions of a number of stars close to the centre of our galaxy. From 2002 onwards[8], observations were undertaken with the VLT and two adaptive optics instruments, NACO and SINFONI, the integral field spectrograph covering the spectral range of 1.1–2.45 μm.

[8] Before that observations were also conducted with the 10-metre Keck telescope on Mauna Kea, Hawaii.

The observations revealed stars orbiting an invisible central object, associated with the radio source Sgr A*, in a very particular way. Out of the 28 stars observed, one, known as S2, completed its orbit in 15 years, i.e. within the period of the study. At a certain point in its orbit it comes as close as 17 light-hours to the central object. The observations enabled the astronomers to calculate the mass of the central object and it was found to be no less than four million solar masses. The conclusion is that our galaxy features a massive black hole at its heart. The study of this region has, if anything, intensified. Readers are advised to stay tuned for exciting updates!

In 2008, Genzel was awarded the Shaw Prize for his research on the black hole in the Milky Way[9].

The central region of the Milky Way seen with the VLT.

[9] Followed in 2012 by the Crafoord Prize, shared with Andrea Ghez from UCLA.

The Milky Way, our cosmic home, is undeniably an exciting place. One topic that scores high on the list of fascinating topics is, naturally, the study of planets outside the Solar System. In Chapter III-7 we discussed the breakthrough in exoplanet research in 1995 with the pioneering work of Michel Mayor and Didier Queloz. But the detection of extrasolar planets by means of radial velocity measurements was an indirect one. Actually seeing such a planet, i.e. obtaining a direct image, had eluded the astronomers. In fact, many considered this to be impossible with existing telescopes. Nonetheless attempts were made and some early claims of success reported, although they could not be confirmed. In Chapter III-14, we mentioned such a case. However, in 2004, ESO published the first direct image of an exoplanet obtained with the VLT. It was based on a research paper by a group of mainly French astronomers[10] led by Anne-Marie Lagrange and with Gael Chauvin, a young astronomer working at ESO Paranal, as first author (Chauvin *et al.*, 2004). Two main problems arise in connection with directly imaging such objects: the small angular separation of the planet from its parent star and the enormous difference in brightness between the star and the planet, which reflects only a tiny amount of the starlight. It is a little like detecting the radiation from the face of the lighthouse keeper in the glare of a lighthouse beam from a great distance[11]. Consequently, the search for exoplanets has focused on intrinsically faint stars, of the kind that astronomers call brown dwarfs, where the difference in brightness would be within the range that could be managed, if still requiring extraordinarily good telescopes and advanced imaging systems. In this case, the object was surmised to be in orbit around a brown dwarf star, known as 2M1207, 230 light-years away in the southern

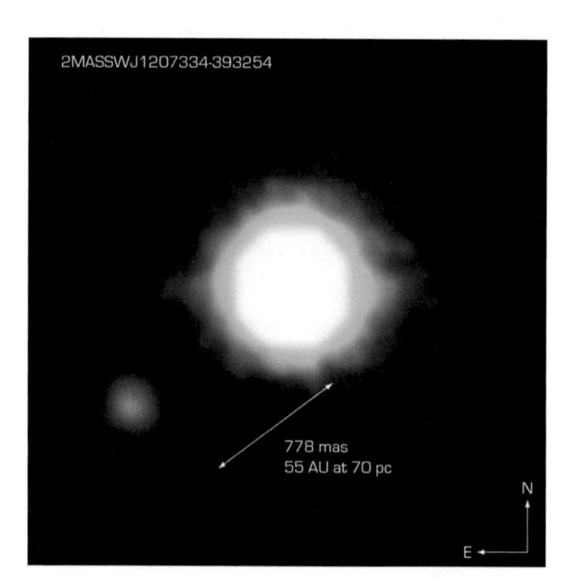

The brown dwarf and its "Giant Planet Candidate Companion", as the astronomers chose to call it.

[10] The team also included three American astronomers.

[11] Planets in an early phase of formation may emit infrared light and thus be detectable in this wavelength domain. This was the case for the object associated with 2M1207.

constellation of Hydra. The planet itself, 2M1207b, was separated from its star by less than 0.8 arcseconds, corresponding to 55 AU at the distance of the star. It is therefore not surprising that it required adaptive optics to detect it. The image was obtained on 27 April 2004 with the NACO adaptive optics camera mounted on the fourth VLT Unit Telescope, Yepun. Alhough ESO had not made such a claim before, given the past history of first images of exoplanets, it was cautious in its first press release, which carried the headline "Is This Speck of Light an Exoplanet?" (ESO/CNRS press release 10 September 2004).

The question now was whether the observed object was gravitationally bound to the star. Subsequent observations with the Hubble Space Telescope followed too soon to provide the answer. However, based on new observations with the VLT and NACO in 2005, confidence grew, leading to a much more assertive science paper and press release ("Yes, it is the image of an Exoplanet", ESO press release 30 April 2005, coordinated with UCLA). Five times as heavy as Jupiter, the largest planet in the Solar System, this distant exoplanet is a cold place, suggested by the presence of water molecules detected by means of spectroscopy. Later observations have confirmed the initial findings, although there is still some argument about whether this is a real planet, because it is likely to have formed in a different way from the planets in the Solar System (i.e. from fragmentation of the initial cloud rather than by accretion). NACO has been used to image other large planets, including the one in orbit around the star Beta Pictoris, as already mentioned in Chapter III-7[12].

Looking at the hundreds of exoplanets that have been found since 1995 it is striking, if not unexpected, that most of them are heavy objects[13]. Some find themselves in extremely close orbits around their parent stars. Others, as the example above suggests, may have formed in a different way from "normal planets". Indeed, few planets come near to what Earthlings would consider normal in terms of looking like the terrestrial planets of the Solar System and being able to host any form of imaginable life. So is Planet Earth — and possibly even life — really an exception? Most astronomers would argue that the reason for the difference between what we see in the Solar System and what we find elsewhere in the Milky Way is simply due to the fact that our current search methods introduce a natural bias. Today's telescopes

[12] Understandably, there was a race to obtain the first image of an exoplanet. A group of scientists at the Friedrich-Schiller-Universität Jena, using NACO, identified an object in orbit around the star GQ Lupi. There was much uncertainty about the nature of this object, so the image of 2M1207 is still regarded as the first direct image of an exoplanet.

[13] This is a natural outcome of the still limited observational capabilities at our disposal.

have managed the incredible feat of confirming that other planets do exist, but they are not able to let us study a representative sample of the planet population to make definitive statements about the distribution of different classes, let alone to study the chemical composition of individual objects. This is a task for the next generation of telescopes that may allow us to provide such answers. But, as we have seen, the current studies have also opened up new questions about planet formation itself. Do all planets form in the same way? Or do we have to revise our current ideas of planet formation? This is perhaps largely a rhetorical question, for such a revision is ongoing as we collect ever more information enabling us to challenge and test these ideas.

We will finish this section on science with the VLT with an example, perhaps not of the most spectacular kind in terms of new science, but which nonetheless provides an impression of the awesome capabilities of the VLT. As we know, cometary nuclei are small bodies in the Solar System, comprised of ice and dust. The American astronomer Fred Whipple famously described them as dirty snowballs. When they get close to the Sun they heat up and gas and dust grains are ejected into the surrounding space forming the comet's coma and the tail. These are the parts that we occasionally see in the sky when a bright comet passes by — such as Comet West in 1976, Comet Halley in 1986, Comet Hyakutake in 1996, Comet Hale-Bopp in 1997 or Comet McNaught in 2007. The spectacular fly-bys of cometary nuclei by ESA's Giotto spaceprobe stimulated the appetite for space missions to comets and in the early 1990s, ESA decided to send a probe, named Rosetta, to intercept a comet and to land on it to extract samples of the comet's pristine material. The target comet selected was Comet 46P/Wirtanen, and to prepare the mission and following a proposal by a joint team from ESO and the Space Science Department of ESA, VLT observations were carried out to study the comet. On 17 May 1999, while the telescope was

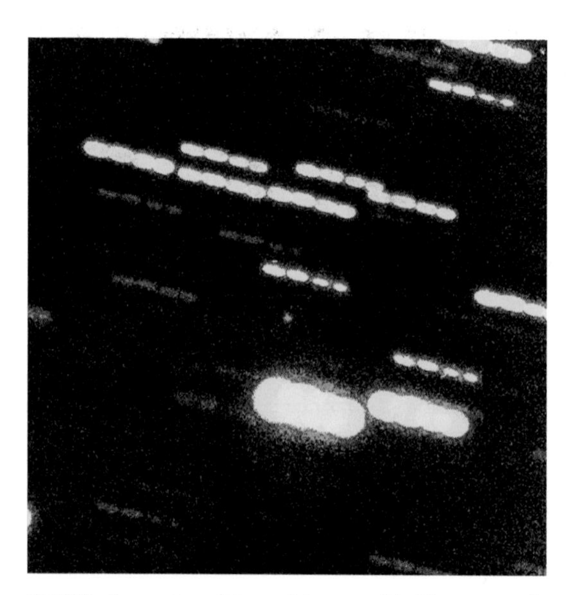

FORS2 observation of Comet Wirtanen (the blue spot in the middle). At the time the comet was about six million times fainter than can be perceived with the unaided eye.

still in its commissioning phase, a series of short exposures was obtained with the VLT Test Camera mounted on Kueyen. At the time the comet was 600 million kilometres (4 AU) from the Earth and 745 million kilometres (5 AU) from the Sun. As the launch date came closer, further observations were made.

On 9 December 2001, several new exposures were made. The comet was now at about the same distance from the Sun as at the time foreseen for executing the landing, i.e. at a distance of 435 million kilometres. Since, by the time of the observations in 2001, the comet was low in the sky and — to be seen at all — had to be observed in twilight, the observations were highly demanding. However, the images, obtained with FORS2 on Yepun, provided a good basis for the mission planners, both by determining precisely the size — barely more than one kilometre — and shape of the target. They also revealed a reassuringly low level of activity on the comet's surface, suggesting that this was not likely to create problems for the lander. Alas, due to launcher problems and the resulting delay, ESA eventually had to move to another target[14], but the example is yet another illustration of the how space missions and ground-based activities can complement each other.

The examples given here offer a kaleidoscopic, if certainly limited and snapshot-like, overview of a few of the important results emanating from research with the VLT. A quantitative assessment is provided by the publication data, collated and analysed by the ESO library and provided every year in ESO's Annual Reports. In 1999, the first publications from the VLT began to appear in the scientific journals. Twenty-nine articles based on VLT observations were registered that year. The number almost doubled during the following year and continued to grow, reaching a peak in 2007 with 494 publications. These numbers must be seen against the total number of publications based on observations at ESO — 348 in 1999 rising to 720 in 2007, where it reached an interim peak (Grothkopf, 2010). After a slight decline in 2008–9, the number of VLT-based papers and ESO-based papers rose again, reaching 783 in 2011.

Let us try to approach the question of the influence of ESO and the VLT on European astronomy in a different way. We will remember that one of the aims was to further extragalactic studies in Europe. For Observing Period 62 (ending March 1999), the last period before the first VLT Unit Telescope was opened to the users, ESO

[14] Instead, ESA selected Comet 67P/Churyumov-Gerasimenko, which was imaged by the ESO NTT on 26 February 2004.

received 250 applications for observing time in the relevant areas (cosmology, galaxies and active galactic nuclei). This was a typical tally in those days and constituted roughly half of the total number of applications. In Period 63, this number rose to 289, in Period 65 to 343. Since then, it has remained at that level, albeit with the usual fluctuations from period to period. These numbers, however, are for ESO as a whole. It is hardly surprising, though, that the overwhelming majority of applications are for VLT time. This is an expression of the attractiveness of the VLT of course, but it is also a result of the lower capacity, and capability, of La Silla. Thus about ¾ of all applications are for the VLT. Perhaps surprisingly, there is a rise in applications for observations in the classical areas of stars and the interstellar medium, in some periods reaching twice the numbers for galaxies and cosmology. The VLT is beyond doubt the observing facility of choice for the ESO community of astronomers.

Some of the examples above have illustrated the importance of coordinating and combining observations with the VLT with those from space observatories or space missions to enable a multi-wavelength study of a specific target. The Great Observatories Origins Deep Survey (GOODS) survey is perhaps the ultimate example in this respect. It covered two fields, each of 150 square arcminutes, around the so-called Hubble Deep Field[15] North and the Chandra Deep Field South (CDFS). We will remember the NTT Deep Field image, produced in 1991. In December 1995, the Hubble Space Telescope was used for ten days to observe a tiny patch of sky in the constellation of Ursa Major with a total of 150 hours of exposure time and combining 342 individual exposures. Deep fields, it seemed, were in vogue. Later, in 1999–2000, after his departure as Director General from ESO, Riccardo Giacconi led a team that produced the Chandra Deep Field in Fornax, a small southern sky constellation, based on a staggering total exposure time of more than 277 hours (one million seconds). The observations were conducted with the Chandra X-ray Observatory, which had been launched in July 1999. These were augmented by optical observations from the VLT, the NTT and the MPG/ESO 2.2-metre telescope at La Silla to enable optical identification of the X-ray sources. Other observatories later carried out further deep observations and so GOODS became possible, as a coordinated attempt to survey the most distant parts of the Universe to detect and study the faintest objects possible and covering a wide range of wavelengths across

[15] Initiated by STScI Director Bob Williams.

the electromagnetic spectrum[16]. In a sense, the GOODS project unified most of the major observational facilities in a common scientific project. Astronomy, perhaps the most international of all sciences, had not only mustered facilities worldwide, but also brought the full multi-wavelength arsenal available to astronomers to bear on a major scientific question.

[16] GOODS brings together observational data obtained with the Hubble Space Telescope, the NASA satellites Chandra and Spitzer, ESA's XMM-Newton satellite as well as the VLT on Cerro Paranal, the Keck and Gemini telescopes on Hawaii and the Very Large Array Radio Telescope in Socorro, New Mexico and others. It was completed in 2009.

On 9–10 December 2004, the President of the Republic of Chile, Ricardo Lagos, and his wife visited Paranal. During this private visit, the President took the time to stay overnight and seized the opportunity to observe with the VLT. In the early 1990s, as Minister of Education, Lagos had been involved in the political turmoil surrounding ESO. Now, he had a chance to familiarise himself with the ESO staff and to see for himself the improvement in relations between the host state and ESO. Here, the President is seen with Jason Spyromilio, at the time Director of Paranal.The picture is a strong testimony to the change in atmosphere. (Photo: Hans Hermann Heyer)

Chapter IV-6

A Love Affair

"The commitment of ESO is and will continue to be the development of astronomy and scientific culture in the country hosting our observatories."

Catherine Cesarsky, on the occasion of the celebration of the 10th anniversary of the Interpretative, Supplementary and Amending Agreement between ESO and Chile.

In the previous chapter, we have looked at what the VLT brought both to ESO's Member States and to the science of astronomy. But the VLT also impacted relations between ESO and its South American host state. Back in 1963, Heckmann had been warmly welcomed in what to Europeans appeared to be a remote country. *"The friendliness with which we were received was unique,"* he wrote (Heckmann, *ibid.*).

Naturally, over the years, both ESO and Chile underwent great changes and it was unavoidable that this would occasionally result in diverging interests. The crisis in the first half of the 1990s, however, proved a healthy shock, forcing both parties to review their partnership and create a new framework in which their legitimate interests could be brought together constructively. There is no doubt that the Interpretative, Supplementary and Amending Agreement of 1996 has played a crucially important role in fostering continued good relations between ESO and Chile. As we have mentioned in Chapter III-10, the agreement included guaranteed access to the ESO telescopes for Chilean astronomers. To enable them to fully exploit this, it also established a fund overseen by a joint ESO–Chilean committee, which simply became known as the Comité Mixto. It disbursed around 400 000 euros per annum. The money was used to support postdoctoral fellowships and professorial positions at the universities, some technical equipment and other smaller items. It also enabled support for the primary and secondary education system, including teacher-training activities, student scholarships and general outreach. In addition, through the ALMA project, Chile receives more than 500 000 US dollars per year for regional and scientific support activities. This support was bound to have a considerable impact on Chilean science and academic life. On 19 June 2006, high-level representatives of ESO and Chile came together in Santiago to take stock of the effects of the Agreement.

Between 1985 and 2005 the number of Chilean astronomers doubled. However, the number of publications increased eightfold and Chilean astronomers were able to use the full amount of telescope time granted. Several concrete examples illustrate the role of the Agreement. At the Universidad de Chile five of the six faculty members received substantial support. Grants enabled lively postdoctoral activity at the institute, as well as supporting the establishment of an instrumentation laboratory for students at the Cerro Calán Observatory. At the Pontificia Universidad Católica de Chile, also located in the capital, Committee support enabled a new faculty position and part of the postdoctoral programme. Funds were furthermore made available for a collaboration in the field of scientific instrumentation with the Department of Astronomy and Astrophysics and the Department of Electrical Engineering. The Católica del Norte University in Antofagasta, located in the same region as the Paranal Observatory, also benefitted from ESO's presence. Firstly, the university used an observing site near Cerro Armazones, prepared by ESO. Secondly, funds allocated by the Comité Mixto enabled the university to engage three academic staff and to conduct an extensive outreach programme. In the southern part of the country, the Universidad de Concepción was able to hire academic staff, acquire up-to-date IT equipment and create a postdoctoral programme of its own. Finally, the Universidad de La Serena had benefitted from the programme, in this case focussing on the training of teachers (ESO/Government of Chile Joint Committee for the Development of Astronomy, 2006).

Chilean astronomy was clearly blossoming. At the same time ESO was increasing its own scientific presence in Santiago. As major administrative and scientific functions returned to the capital in the mid-1990s and the VLT operations got underway towards the end of the decade, the scientific staff at the Vitacura office doubled. The development of an astronomical centre was first overseen by Bo Reipurth, then by Danielle Alloin and, later, by Felix Mirabel[1]. With such a concentration of astronomers, both at Vitacura and at the Chilean institutes, joint seminars and workshops were organised, contributing further to creating an integrated scientific community. The first event was a three-day workshop entitled "Astronomy in Chile" held in

[1] The current leader is Michael West.

From the press meeting in Santiago on 19 June 2006, on the occasion of the tenth anniversary of the Supplementary Agreement. Seen from left to right: Felix Mirabel, who succeeded Daniel Hofstadt as ESO Representative in 2004, Catherine Cesarsky and Leonardo Bronfman, Professor at Universidad de Chile. Speaking is Ambassador Luis Winter, Director of Special Policy for the Chilean Ministry of Foreign Affairs.

April 1995 (Reipurth, 1995). Another example was the Interferometry Week in January 2002. Vitacura became a centre of attraction far beyond ESO, just as the VLT attracted astronomers from all over the world. More than ever, Chile had become a mecca for astronomy. Let us, however, dwell a little more on the importance of the VLT for Chile. The VLT confirmed the outstanding quality of the observing sites in northern Chile, while showing how the substantial logistical challenges could be addressed. Thus, in the relationship between ESO and Chile, the VLT also cast its shadow forwards, leading to the decision in 2010 to place the next-generation optical telescope the E-ELT, on Cerro Armazones, in sight of the VLT.

≈

As the relationship grew in intensity, some began to see it not simply in terms of European scientists using the observational facilities to do their own science, but to value the intellectual exchange and ESO's role as a cultural bridge between Chile and Europe. Of course, in the contemporary Chilean discourse, ESO often stood as a symbol of Europe, but sometimes it was the other way around. In 1999, the year after the VLT first light, Chile was chosen as the international partner country for the Hanover Fair in Germany. This is the world's largest industrial fair with more than 300 000 visitors strolling through 30 large exhibition halls. Being chosen as the partner country is a special honour and Chile understandably wished to use this opportunity to showcase the very best of the country, its companies and products and its attractiveness in terms of foreign investments and tourism. But the Chilean government also wished to highlight cultural and scientific aspects and therefore kindly invited ESO to fit out the entrance part of the 1700 square metre stand. The author worked on this project, together with representatives of the Chilean Ministry of Foreign Affairs and the Chilean Foreign Trade Organisation (ProChile). The result was a 26-metre-long entrance corridor with an inclined floor, leading up to a first-floor cinema area. Giant pictures of the ESO facilities graced one wall and and on the opposite side there were pictures and texts describing the climate and geology of Chile, and relating these to their effects on observational conditions. Three-dimensional exhibits included a model of the VLT. Videos provided information about the telescope and astronomy in general. Visitors entered this corridor and walked beneath a huge backlit mosaic photograph of the Milky Way, before leaving this part of the exhibition. The Chilean pavilion was formally opened on 19 April by the President of Chile, Eduardo Frei. An engineer by background, and as we have seen, personally familiar with the VLT, he clearly took pride in explaining the project to his German hosts.

From the opening of the Chilean Pavilion at the Hanover Fair on 19 April 1999: (From left to right) Rodrigo de Castro, President Eduardo Frei Ruiz-Tagle, Marta Larraechea Bolívar and Massimo Tarenghi. Photo right: The exhibition corridor. (Photos by the author)

The Chilean pavilion presented a modern, forward-looking country, eager to interact with the rest of the world. And it is true that regular visitors to Chile cannot but be deeply impressed by the progress that has occurred in that country, especially since 1990. With an average economic growth of 5.6% per annum, Chile outpaced all other countries in Latin America during the period 1990–2006. The boom created a new middle class and cut poverty, the scourge of most third-world countries. Whilst in 1990, 40% of the population lived below the poverty line, by 2006, it had fallen to 13% (Muños, 2008) This notwithstanding, it is apparent that historically, large social differences — with their potential for tension and conflict — have had a huge impact on the societies of South America. Chile has had its share of problems. It is beyond the scope of this book to deal with these issues, but it is worthwhile looking at the interaction between ESO and its staff and Chilean society, as it evolved over the years, and particularly over the last one and a half decades.

≈

Many European ESO staff members have come to see Chile as their second homeland and in quite a few cases it has actually become the first. While admiring the progress the country was making, they could not ignore the need for help in some places. In the early 1990s, a group of ESO staff came together to organise a charity effort. About 35 ESO personnel agreed to voluntarily support a charity scheme, which set itself the goal of providing help, especially to street children and orphans in Chile. The main beneficiaries of this effort were the SOS Children's Village in Antofagasta, the Kindernothilfe projects in Santiago and the Hogar Santa Clara, a charity in Santiago that supports HIV/AIDS victims. Contributions from staff members, together with money raised through a number of activities, have over the years netted a sum of about 100 000 euros in support of these organisations. To this should be added voluntary work by ESO staff in Chile, not least in connection with the relief efforts after earthquakes, a curse that relatively often strikes this country with force. A report by Arno van Kesteren about the work of the group after the 27 February 2010 earthquake, which reached 8.8 on the Richter scale and hit the area south of Santiago, serves as an illustration of the dedication and vigour of those involved: *"In the village of Cumpeo (located 300 kilometres south of Santiago), six families who had lost their homes have been 'adopted' by our group. The primary purpose was to build new houses. We first selected and bought the construction material. Then a group of ESO volunteers generously donated several of their supposedly 'off-shift weekends' to initiate the construction work. At the time of writing, four families have already moved into their new houses while two others are about to do so. In the area of Doñihue (located 120 kilometres south of Santiago), another group of ESO volunteers worked to improve the thermal insulation of the temporary houses in April. Some complementary construction material has been recently distributed to persons in need."*

Thus what began in 1961 with Jürgen Stock's colour slides and his enthusiasm about the observational qualities of the Chilean desert sites has evolved into a multifaceted relationship with strong bonds of friendship at the professional[2] as well as at the personal level. For some, it has even become a love affair.

[2] As mentioned, in 2005 Daniel Hofstadt was awarded the Order of Bernardo O'Higgins. In 2011, the same honour was bestowed on Hans-Emil Schuster.

Chapter IV-7

A Growing Organisation

"Now being an astronomer in Europe, it is a must to cooperate with ESO, as it is a must for particle physicists to cooperate with CERN, as it is a must for a space scientist to cooperate with ESA. We needed some organisations in Europe, we now have this and we are very happy to see how efficient they are...."

Hubert Curien, French Minister of Science and Technology, speech at the 1990 NTT Inauguration Ceremony in Garching.

In Chapter I-5, we saw how ESO began to organise itself in order to serve the scientific community. Among other things, this implied the establishment of a coherent system of governance. The Council and the Finance Committee were both stipulated in the Convention and the Financial Protocol. These important committees were supplemented by the Scientific Technical Committee, the Observing Programmes Committee and the Users Committee. The latter two committees serve the Director General and the STC is linked to Council as a parallel body to the Finance Committee. In addition, over time many *ad hoc* committees and working groups have dealt with particular issues, mostly related to specific projects. All this has provided a robust system from which ESO and its user community has profited ever since. But adjustments have had to be made along the way, both as regards operations and governance. In a sense, one can say that ESO became a victim of its own success. In such cases, reform is called for.

The growth in the number of telescopes and their attractiveness to an increasingly vibrant scientific community brought with it a growth in the number of applications for telescope time. In the late 1980s, the number of observing applications had reached about 350 per semester, too many for the members of the Observing Programme Committee to deal with. Like the other ESO Committees, the OPC comprised representatives of the Member States, but in 1988 additional members were appointed, described as members at large, to alleviate the pressure on the proposal reviewers. The respite did not last long. By 1993, the number of proposals had reached 500. Thus, under the chairmanship of Joachim Krautter, it was decided to introduce a two-step selection process with discipline-oriented panels to evaluate the proposals before they were discussed at the OPC. Six panels were established, covering the main categories of research — *"(i) Galaxies, clusters of galaxies, and cosmology; (ii) Active galactic nuclei and quasars; (iii) Intergalactic and interstellar mediums; (iv) High-mass and/or hot stars; (v) Low-mass and cool stars; and (vi) Solar System."* (ESO Annual Report, 1993). This classification system, as well as the procedure, has remained in force ever since. Not surprisingly, with the advent of the VLT, the number of observing proposals has kept rising, and is now at a level of 1000 proposals per semester. It would seem as if the thirst for observing time of astronomers is impossible to quench.

The pressure factor on the telescopes, expressing the demand for time relative to the available facility time, is substantial and the Users Committee have kept ESO's focus on expanding its telescope park and on securing smooth operations. But no organisation is stronger than its weakest link. To ensure that ESO could live up to the demands for excellence across the board, Giacconi, introduced the idea of an international visiting committee[1], comprised of scientists from ESO Member States and beyond. They were charged with reviewing all aspects of the organisation. The first Visiting Committee was appointed in December 1994 with Alec Boksenberg (Royal Observatories), Claes Fransson (Stockholm Observatory [Stockholms Observatorium]), Ken Freeman (Mount Stromlo and Siding Spring Observatory), Johannes Geiss (Universität Bern), John Huchra (Harvard Smithsonian Center for Astrophysics), Rolf Kudritzki (Universitätssternwarte München), and Guy Monnet (at the time from Observatoire de Lyon). George Miley (Leiden Observatory) was asked to chair the committee. Since then, with varying composition, the Visiting Committee has become a recurrent feature in the oversight of the organisation, providing important input to the Council, the ESO management and its staff.

We described how the now familiar operational structure of ESO had emerged in the mid-1970s: the European headquarters with strong technical and scientific competences and the observatory in Chile. The facilities in Santiago were largely dormant for twenty years. But times change.

Once the decision to develop a second observing site at Paranal was taken, the question of the role of the Santiago office was re-opened. Most of ESO's activities in

[1] This was in itself not new in the academic world, but it had not been used at ESO before.

Chile, including the administration, had been concentrated at La Silla since 1975. Would it make sense to run Paranal from La Silla in the long run? And, in view of the troubled relationship between ESO and Chile in the early 1990s, would a strengthened presence in Santiago help? In 1994, ESO decided to move the administration back to Vitacura. During September 1994 the Astronomy Support Department also moved its offices from La Silla to Santiago, breathing new life into the Vitacura premises. As the number of staff members grew in connection with the VLT, the former astronomical workshop in Vitacura was refurbished and converted into offices for astronomers. By 2002, the scientific staff in Vitacura comprised approximately 35 staff, 15 fellows (postdocs), five paid associates and ten PhD students/co-opérants (Alloin, 2001). As we discussed in the previous chapter, this had very positive repercussions for Chilean astronomy and illustrates the catalytic effect that ESO's presence, together with that of the American observatories, had in the region[2]. Yet, from ESO's historical perspective the return to Santiago represented something of a *volte face*. Sometimes, it seems, such a *volte face* can be a healthy development. In any event, the old tension between Santiago and "Europe" has not reappeared and both centres exist in a fruitful symbiosis today.

The next stage of the reorganisation followed logically from the relocation of major activities to Santiago, combined with the shift in the centre of gravity that occurred as the VLT moved into pole position as ESO's prime facility. In February 2005, the two observatories, La Silla and Paranal, merged to become one observatory in two sites, under the name the La Silla Paranal Observatory — streamlining both management and operations. In the 1980s and 1990s, the development of the Paranal site had leaned heavily on La Silla for logistical support. Now, Paranal had grown strong and could provide support and help for La Silla, when needed.

≈

Moving now to issues of governance, as the organisation took in more Member States and embarked on new projects, the budget grew accordingly. The main income for ESO is the annual contribution by its Member States. The individual contributions are determined by the size of the economy of the respective countries. This means that, in principle, every country contributes its fair share. The 1962 Convention has, however, built in an upper limit for the contribution of a single country. This was

[2] The Vitacura Centre was further strengthened when the new ALMA office opened in 2010 on the premises.

originally set at one third of the budget[3], but has, by Council decision, been lowered first to 27.6% and periodically to somewhat less. The reason for the upper limit is to avoid a situation in which a single country becomes too influential in the organisation. Historically Germany, Europe's largest economy, has continuously hit the contribution limit, as has France for long periods of time. Despite this exception, the "fair share" principle applies. Taken in absolute numbers, however, the contributions of the different countries vary significantly, from Germany and France at the very top, to Finland, Portugal and the Czech Republic as the smallest contributors among the current Member States. Based upon the idea of the "fair share", Article V, Paragraph 4 states that *"each member state shall have one vote in the Council"*. This ensures equal treatment and equal influence on decisions and also extends to decisions on the budget. But even if there are great similarities between the Member States, their national economies are not necessarily in step. This can lead to different views on the budget and, in some cases tension between the few large countries and the more numerous smaller ones. This problem became apparent in the course of the 1990s under the financial strain imposed by the budgetary belt-tightening of many countries, but also in the face of increased interest of new countries in joining the organisation. In 1995, Council reached a compromise solution, implementing the principle of weighted voting on a trial basis. The idea would be that while the small countries should not be able to outvote the two largest contributors, the latter should not be able to impose their will on the smaller countries either. At the same time, however, a stalemate should also be avoided, to ensure that the organisation could function.

This implied that endorsement of the annual budget would require a "double majority", meaning a majority both among the Member States and the yes votes securing contributions of at least 55% of the funding. This was implemented for the Finance Committee, which *recommends* the budget. The *decision* on the budget, however, is the prerogative of the Council, for which the Convention continues to apply. Initially, the decision was for a trial period of three years, but it remained in force until the end of 2003. In 2004, it was modified somewhat, but the basic idea was retained. According to the revised practice, *"A double majority procedure shall apply to the Finance Committee recommendations to Council.... The percentage of contributions necessary to reach the financial majority under this procedure is set at a level of more than the total contributions of all Member States (100%) reduced by the sum of the percentages of the two largest contributors, according to the scale of contributions in force at the time the vote takes place."*

[3] Article VII, para. 1 *c)*.

≈

Membership of the STC was originally based on the idea of national representation, as it was for Council and Finance Committee. In 2005, however, the composition of the Scientific Technical Committee was changed *"to ensure broad coverage of all the range of scientific disciplines required for VLT, VLTI, ALMA, and ELT"*[4] (ESO Annual Report, 2005). For a while members would be nominated not by their national constituencies and as representatives of those, but by a dedicated nomination committee, with the membership confirmed by Council[5]. It seemed like a natural follow-up to the changes in the OPC, reflecting the growing complexity of ESO's programmes, but also, in a deeper sense, revealing the growing perception of a unified European scientific landscape, rather than a group of national ones[6].

---

[4] Following a recommendation by the 2004 Visiting Committee.

[5] This procedure was later simplified, but the focus on securing the widest possible set of expertise was retained. Whilst national *nomination* was reintroduced, the selection is made by the Director General, subject to confirmation by the Council.

[6] It was the same thinking that enabled the establishment of the European Research Council as a pan-European funding mechanism for scientific research, based on scientific excellence, only, and without any national selection mechanism.

Chapter IV-8

The Surge

"The VLT changed everything."

Gerry Gilmore, Speech at Lancaster House on 8 July 2002
on the occasion of the UK accession.

There were five founding countries of ESO — Belgium, France, Germany, the Netherlands and Sweden. In 1967, Denmark joined. Many years passed before the next countries joined — Switzerland and Italy, both in 1982[1]. Once more, a long period would pass with ESO in a stable situation as regards Member States, but from 2001, things changed dramatically. Between 2001 and 2008 ESO took in six new Member States, in the following order: Portugal (2001), the United Kingdom (2002), Finland (2004), Spain (2006), the Czech Republic (2007) and Austria (2008)[2]. There were a number of reasons for this surge in membership. Undoubtedly, the success of the VLT played a major role, as the quote by Gerry Gilmore suggests. Aside from the research perspectives opened up by the VLT, it also accorded ESO credibility as regards the future. It seems that ESO's involvement in the ALMA project was equally important, and helped the entry of both the UK and Spain. ESO's future projects were of great interest both to scientists and industrialists in many countries and stimulated a strong desire to become part of them. But there was more to this surge. The early years of the first decade of the twenty-first century saw an unprecedented move towards European integration. This went far beyond the issue of science, but it was accompanied by a growing recognition among governments that science and technology played a central role in the creation of wealth. Among other things, this manifested itself in many countries by an increase in their investments in research

1 To join ESO a country must accede to the ESO 1962 Convention and its associated protocols. This obviously requires a decision by the Government in question backed up by a confirmation by the parliament of the country, in the process known as ratification. Only when the ratification documents have been presented (to the French Ministry of Foreign Affairs, which holds the original of the Convention), will the country formally become a member.

2 During that period early, informal discussions also took place with almost ten additional countries, some of which may, ultimately, lead to membership.

activities[3]. This in turn became an important element of the overall political development towards the creation of the European Research Area. As regards ESO membership, it is probably not possible to assess how much the ERA development actually meant *vis-à-vis* the other factors, but there is no doubt that it led to an overall political understanding of the value of European collaboration in research and therefore also provided important arguments in the political decision process that underpins government actions. The tailwind for science was notable. But to obtain a full appreciation of what this development meant for ESO, we once more need to acknowledge the human factor. Again and again we see how young scientists — from outside the original Member States — became familiar with ESO and its projects, and how seeds were planted that would ultimately blossom, sometimes twenty years or more later. This is a fact of crucial importance, for it takes dedicated people to join up the technological, scientific, political and strategic dots that become apparent in the evolution of nations. It is therefore not surprising at all that we shall see a number of individuals play decisive roles in helping their countries join the organisation.

≈

Article VII, paragraph 3 of the ESO Convention proscribes that *"States becoming Members of the Organization after the date on which the Convention comes into force shall be required to make a special contribution representing their share in capital investment and fitting-out cost already incurred."* As the observatory grew and the sums representing the capital investment became considerable, this created a major hurdle for countries wishing to join. ESO's policy deviated from most other organisations in this respect, but it is important to remember that when a country joined, the community of scientists obviously also grew. It was therefore important for the organisation to have the necessary resources to serve the enlarged community properly. Thus we remember that the special contributions made by Italy and Switzerland were used to pay for the NTT. On several occasions Council reaffirmed two important decisions: that the special contribution would be used to increase the budget (rather than lowering the contributions of the existing Member States), and, importantly, that a calculation of the capital investment would only comprise those facilities that were of real scientific significance at the moment of accession. Furthermore, ESO's facilities would also depreciate with time and this would be accounted for in

[3] Thus, in 2003, the European Council — the meeting of Heads of Governments and States of the EU — agreed to raise their spending on research to 3% of the Gross Domestic Product (GDP) to be reached by the year 2010. This became known as the Barcelona Decision. Whilst not achieved within the foreseen timeframe, it has been retained in the Europe 2020 Strategy.

the calculation of the capital investment considered. In round numbers, during the first decade of the new century, the relevant capital investment was around one billion euros. The special contribution was thus based on this sum, but the amount to be paid depended on the GDP — or to be precise the Net National Income (NNI) — of the candidate country relative to the equivalent number for the existing Member States. Joining ESO constituted a major effort on the part of the candidate country, and required the mobilisation of political will.

We will now try to trace the main steps towards membership that led to the 2001–2008 surge (the accessions of Switzerland and Italy have been described briefly by Woltjer in his 2006 book). Although the membership increase happened at quite a pace, it was built on a long — sometimes decades long — sequence of contacts, informal exchanges and discussions and, from a certain point in time, more formal interaction. Each of the stories is a story of hope, frustration, sustained lobbying and, mostly, the attainment of the coveted goal. But this was not always the case. The first example would, in fact, appear to be a failure.

Australia

Historically, the main observatories in the southern hemisphere have been located in South Africa and in Australia. It was only in the 1960s that Chile, with its superior observing sites, was added to the list. Modern astronomy in Australia was vibrant. The scientists had advanced optical telescopes[4] and some of the most impressive facilities for radio astronomy as well, including the Australia Telescope, which was opened in September 1988. The success of Australian astronomy was not only rooted in the historical fact that the continent had been the home of observational facilities since the late 18th century, but also because in the second half of the 20th century, there had been a tacit understanding among the twin communities of optical and the radio astronomers that large investments would alternate between the groups. Standing together had helped both communities secure the necessary funding. In the 1960s, the 64-metre Parkes Radio Telescope had been the pride of the community, in the 1970s, the Anglo Australian Telescope had been built and in the 1980s, it was the turn of the Australia Telescope, as already mentioned. However, the two main sites for optical telescopes, Mount Stromlo near Canberra and Siding Spring

[4] Including the 3.9-metre Anglo Australian Telescope, commissioned in 1974, two years ahead of the ESO 3.6-metre telescope and now part of the Australian Astronomical Observatory.

in the Warrumbungles (New South Wales), were far from ideal, and so, as the next generation of very large telescopes — the 8–10-metre-class telescopes — were being planned in the US and in Europe (but to be placed in Chile), Australian astronomers became interested in participating in one of these projects[5]. Their interest coincided with ESO's interest in expanding its membership base, and in reinstating the VLTI programme that had been postponed in 1993. And so, towards the end of that year, Giacconi contacted Jeremy Mould, then Director of the Mount Stromlo and Siding Spring Observatories.

In early 1994, Giacconi paid a visit to Australia. Aside from seeing the Australian astronomers at Mount Stromlo, he met the Federal Minister for Science, Christopher Cleland Schacht, at Canberra's conspicuous Parliament House. Schacht appeared to be favourably disposed towards an ESO bid[6], and, given this head start, Australian astronomers began preparing their case for joining ESO. This coincided with an ongoing one-off competition for extra funding among all major national research facilities and across disciplines, resulting in an investment plan entitled Innovate Australia. The key scientists behind the ESO proposal were Ron Ekers, Director of Commonwealth Scientific and Industrial Research Organisation's (CSIRO) Australia Telescope National Facility, Lawrence Cram of the University of Sydney and Jeremy Mould. In June 1995, Jeremy Mould attended an ESO Council meeting at Garching and presented the Australian ideas, leading to a green light from Council to continue the process. To further assist in the efforts, an ESO team comprised of Peter Shaver and the author visited Australia in July 1995. They took part in a policy event in Canberra and a public exhibition was held at the national Australian science centre, Questacon. At the ESO Council meeting held in Milan on 28–29 November, the issue was again on the agenda.

Before we continue with the chronology of the Australian attempt to join ESO, let us quickly address an obvious question that uninitiated readers may well pose: for all its magnificent qualities, Australia can of course hardly be considered to be part of Europe. Could a non-European become a member of a "European" organisation? Although the Convention determines the name of the organisation (containing the word European), perhaps somewhat surprisingly, it does not specify any

[5] In an Australian Science and Technology Council (ASTEC) report of 1992, possible participation in either the Gemini or the Magellan Telescope Project with two 6.5-metre telescopes at La Campanas in Chile was mentioned.

[6] Unfortunately, the minister was replaced shortly afterwards and his successor seemed much less enthusiastic about ESO membership.

geographical limitations for the Member States. Discussions had already taken place about a possible Chilean membership. Council was, in other words, open to the idea of non-European Member States. This was re-confirmed at the Milan meeting, in which Council adopted a resolution stating that *"The admission of ... non-European states is desirable provided that the European identity is not altered."* It appeared as if Australia might be on track to join within a short space of time. Alas, towards the end of the year, although highly ranked in the peer review process, the proposal by the Australian astronomers was not selected by their government. Then a somewhat ambiguously phrased letter, dated 14 December 1995, arrived from the Australian government. The government had decided to fund an upgrade for the Australia Telescope in Narrabri, the letter said, *"but not to provide additional funding for participation in the ESO VLT project immediately"*. Nonetheless the minister had appointed a negotiating team to meet with ESO. The ESO Council responded by setting up its negotiation team and a tentative date of 5 February 1996 was set for the first meeting, but it never took place. And then, in March 1996, parliamentary elections led to the ousting of Paul Keating's Labour government. On 23 May 1996, the incoming government of John Howard finally withdrew Australia's bid, citing budgetary problems. In the end, Australia joined the Gemini Telescope project with twin 8-metre telescopes, one located in Hawaii, the other in Chile.

Despite disappointment on both sides, a link remained: in June 1998 Australia was awarded a contract for the delivery of the fibre positioner (OzPoz) for the VLT, mentioned in Chapter III-2. Even if the bid of 1995 failed, interest in Australian membership has remained. This book is about ESO's past, not its future. As with other countries that have wished to join, the road can be long and difficult, and even at times appear to have reached a dead end, only to be opened up again at an opportune moment.

Portugal

With the Carnation Revolution of 1974, which led to establishment of the Third Republic, Portugal began a major political reorientation. The African colonies were given independence and in 1986, Portugal joined the European Communities. European integration brought with it rapid economic growth, due to support from the European Communities, but also because Portugal, with its low wages, appeared to be attractive for low-tech industrial production. On its own this was hardly a sustainable model for growth and a cohesive society. The reorientation therefore also aimed at freeing the country from the stagnation of the past regarding investment in scientific and technological infrastructure and in highly qualified human resources.

As a seafaring nation, Portugal could look back on a long astronomical tradition. The Observatório Astronómico de Lisboa was established in the 19th century, inspired by the observatories in Helsinki and St. Petersburg. Yet astronomy — along with other sciences — had suffered from serious underinvestment and lack of perspectives for decades. By 1986, Portugal had only three professional astronomers at PhD level. One of them, Teresa Lago, decided to make a bid to change this. She started to develop astronomical studies at the Universidade do Porto in 1984, and as part of this initiative, she contacted José Mariano Gago, who at the time headed the Portugal National Board for Science and Technology (JNICT)[7]. He was receptive and encouraged Lago to develop a formal proposal for how Portugal could develop a long-term programme for astrophysics. Together with her colleagues and the support of JNICT, she initiated a survey of the existing activities and subsequently developed a five-year development plan. And, importantly, she also set up a small team of international advisors with Françoise Praderie, Jean Heyvaerts and Alec Boksenberg as members. Following an international review in 1987, the proposal was approved by Luis Valente de Oliveira, then Minister for Planning and Science and Technology[8]. The plan provided a framework for the further evolution of Portuguese astronomy. Its proposals included initiating a dedicated formal training programme at doctoral level, the creation of positions for postdocs and support for research institutes. But it also recognised the need to establish close links with international facilities, notably the observatories in the Canary Islands or ESO. In 1987, Lago visited Woltjer and Giancarlo Setti, then Head of the ESO Science Division, at ESO. Woltjer, in his final year as Director General, was sympathetic, but referred the issue to his successor Harry van der Laan. Van der Laan was equally positive and set up a joint astronomy working group with Setti and Marie-Helène Ulrich for ESO and Teresa Lago and João Dias de Deus for Portugal as members. The charge to the working group was to *"survey ... potentialities for the development of Portuguese astronomical research in view of a possible membership of Portugal"* (Setti *et al.*, 1988). Meanwhile, on 30 November 1989, the Portuguese Secretary of State, José Pedro Sucena Paiva, together with Lago, visited the ESO Headquarters. The visit bore fruit in terms of securing the full support of the Portuguese government, whilst the working group report

[7] Junta Nacional de Investigação Cientifica e Tecnológica.

[8] A significant step was the establishment of the first research centre in astrophysics in Portugal, the Centro de Astrofisica (CAUP) in Porto in 1988.

demonstrated the growth potential. The ESO Council therefore agreed to consider a special arrangement that would support the development of Portuguese astronomy, but which also aimed at membership of the organisation within a period of ten years. On the suggestion of Sucena Paiva, Lago made contact with Fernando Bello[9], at the time Director of Projects and Planning Department at JNICT. Together, they formed the driving force behind Portugal's bid to join the ranks of highly developed science nations as regards astrophysics.

In December 1989, the ESO Council gave its blessing to the agreement, and so, on 10 July 1990, a Cooperation Agreement between ESO and Portugal was signed by Sucena Paiva and Harry van der Laan. The Portuguese government committed itself to support the development of its astronomical potential. In practical terms the government would invest a steadily increasing amount in national efforts that would essentially be equal to what its annual subscription to ESO would have been. In return, Portuguese astronomers would gain immediate access to ESO facilities and Portugal would be able to observe ESO committee meetings[10]. In a sense, this amounted to an associate membership of ESO for Portugal. It was a historic deal, both for Portugal and for ESO. In fact, the ESO Council has traditionally declined any suggestion of associate membership, with the exception of this particular case. To mark the agreement, ESO mounted a major exhibition in September 1990 in Porto and subsequently in Lisbon.

To monitor the implementation of the agreement, a joint Portugal/ESO Consultative Board was established, with Peter Shaver and Richard West representing ESO, Teresa Lago and Fernando Bello, Portugal. The effect of the agreement was quickly noticeable, with new research projects, fellowships and infrastructure investments. Since the late 1980s, Portugal had made huge efforts not only to strengthen its science base, but also to foster the public awareness of science[11]. In 1994, the year that Lisbon was European Capital of Culture, CERN and ESO, in partnership with the Portuguese authorities, created a 2000 square metre exhibition under the name Infinitus at the former Tejo power station. The exhibition was opened by the Portuguese Prime Minister at the time, Aníbal Cavaco Silva.

[9] Bello would later act as negotiator for Portugal *vis-à-vis* ESO and subsequently as Portuguese delegate to the ESO Council, together with Teresa Lago.

[10] The first Council meeting attended by a Portuguese observer was the landmark meeting in December 1990 where Paranal was selected as the site for the VLT.

[11] This led to the creation a dedicated organisation, Ciência Viva, which later became an effective partner for ESO's outreach activities.

From the opening of the Infinitus Exhibition: Richard West, the Portuguese Prime Minister, Vitor Constâncio (President of Banco de Portugal and of Lisbon European Capital of Culture) and Peter Creola passing by a 1:50 scale model of the VLT. Also seen is Werner Kienzle from CERN. (Photo by the author)

The following year, a mid-term report prepared by JNICT was clearly encouraging: Portugal had spent more than its "ESO contribution" during the first five years and doubled the number of people in astronomy. By then CAUP alone had 21 PhD students being trained abroad and five in Porto. Furthermore four PhDs had been concluded abroad and one in Porto. It also confirmed that cooperation with ESO was seen as *"fundamental for the future of astronomy in Portugal"*[12]. Portugal was gradually moving towards a situation where it could benefit from full ESO membership[13].

So, towards the end of the decade the time had come to discuss the full membership. First discussions between Giacconi and the various Portuguese stakeholders had already occurred in 1997, but, in view of the impending expiry of the Cooperation Agreement, they intensified in 1999. A meeting on 13 April 1999 led to the signing of a Statement of Intent by Gago and Giacconi. In parallel a series of events were organised. These included a workshop entitled Portugal — ESO — VLT at the Lisbon Museum of Science, mainly for young Portuguese scientists; a public exhibition and lectures and a press conference with the Minister of Research and ESO's Director General. The Minister was now José Mariano Gago, who had not only

[12] Report by the Portuguese observer, Fernando Bello, to the ESO Council.

[13] In fact, over the ten-year period, the number of professional astronomers increased from three to about sixty.

worked hard to enable his country's scientific community to catch up, but was also a driving force on the European science policy stage as one of the key figures behind the idea of a European Research Area, which had been endorsed by the European Council only a few weeks earlier[14].

Discussions resumed in Lisbon in November, now between Catherine Cesarsky, as new Director General, and Gago. A further meeting happened on 7 March 2000, when a basic mutual understanding about the terms of accession was reached. With the successful conclusion of the negotiations, Gago and Cesarsky signed the formal accession agreement on 27 June at a ceremony at the ESO Headquarters. To mark the new membership the ESO Council held its summer meeting in 2001 on 18–19 June, in Porto, at Teresa Lago's University.

The Accession Agreement was followed up by two additional agreements. An agreement with the Portuguese Ministry of Science and Technology foresaw a continuation of the Portugal/ESO Committee for several years. The other, with the Portuguese Innovation Agency, allowed Portugal to send young "technical graduates" for a period of one to two years to work at ESO. A similar agreement had been concluded with CERN a few years earlier. These agreements have proven to be very beneficial both for the organisations and for the continued Portuguese efforts to catch up with some of the technologically more advanced countries on the continent.

United Kingdom

The connection between ESO and the United Kingdom began, as we have already seen, before the organisation came into formal existence. Harold Spencer Jones, the Astronomer Royal, took part in the initial discussions about the common southern observatory. However, with his retirement in 1955, British interest faded. Influential British astronomers such as Fred Hoyle, Richard Woolley, Roderick Redman and others saw greater prospects for UK astronomy in a Commonwealth project, which led to the establishment of the Anglo-Australian Observatory in New South Wales, Australia. Few countries in Europe can look back on a history of astronomical research more impressive than that of the UK. The early loss of the UK as a partner in ESO was therefore undoubtedly painful, but somehow as we saw in Chapter I-2

[14] Under the Portuguese EU Presidency.

it was water under the bridge. In any event, the UK went its own way and expanded its engagements not just in Australia but also in Hawaii and in Spain[15].

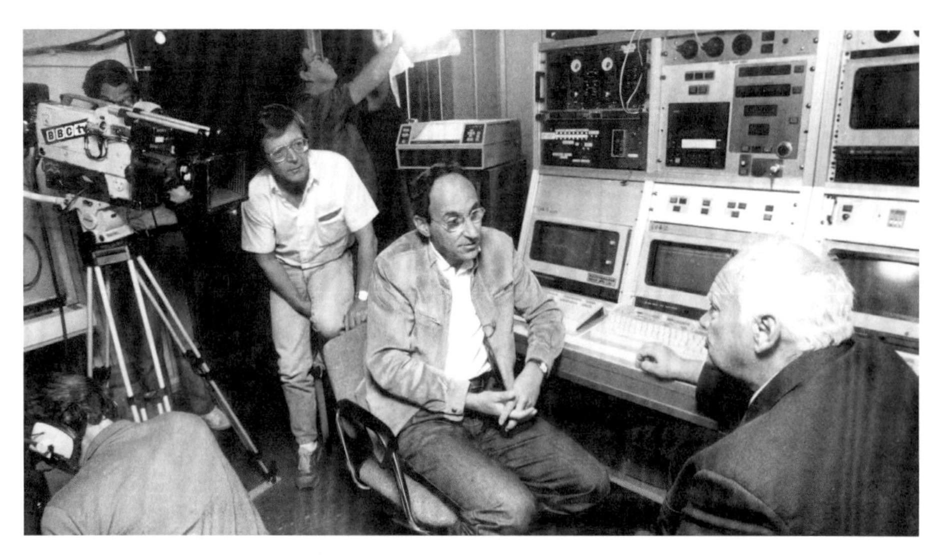

In 1989, Sir Patrick Moore, possibly the most famous astronomy populariser ever, visited La Silla to make two programmes for the BBC series The Sky at Night, *as well as a film for ESO. Here, he is doing an interview with Jorge Melnick in the 3.6-metre control room. The man behind is the long-term producer of* The Sky at Night, *Pieter Morpurgo. (Photo by the author)*

By the mid-1980s, as the new generation of 8–10-metre class was being planned, an "8-metre task group" had been established in the UK. Towards the end of the decade, Ian Corbett, who at the time served as Head of the Astronomy Division at the UK Science and Engineering Research Council[16], was involved in the discussions together with the other members of the task group. It seemed clear that, unlike in the days of the 4-metre-class telescopes, the UK would not be able to build its own telescope and therefore the attention was focussed on possible partnerships. Corbett had informal contacts with ESO, first with Woltjer, then with his successor van der Laan, but these contacts did not lead anywhere. Instead, it appeared that, given the financial constraints, the UK would not be able to afford membership of ESO and, in view of the many telescopes accessible to UK astronomers on La Palma and Hawaii,

[15] A link between the UK and ESO was retained, however, in the shape of the joint survey of the southern sky, mentioned in Chapter I-7.

[16] Corbett would later become Head of the ESO Administration and, subsequently, Deputy Director General of ESO.

it would be difficult to justify buying into La Silla. The two alternatives for the UK that appeared to be viable were participation in an American telescope project (which evolved to become the Gemini telescopes) or in the very large telescope project in Spain that ultimately became the GTC (to which we shall return shortly). In the end, in 1994, the UK became a partner in Gemini. But British astronomers never left the much more powerful VLT out of their sights. By that time, the VLT was progressing and, to Corbett and many of his colleagues, it appeared increasingly attractive, not the least because of the strategic thinking that underpinned ESO's project. Once again, Corbett undertook informal conversations with ESO, now under the leadership of Giacconi, and once more, it seemed that they failed to find a basis for more formal talks.

Then, almost simultaneously two important developments occurred that would change the relationship between the UK and ESO.

The first was to be repeated shortly afterwards in countries like Finland and Spain: the effect of the emergence of the joint European/American millimetre-wave project LSA/MMA (which evolved into ALMA), in which the radio astronomy community of the UK had a strong interest. When it became clear that the two projects would merge and that ESO might play a leading role at least in the European context, this contributed to stimulating interest in joining the organisation. Thus ESO, traditionally seen as the optical astronomers' organisation, became attractive to the wider community. In April 1998, Corbett — now as Deputy Chief Executive — sent a letter on behalf of PPARC, the successor organisation to the SERC, formally declaring the interest of the UK in participating in *"a mm/submm array project"*[17]. This led to a meeting on 13 May between Ian Corbett and Paul Murdin, representing PPARC, and Giacconi, Richard Kurz, Peter Shaver and Norbert König[18], in which *"the meeting concentrated on PPARC participation* through *ESO."*

The second development had its origin, not in astronomy, but in politics. In 1997, Tony Blair had taken over the reins of government in the UK. An important goal for the new Labour government was a stronger, more constructive participation by the UK in European matters. Furthermore, it was open to the idea of increasing the spending on science and technology, based on *"several reports [that] found that a shortage of state-of-the-art research equipment was making it difficult for universities*

[17] Letter to the ESO Director General, 29 April 1998.

[18] Head of the ESO Administration at the time.

in the UK to remain at the forefront of international research." (*Physics World*, 1999). Under the heading "Making UK science fit for the Millennium", the 1998 Comprehensive Spending Review declared that *"the Government will invest a total of more than GBP 1.1 billion to transform the science base, after years of under-funding, in a unique partnership with the Wellcome Trust."* (UK Government, 1998). And so — in the year of the VLT's first light — the government, together with the Wellcome Trust, the world's largest biomedical research charity, established a Joint Infrastructure Fund (JIF). The participation of the charity might suggest a certain bias towards the life sciences, but the fund actually supported a broad range of projects. In fact, one of the projects that received funding during the first round was the VISTA project (see Chapter IV-3). The UK participation in ALMA and the VISTA telescope, which would soon be looking for a home, created a unique opportunity for a new approach to ESO. At PPARC both Ken Pounds, its chief executive officer, and Corbett saw this. Pound's successor Ian Halliday also gave strong support to the idea. This time things looked more promising. The proposal received a strong boost when Corbett and others from PPARC, riding on a Eurostar train back from an ESA Council meeting at ministerial level in May 1999, had the opportunity to broach the idea with the Undersecretary for Science, Lord Sainsbury of Turville. In a follow-up meeting with Sainsbury and John Taylor, a senior civil servant, PPARC received a green light to explore the conditions for UK membership. Thus, on 22 November 1999, Halliday and Cesarsky met at ESO to discuss possible membership. The ball had begun to roll.

British astronomers, who still had to wait another two and a half years before the 8.1-metre Gemini South saw first light, were not the only ones who had noticed the initial success of the VLT. The first results obtained with the VLT had also created echoes in the world of politics. Thus, in March 2000, one year after the high-profile inauguration of the Paranal Observatory, a debate about ESO membership took place at Westminster in the House of Lords. In a passionate speech, Simon Brooke Mackay (Lord Tanlaw) said: *"If this country fails to subscribe [to ESO] ... this will mean that astronomy will hardly be worth pursuing as a career in Britain in the new millennium. Does the Minister agree that this will be a most unsatisfactory situation, especially in view of what the Government says about the need to encourage science in our schools coupled with the keenness to participate in Europe?"*[19]

[19] As in other cases, ESO's information service was active in support of UK membership. Public activities included presentations (with an audience of almost 1000 people) at the annual Astrofest event at the Kensington Town Hall, presentations at the British Astronomical Association and close links with the UK media, not least the BBC.

On 24 May 2000, a first formal meeting took place between the UK and ESO in Garching. At this stage, it was still an exploratory discussion. The core of the proposal was that the money for the UK participation in ALMA would be spent towards the special entrance fee. Furthermore, the VISTA telescope would be delivered to ESO as an in-kind contribution. Since it was clear that the survey telescope would only be ready much later, deep infrared sky survey data obtained with the Wide Field Camera at UKIRT[20] (on Hawaii) was included as a UK in-kind contribution.

Further informal conversations followed at the 24th IAU General Assembly in Manchester in August of that year. While it was obvious that the details would take some time to work out, the UK government felt confident enough in November 2000 to officially announce its intention to join ESO[21]. Shortly afterwards, Taylor sent in the formal application for membership. As in the case of the other acceding countries, months of negotiations followed, but by the autumn, a deal had been reached. The UK formally declared its agreement in November 2001 and the ESO Council approved it at its meeting on 3 December of the same year. The accession, which became effective as of 1 July 2002, was marked by an ESO Council meeting in London, including an evening reception at Lancaster House as well as a press event in the historic Octagon Room of the Royal Greenwich Observatory. This was just three

From the press meeting at the Royal Greenwich Observatory. From left to right: Roy Clare, Arno Freytag (President of the ESO Council), Lord Sainsbury, Gerry Gilmore, Ian Halliday (Chief Executive Officer of PPARC), Catherine Cesarsky and Pat Roche. (Courtesy: PPARC)

[20] United Kingdom Infrared Telescope.

[21] UK Announces Intention to Join ESO (ESO press release 22 November 2000).

months and four days short of 40 years after the signing of the ESO convention in Paris. Probably for that reason, at the press conference, a journalist posed the unavoidable question: *"If ESO is the answer, why did it take 40 years for the UK to ask the question?"* Gerry Gilmore provided the answer: *"The real change that has happened in the last few years was the VLT.... Everybody said: Wow! They've just jumped ahead of us. These guys are good. We better get on-board!"*

Finland

Finland became the third country in the surge. Finnish astronomy had a proud astronomical tradition linked to the Helsinki Observatory (Helsingin yliopiston Observatorio). In fact the first observatory was built in Turku (Åbo) in 1819, but after a fire in 1827, the university was moved to a new location in Helsinki (Helsingfors) and a new observatory was built. When it opened in 1834, it was a state-of-the-art observational facility. The most famous astronomer at that time associated with this observatory, was Friedrich W. A. Argelander, but perhaps the most conspicuous contribution to observational astronomy was its participation in producing the Astrographic Catalogue in the late 19th and early 20th century, which itself was part of the famous Carte du Ciel project. Besides Helsinki, astronomy was revived in Turku in the 1920s and later also taken up in Oulu in the north of the country. Significantly, Finnish astronomers had also become involved in millimetre-wave astronomy with the 14-metre "radio" telescope, operated by the Helsinki University of Technology (Teknillinen Korkeakoulu) at the Metsähovi Radio Observatory, established in 1974, and located at Kirkkonummi, outside the Finnish capital (Mattila *et al.*, 2004). With growing activities in Turku, Oulu and Helsinki, a new generation of astronomers was emerging.

One of them was Kalevi Mattila, who in 1980 was involved in a research collaboration with Gerhard Schnur, a German ESO staff astronomer. Their goal was to measure the optical extragalactic background light, and as part of this project, Mattila visited ESO's science group in Geneva to scan ESO Schmidt plates. He was also introduced to the IHAP image processing system that we have mentioned in Chapter I-10. This was new and Mattila was duly impressed. Soon afterwards he was appointed professor of astronomy at the University of Helsinki (Helsingin yliopisto). In this function he chaired a working group that considered *"Finland's possibilities to participate in international astronomy projects."* (Mattila *et al.*, 1982). The cover of their report featured a picture of the globular cluster Omega Centauri, obtained with the

Danish 1.54-metre telescope at La Silla[22]. The scope of the report was shown by the mention of two international projects of interest to Finland: the proposal for a 2.5-metre Nordic Optical Telescope to be placed at La Palma and the new project for a submillimetre telescope driven by Swedish astronomers, originally conceived as a Nordic project but eventually becoming the 15-metre SEST. Aside from the scientific potential, Nordic cooperation in general was in high political standing in Finland, since it had been one of the options open to this country, which for many decades had lived uneasily in the shadows of its mighty neighbour, the Soviet Union. While, as we have mentioned, Finnish astronomers had already become deeply involved in millimetre astronomy, the optical telescope offered interesting opportunities for a Finnish in-kind contribution in terms of the optics for the NOT. However, the Finnish involvement in the NOT required more resources than originally thought, and thus Finland had to restrict its involvement in the SEST to a bilateral deal with Sweden, giving Finland a 5% stake in that telescope. Perhaps not a lot, but given that the SEST could be operated 24 hours a day, Finnish millimetre astronomers found themselves with plenty of observing time. This involvement also increased the interaction with ESO scientists and technicians considerably and raised interest in joining the organisation. Finnish optical astronomers opted for NOT participation, feeling that fully-fledged ESO membership would be unrealistic at the time.

There is a somewhat curious aspect to this: whereas interest in ESO activities was normally very strong among the community of optical astronomers in most countries, it has been rather low among radio astronomers. However, in the case of Finland, as its entry into ESO went mainly through the SEST participation, it had strong support from the growing field of millimetre-wave astronomy. Besides, many of the astronomers were working in both wavelength regions. Even so, it would take another 20 years, before membership became possible.

In the early 1990s Finland experienced a near collapse of its economy. Deregulation of the banks created a bubble economy based on foreign debt. When the communist Eastern Bloc fell apart, this meant the loss of major trading partners for Finland. In those dark days, the Finnish government took the conscious decision to invest heavily in education and science as a way towards economic recovery. Since then Finland has seen the highest spending on R&D in relation to its GDP among all the European countries, and in fact in the world.

[22] This had a Finnish connection since the corrector lens for that telescope (to increase the field of view) had been produced by the Tuorla Observatory.

As the economy improved, the Tuorla Observatory became interested in participation in the project to build the 10-metre telescope on La Palma, the GTC. In response, however, the Finnish Ministry of Education set up a working group in September 1998 to look in a more general way at *"megaprojects in astronomy"*. The group was chaired by Mirja Arajärvi of the Ministry of Education. Other members of the group included Mattila, Mauro Valtonen (from the Tuorla Observatory), Merja Tornikoski, from the Metsähovi Radio Research Station, representatives of the space science community, of TEKES (the Finnish Agency for Technology and Innovation), the Academy of Finland and finally Risto Pellinen, who represented Finland in ESA.

The report of the working group was published in 1999 and concluded that *"Finland [should] launch negotiations about membership in the European Southern Observatory (ESO)."* However, it continued, *"should this fall through, Finland should explore possibilities to join the Canary Island GTC project and the ESO LSA/MMA radio telescope projects."* (Finnish Ministry of Education, 1999). Clearly, a major advantage of ESO membership, from the Finnish perspective, was that it would offer both the optical and the millimetre-wave astronomical communities interesting perspectives. In addition it was found to bring synergies with ESA, of which Finland had become a full member in 1995. The proposal fell on fertile political ground. Finland had also joined the European Union in 1995 and the overall attitude was favourable towards international and especially European cooperation. Finland had furthermore followed the discussions in the framework of the OECD Megascience Forum and appreciated the value of participating in major research infrastructure projects. Mirja Arajärvi herself was a staunch supporter of the bid by the astronomers, patiently and skilfully paving the way through the political system during the time to come. We are yet again reminded of Blaauw's comments regarding the importance of broadminded science policy makers.

Following an informal contact, the ESO Director General, Catherine Cesarsky, together with Massimo Tarenghi, visited Helsinki in November 1999 to present the activities of organisation and discuss the formal process that would be necessary for the accession to the ESO Convention.

Finland was now working towards a formal application for membership, but it would still take some time, because, among other things, it was waiting for the recommendations of an international evaluation panel in 2000. This report endorsed Finland's aspirations to join ESO, but for a while the process seemed to slow down. In February 2002, another meeting took place, this time between Mirja Arajärvi, Pentti

Pulkkinen (from the Academy of Finland) and two ESO representatives, Richard West and the author. The meeting led to an information seminar at the Academy of Finland, held in May 2002. It may also have slightly boosted the formal application to ESO for membership, which was submitted by the Finnish Government only a couple of weeks later.

As in the case of the United Kingdom, Finland was interested in paying a part of the accession fee as an in-kind contribution. Such an approach has often been used, partly to alleviate the financial burden, partly to give the institutes and industry of the candidate country a headstart in ESO. But it has its drawbacks. ESO is through and through based on the idea of excellence through competition. According a kind of protected status for goods and services, if only for a limited period of time, runs counter to this principle. Furthermore, there may be questions of liability and product guarantees, etc. But since the accession of a country is a political process, there is a certain scope for compromise, as we saw it in the case of the UK. For Finland this meant that a limited part of its special entrance fee was a three-year study of the future needs for the ESO community in the area of data reduction and analysis

Catherine Cesarsky and Tuula Haatainen exchanging the accession agreement documents on 9 February 1994 with Ian Corbett, now as Head of the ESO Administration, overseeing the procedures. (Photo: Hans Hermann Heyer)

environments, called Sampo[23]. This study dovetailed with the efforts regarding the Astrophysical Virtual Observatory and also with e-science activities in the UK under the name AstroGrid. The in-kind negotiations unavoidably took time. Meanwhile also parliamentary elections and the formation of a new government occurred in 2003, but ultimately an agreement was reached and the new government was ready to take the decisive step.

The ESO Council endorsed the agreement with Finland on 30 January and on 9 February 2004, the membership agreement was signed by the Finnish Minister of Education and Science, Tuula Haatainen, and the ESO Director General at a ceremony in Garching. Membership took effect on 7 July of that year with Mirja Arajärvi attending the ESO Council meeting on that date.

Spain

Like its neighbour on the Iberian peninsula, Spain had lived in relative isolation from the rest of Europe under its Falangist rule since the Second World War. However, with the death of the leader of the government in 1975, Francisco Franco, things began to change. Spain had already applied for membership of the European Communities in 1977. After a somewhat precarious transition period, democracy had been firmly established in the early 1980s and in 1986 Spain became a member as part of what is commonly known as the third enlargement round[24]. As we saw in the case of Finland, joining Europe had become a political imperative. At the same time, Spanish astronomy was developing fast. The observatories in the Canary Islands, at El Teide in Tenerife and Roque de los Muchachos at La Palma, were expanding their park of telescopes, owned by national institutes in a number of European countries, notably the UK, the Netherlands, Germany and Sweden. The observatories were operated under a set of agreements, including one between the user institutions and

---

[23] Sampo is a mythological artefact in Finnish folkore and the national epos *Kalevala*, which has strong links to the sky and origins of the world.

[24] Not for the first time do we see that science and scientific cooperation can precede (and even run counter to) political developments. Spain had joined CERN in 1959, but withdrew ten years later. In 1983 it rejoined the organisation. In 1964 Spain was a founding member of ESRO, one of the two parent organisations that then merged to form ESA in 1975. The agreements for the establishment of the Observatorio de Calar Alto date back to the early 1970s, and international collaboration in astronomy in the Canary Islands started in the 1960s, with the Observatorio del Teide being established in 1964. Nonetheless, the cases of Spain, Portugal, the Czech Republic and Finland demonstrate that an overall political agenda can help to boost international cooperation in science considerably.

the Instituto de Astrofísica de Canarias under the leadership of its influential director, Francisco Sánchez Martínez. On the mainland, the German–Spanish Astronomical Centre (Centro Astronómico Hispano–Alemán) had been established at Calar Alto, in the Almería Province, with a 3.5-metre telescope as its largest facility[25]. The Instituto Geográfico Nacional (IGN) had built a 14-metre millimetre-wave radio telescope in Yebes (near Guadalajara, in central Spain) in the mid-1970s and through the participation of IGN in IRAM, radio astronomers enjoyed access to the 30-metre millimetre/submillimetre telescope on Pico Veleta in the Sierra Nevada as well the IRAM array in France[26]. Spanish astronomers enjoyed guaranteed observing time at these telescopes, but some also found their way to La Silla, often participating in collaborations with scientists in the ESO Member States. Astronomers who had worked with ESO in some capacity included José Miguel Rodríguez Espinosa, Josefa Masegosa, Montserrat Villar Martín, Carme Gallart, Eduardo Martín and María Rosa Zapatero Osorio. A first push for Spanish membership of ESO, however, was to come from elsewhere.

Among the early observers at La Silla was Alvaro Giménez Cañete, who later became ESA Director of Science and Robotic Exploration. Here he is seen working at the Danish 0.5-metre telescope in 1982. (Photo by the author)

In 1987, Jon Marcaide proposed to the Spanish Secretary of State, Juan Rojo Alaminos, that Spain should join ESO. Marcaide, a radio astronomer, had returned to Spain after stints at Massachusetts Institute of Technology and the Max-Planck-Institut für

[25] Another observatory, the Observatorio de Sierra Nevada near Granada, started operating in 1981 as an initial collaboration between the Institute of Astrophysics of Andalucía (IAA), the Royal Greenwich Observatory, and the Observatoire de Nice.

[26] Spain also hosted the Villafranca satellite tracking station near Madrid, which became the operations centre of the International Ultraviolet Explorer Satellite and later also the European Space Astronomy Centre.

Radioastronomie in Bonn. Having also lived for a while in Munich, he was familiar with ESO. Furthermore, he knew Daniel Ponz, who worked at ESO in Garching. With his international experience, Marcaide realised the importance for Spanish astronomers of becoming more integrated in the international scientific community and involved in European-level collaborations. He also saw a need to establish a grass-roots framework for its growing astronomical community. Marcaide was of course not the only Spanish astronomer with international experience, and so he soon found like-minded colleagues, notably Ramon Canal Masgoret, who had worked at the University of Chicago and the Max-Planck-Institut für Astrophysik in Garching and Eduard Salvador-Solé, who obtained his PhD at the Institut Astrophysique de Paris (IAP) in Paris. In 1991, they took the initiative to establish the Sociedad Española de Astronomía (SEA) with Canal as its first president. Marcaide's initial attempt regarding ESO was not successful. Rojo answered that although ESO membership was only believed to cost 7% of the CERN membership, *"Seven percent of an astronomical figure is still an astronomical figure!"* (Marcaide, 2011). Around the same time, ESO was invited to mount an exhibition at the new planetarium in Madrid. The exhibition opened on 1 March 1988 with a major social event including public talks and the gathering of many prominent astronomers in Spain. In the following years, Marcaide, who had become perhaps the strongest promoter of ESO membership found allies and so, in October 1996, ESO's Director General accepted an invitation to attend the annual meeting of SEA, which took place in San Sebastián. Giacconi's presentation had a considerable effect. Understandably, Spanish astronomy was strongly associated with the Canary Islands observatories, which harboured plans for an 8–10-metre telescope. Even so, the meeting led to a resolution with a double recommendation: *"The SEA is committed to joint action on the entry into ESO and construction of the 10-metre Spanish telescope. It considers both aspects together will not only allow to maintain the scientific level reached by the astronomical community in recent years but will give access to even higher steps within the worldwide astronomical landscape."*[27] Importantly, the outcome was communicated to Gonzalo León Serrano, then Director General of the Office for Science and Technology of the Spanish Government. León supported the idea. Among the participants at the San Sebastián meeting was a prolific journalist from *El País*, Alicia Rivera, who became a staunch

[27] *"La SEA apuesta por una acción conjunta sobre la entrada en ESO y la construcción del telescopio español de 10 m. Considera que ambos aspectors al unísono permitirán no solo el mantinimiento del nivel científico alcanzado por la comunidad astronomía en los últimos años sinó también acceder a escalones aún más destacados dentro del panorama de la astronomía mundial."*

supporter of Spanish ESO membership[28]. With the Spanish economy expanding, all of this activity suggested that Spanish membership might be possible. The ESO Council saw this positively and was favourably disposed towards Spanish membership, but at the time ESO had its own problems and Council struggled to develop a clear vision for what it wanted *vis-à-vis* Spain, beyond straightforward membership. ESO was still shaken by the land dispute in Chile and was interested in alternative sites, including those in the northern hemisphere, also for scientific reasons. At the same time ESO was facing serious cash flow problems relating to the VLT. One idea was to use the Spanish contribution to finance one of the VLT Unit Telescopes and place it in Spain. This idea did not find many supporters, however. For Spain, the challenge was to balance the understandable interests of its major observatory at La Palma with the demands arising from an accession to ESO. Even so, on 28 October 1996, an ESO delegation led by Riccardo Giacconi and comprising Jean-Pierre Swings, Arno Freytag and Norbert König[29] met with Gonzalo León and Francisco Sanchez. The meeting did not lead to any result. It was clear that the Spanish negotiators wished to give priority to their large telescope, then thought to be an 8-metre telescope, at La Palma, and ESO was not willing, after all, to part with one of its UTs.

However, the San Sebastián meeting had also been attended by Paul Vanden Bout from NRAO and Michael Grewing from IRAM, who spoke about the plans for a large array of millimetre-wavelength telescopes in the southern hemisphere; the project that eventually became ALMA. This project also met with strong interest from Spanish astronomers and they managed to win the support of the government in their desire to become part of it. So, in the spring of 1997, the government declared its interest in joining ALMA, by means of a telefax to Giacconi. The telefax arrived shortly before Giacconi's meeting with Paul Vanden Bout that we mentioned in Chapter IV-1, and strengthened Giacconi's position, although it did not imply Spanish membership of ESO itself.

[28] ESO carried out several outreach activities in those years with Spanish participation. This included further exhibitions in Barcelona and at EXPO 1992 in Seville, the Future Astronomers of Europe contest in 1993, its support for the European Association for Astronomy Education (EAAE) with its high-profile outreach activities, support for the Spanish Física en Acción programme and finally an event for Spanish industry in relation to the Spanish participation in ALMA. Conversely, Spain began to support the ESO/EAAE outreach programme Catch a Star.

[29] As well as a representative from the French Embassy in Madrid.

The case for joining ESO, however, was boosted by a number of events that occurred around the turn of the century. The first was a study instigated by SEA and led by Xavier Barcons about *"the status of Astronomical Research in Spain (1999–2002)"*.

This was followed up by ESO's next Director General, Catherine Cesarsky, in 2001 with meetings in July and in November of that year, in Garching and Madrid respectively. The November meeting was with the Secretary of State for science policy, Ramón Marimón. Although the subject of the discussions was Spain's participation in ALMA, full Spanish membership of ESO was discussed, but it was decided to postpone this[30].

The third event was the decision in December 2002 by the Spanish government to participate in Phase Two of the ALMA project, the construction. As we have seen, there was strong Spanish interest in this project, not the least from José Cernicharo, from CSIC[31], as the main driving force behind the participation of Spain.

A recommendation from the National Astronomy Committee to the Spanish government to start negotiations to join ESO followed in March 2003. The membership of this body included the President of CSIC and the Director General of IGN, as well as the directors of all the major astronomy institutes in Spain.

Finally, in November 2003, the Spanish government published a comprehensive National R&D&I Plan for the period 2004–2007, in which membership of ESO was explicitly recommended. In the same month, the government sent a formal letter to ESO requesting the opening of negotiations.

The first meeting of the negotiating parties took place on 4 March 2004. This was followed by an "in-kind working group" meeting in Garching in April, co-chaired by Gerry Gilmore and Xavier Barcons on behalf of ESO and Spain, respectively. As part of its special contribution, Spain wished to offer access to the 10-metre GTC telescope at La Palma. This was an interesting proposition for ESO. Firstly it would give access, albeit temporarily, to a powerful northern hemisphere facility. Secondly,

---

[30] Independently, in 2001, speculations emerged about a possible merger between ESO and the Canary Islands observatories, which was occasionally also described as the ENO, the European Northern Observatory. This did not seem an attractive option to ESO, as it might have diverted resources away from ESO's new main projects, on which the future of European astronomy depended. Instead ESO reiterated that Spain would be welcome as a member of the organisation.

[31] The Consejo Superior de Investigaciones Científicas.

the primary mirror of GTC is a segmented mirror, rather than a monolith. GTC thus employs the same technology as the Keck telescopes in Hawaii. A segmented mirror was also planned for ESO's next giant optical telescope, and access to the GTC would enable ESO to gain first-hand experience in this area, which so far was new to the organisation[32]. Despite the obvious benefits, however, the negotiations took quite some time. This was partly because ESO had by now put an upper limit on the size of the in-kind part of the special contribution and putting together a solution that would satisfy both sides was not easy. Furthermore the completion of the GTC itself was delayed, hardly an unusual fate for a large telescope project, as we know. In December 2005, the negotiators had, however, reached agreement and the declaration to join ESO was formally signed by María Jesús San Segundo, Spanish Minister of Education and Science and Catherine Cesarsky, at a ceremony in Madrid on 13 February 2006. The agreement foresaw Spanish membership by 1 July 2006, but the formal ratification procedure lasted into the early days of 2007. Spain thus became ESO's 12th Member State on 14 February 2007. And just eight weeks earlier, another country had announced its readiness to join: the Czech Republic.

The Czech Republic

The first informal discussions about ESO membership for what was then Czechoslovakia occurred on 28 April 1990, on the occasion of a celebration of Jan Hendrik Oort's 90th birthday in Leiden[33]. Among the many guests were ESO's Director General, Harry van der Laan, and the ESO Council President, Per-Olof Lindblad. A young Czech astronomer, Jan Palouš, was also present. This was just one and a half years after the so-called Velvet Revolution, which was part of the political changes that took place across Central and Eastern Europe at the end of the 1980s, and which allowed these countries to re-establish their historical connections with the West. In the field of science, a major step was taken in 1991, when Czechoslovakia joined CERN. Early accession to CERN became possible because the CERN Council had decided to offer membership to a number of the countries that had been under Soviet influence under very favourable conditions, as a welcome gesture. ESO, however, was less forthcoming, in the sense that the ESO Council was not minded

[32] It is also worth noting at the time several Spanish institutes and companies were already actively collaborating with ESO through the FP6-funded ELT Design Study, which started in 2005. We have alluded to this study in Chapter IV-4.

[33] For readers proficient in the Czech language, a comprehensive description of the country's accession process is given by Palouš, J (2009): *Česká republika v Evropské jižní observatoři.*

to make any special concessions to new countries at the time. ESO's insistence on the payment of a substantial entrance fee has been the subject of debate over time, but, as we have seen, it has provided for a fresh inflow of capital and thus enabled the organisation to undertake new projects and also to better cope with the increase of its user community every time a country joined. The first discussions between van der Laan and Palouš therefore focussed on how Czechoslovakia could raise this amount, which was not insignificant for a country that had just emerged from the shadows of the Soviet system with its regulated economy. At the time, it was thought that financial help might be obtained from the European Community, but this option never materialised. By July 1992, the Czech Ministry was seriously interested in joining and requested a report from the Czechoslovak Academy of Sciences about the implications of Czechoslovak membership of ESO. But then the process was interrupted by yet another political earthquake: the country split in two — into the Czech Republic and Slovakia — and in the wake of this, the Czechoslovak Academy naturally ceased to exist. Things only began to move again in 1999. By that time, Palouš had become chairman of the Czech National Committee for Astronomy and the proposal was relaunched. He invited Peter Shaver from ESO to give a talk about ESO's future plans, especially regarding ALMA, at an astronomy meeting in Prague in April 1999. He also contacted Catherine Cesarsky, who had just assumed her post as the new Director General of ESO. The proposal to join ESO gathered rapid momentum and, in January 2002, the Czech astronomical community sent a formal inquiry to ESO about membership conditions. This was supported by the President of the Czech Academy of Sciences of the time, Helena Illnerová, who knew Catherine Cesarsky well from their participation in the European Research Advisory Board (EURAB), a body set up to advise the European Research Commissioner, Philippe Busquin. Palouš had gained confidence in the project and he decided to aim for August 2006 as the accession time to coincide with the IAU General Assembly, scheduled to be held in Prague.

He also tried to arrange for a stop at ESO in Chile by the Czech Prime Minister, Vladimir Špidly, as he visited the country in October 2003. The visit to ESO did not happen, but in the process, the Czech Ambassador to Santiago, Lubomir Hladík, became a good ally. Although the prime minister did not manage to visit ESO, political attention had been created. In a curious twist of fate, the ESO cause may also have been helped by an incident that itself had nothing to do with it. In early 2005, Czech tourists in Chile caused a major forest fire. The incident became an international issue and the Czech government sent two high-ranking officials to Chile to deal with the matter. Thanks to the good offices of the ambassador, a side visit to the

ESO Santiago office was arranged, adding to government awareness of the ESO question. Later in 2005, things began to speed up. In June the Czech Academy of Sciences made a formal proposal to the government. On 10 November 2005, Catherine Cesarsky and the author visited Prague and met with the Academy President Václav Pačes, representatives of various Czech Ministries and of the Czech astronomical community. Later in the day, they also met with the Deputy Prime Minister, Martin Jahn, who was very supportive. It looked good, but Jahn was about to resign and furthermore, elections had been called for the coming summer. These elections unfortunately resulted in a hung parliament and a long period of political uncertainty.

Nevertheless, the outcome of the informal discussions was substantial. The Czech government agreed to make a preliminary payment of the accession fee in advance and the money was transferred into ESO's account on 14 December 2005.

In February of the following year, the government sent its official application to ESO and nominated a negotiating team. In July of that year, the ESO Council did the same and the two negotiating teams met for one single meeting, on 20 September 2006 in Garching.

Meanwhile, in August, Prague had hosted the XXVI General Assembly of the International Astronomical Union with 2500 participants. The situation was such that the Czechs could not announce their accession to ESO, as they had hoped, but the General Assembly nevertheless played a role. A major topic at this General Assembly was the (re-)definition of what constituted a planet, a topic that generated much heated debate and media interest. When the General Assembly finally decided on the issue, Pluto — which used to be considered the ninth planet of the Solar System — was downgraded to a "dwarf planet". We shall not recapitulate the scientific arguments here, but the debate generated immense public interest worldwide — and of course also in Prague. Astronomy became a topic on the streets. As Jan Palouš later recalled, *"Pluto helped us into ESO!"*

In November a Czech delegation visited Paranal. Here it turned out that the draft agreement reached in Garching had raised additional issues within the Czech administration, linked to the advance payment of the accession fee and what would happen if parliamentary ratification failed. In a swift recovery action, the problems were solved, literally while driving through the desert to Paranal and with some frantic telephone calls to Europe. The accession agreement, with the additional clauses, was approved by the ESO Council on 6 December 2006 and by the Czech government

The Czech delegation visit to Paranal — from left to right: Ambassador Hladík, Jan Palouš, Andreas Kaufer (Director of the Paranal Observatory), Stanislav Štefl and Václav Pačes, president of the Czech Academy of Sciences. (Photo by the author)

on 12 December, with the signing of the document foreseen before Christmas. The Czechs were in a hurry, for the political situation in the country had become increasingly uncertain. Ultimately, the date was set for 22 December, but the time of day kept changing. The small ESO delegation[34] opted to fly in on the previous afternoon, only to learn that the Prime Minister had tendered his resignation that very same day. Luckily, the President had refused the resignation and thus, the responsible minister, Miroslava Kopicová, was still authorised to sign on the following day. The agreement was signed and submitted to parliament for ratification. Only 18 days after the signing, the government finally fell, but at that stage, the ratification process was underway, closely monitored and nurtured by Jan Palouš, and on 13 April 2007, the instruments of ratification were deposited at the French Ministry of Foreign Affairs as required by the ESO Convention.

[34] Associate Director Thomas Wilson and the author.

This was a true cliffhanger, but it was also a remarkably well executed process by the Czech astronomers[35] in the middle of the most difficult political situation that one can think of. It was a country that had only joined the European Union in 2004 and was therefore in the process of deep and rapid adjustment, while having to cope with a complete parliamentary stalemate and a government on the verge of collapse.

The Czech Republic was the first country of the former Eastern Bloc in Europe to become a member. The accession raised interest in ESO in other countries in the region (notably Poland) — and also in a country with close historical ties, Austria.

Austria

Many Austrian astronomers had worked in Germany and had observed at La Silla as "German astronomers". They were therefore well acquainted with ESO. The first suggestion that Austria should join ESO came up in the early 1970s in conversations between Blaauw and Hans Michael Maitzen (Maitzen & Hron, 2008), then a young astronomer working at the Ruhr-Universität Bochum and therefore also occasionally observing with the Bochum telescope at La Silla. Maitzen returned to his native Austria in 1976 and began lobbying for ESO membership, with some initial success. However, in 1980, the Austrian government decided against membership, due to a veto by the Austrian Ministry of Finance. A second attempt was undertaken in the period 1988–1991, rekindled by a visit of the ESO Council President, Kurt Hunger in 1986, and also with support from the Austrian Minister of Science, Erhardt Busek. Formal negotiations began in June 1990, partly preceded by various outreach activities, including an exhibition at the Vienna Planetarium in December 1987 and at the Austria Centre in Vienna's UN City in connection with the 7th EUREKA Minister Conference in June 1989. However, in 1991, the Austrian Ministry of Finance again said no. The ESO project went into hibernation until the year 2000 when the new ESO Director General, Catherine Cesarsky, attending a science meeting in Alpbach in Austria, encouraged the Austrian astronomers to restart the process. Formally the showstopper so far had been the veto by the Ministry of Finance, but other reasons for Austria's arduous path towards ESO membership were disagreement within the astronomical community and lack of support within the bureaucracy.

[35] Occasionally with a little help from friends: the first Czech astronomer to work as an ESO staff astronomer, Stanislav Stefl, took up his duties in 2004. Also in that year, the Venus transit project, led by ESO, held what was possibly the first EU-funded, science-related meeting in the Czech Republic.

Among the astronomers, the opponents of ESO membership feared that funds for Austrian astronomy would diminish if money were diverted to pay for membership. This apparently logical argument has been seen again and again in many countries contemplating to join, but it has seldom been borne out by reality. In most cases — though admittedly not all — the outcome has been the opposite: keen to make the best of membership, governments have actually allocated more funds to the national scientific effort, not less. In the end, the formation of the Österreichische Gesellschaft für Astronomie und Astrophysik (ÖGA^{2}) in 2002 helped resolve the internal differences. The other important element, the need for support by strong, dedicated civil servants is equally clear. Blaauw's comment about *"people in these important positions of science policy"*, quoted in Chapter I-2, bears testimony to this. Conversely, without support from within the administration, things become very difficult. For a long time this appeared to be the case in Austria. To be fair, Austria was far from being an isolated case. And, in any case, by 2006 things had begun to look much more positive. Consensus had been achieved within the scientific community and the bid was backed by a young, dynamic civil servant, Daniel Weselka, and also by his hierarchical superiors, notably Peter Kowalski, the Director General for Scientific Research and International Relations at the Austrian ministry.

So, for the third time, on 28 June 2006, discussions assumed a formal character with a negotiating team appointed by Council paying a visit to Vienna. During the coming months, discussions progressed, although at a very slow pace. The main reason was that Austria wished to pay a part of the entrance fee as an in-kind contribution, requiring long discussions to define the nature and the value of this. However, by June 2007, the main hurdles had seemingly been overcome and the way lay open for a formal agreement. Then, in a surprise move, the Austrian government abruptly retreated from the process, bringing back memories of the situation in 1991. The shock can be felt in an e-mail of 10 July from Catherine Cesarsky to the Council president, Richard Wade: *"Dear Richard: I must inform you ... that, after a very positive and even conclusive meeting on the in-kind contribution, the negotiation with Austria has suddenly collapsed."* The unexpected change of heart caused much commotion, in particular among Austrian astronomers, who once more saw their hopes crushed. Soon after taking over as ESO Director General on 1 September 2007, Tim de Zeeuw therefore met with the Austrian minister, Johannes Hahn, in a last-ditch attempt to salvage the situation. No real breakthrough was achieved at the meeting itself, but it opened the way for continued, quiet discussions between the parties and in March 2008, the government informed ESO that it was ready to resume formal negotiations with the aim of joining the organisation by the middle of the year.

Things were finally falling into place, it seemed. But it meant that negotiations had to be concluded and all documents approved, both by the various Austrian authorities and ESO within three months. On 30 June, the Austrian minister and ESO's Director General signed the accession agreement. The negotiators had reached their goal. It wasn't the only goal scored then as Vienna was one of the hosts for the European Soccer Championship at the time! The speed at which things developed turned out to be fortuitous. Only nine days after the signing of the agreement, the government collapsed and parliamentary elections were called. Ratification of the agreement, thus fell to the new parliament. Luckily, the minister was able to continue in office and, with the able help of Weselka and Hans-Joachim Sorger, an aide to the minister, the process was completed, so that Austria became ESO's 14th Member State.

(Photo: Hans Hermann Heyer)

Chapter IV-9

Born in Europe, at Home in the World

"Sometimes a global approach as well as a global strategy would be appropriate. It would bode well for Council to think of global strategy if not necessarily of global projects."

Richard Wade, ESO Council President, Statement to the ESO Council.

With the commissioning of the VLT, the successful relaunch of the interferometry option (VLTI), engagement in the ALMA project and the E-ELT project on the horizon there is no doubt that ESO has come to be seen as a scientific powerhouse in the world of astronomy. The so-called Senior Report by the National Science Foundation of the US, published in 2006, testified to this by stating that *"there is no U.S. organization that provides a suitable match to ESO. Despite its origin as an optical observatory, ESO was the natural organization to lead European involvement in the ALMA project and, through the international nature of its operations, obtained the resources to match NRAO in constructing and operating ALMA."* (NSF, 2006).

It is also evident that ESO has become the centre of gravity for European astronomy, working in symbiosis with the national institutes and the various European networks that have sprung up since the year 2000 and to which we shall return below.

But ESO has also played its own role in the wider development towards European integration. Firstly, the very wording of the Convention gave the organisation a European remit and perspective, even if it spoke of "member states" rather than "Europe". But in the discussions between CERN and ESO in 1969, we find an interesting comment about the possible wider implications of the foreseen close cooperation between the two organisations. In a letter to the incoming Director General Blaauw, Kees Zilverschoon — then head engineer at CERN — reported from the discussions at the CERN Council about this that *"it was remarkable how almost everyone ... entirely lost sight of the orginal aim, the construction of the telescope, and rather emphasized the scientific importance ... and the political aspect: Formation of a 'communauté scientifique Européenne', in which there would be room also for other organisations for fundamental science."* (Blaauw, *ibid.*). It is in fact a remarkable statement because here we see a very early expression of the ideas that have ultimately led to the creation of the European Research Area. Perhaps even the first? It is in any event noteworthy that the initial ideas at the purely political level were formulated only in 1973 by European Commissioner Ralph Dahrendorf (Madsen, 2010).

≈

ESO began to develop activities outside its Member States in the 1980s. An early example — in September 1987 — was a joint school between ESO and the Astronomical Council of the Academy of Sciences of the USSR, organised at the Byurakan Astrophysical Observatory in Armenia. For a while, it would be an isolated initiative. ESO's primary task is to serve its Member States, but, as we shall see, the organisation soon became much more involved in developments outside its own narrow area, both geographically and thematically.

ESO gradually adopted a broader view of its obligations as a European organisation. One of the early examples was the support programme for the Central and Eastern European (C&EE) countries, which was decided by the ESO Council in December 1992. Let us remember that at the time, Eastern Europe was experiencing its greatest upheaval since World War II. In November 1989, the Berlin Wall had come down, leading to German re-unification in 1990. In 1991, the Soviet Union collapsed and the existing institutional and organisational frameworks there either ceased to exist or were paralysed. This did not only happen in the Soviet Union, but countries that had belonged to the then Soviet sphere of influence were also affected. Science and scientists were affected along with everyone else, but for them it was perhaps even more dramatic, since in the communist system, science was more an integral part of the public system than in the West — a system that for a while seemed simply to evaporate. An example illustrates how desperate the situation was: for a while, the monthly salary at the Abastumani Astrophysical Observatory in Georgia was the equivalent of two US dollars, if it was paid at all. In fact for many months the astronomers did not receive any payment whatsoever. One day, they came across an abandoned lorry that turned out to be loaded with sugar. The observatory confiscated the sugar and distributed it among its employees — as salary. At the observatory in Armenia, the situation may have been slightly better, but instead they only had electrical power for one hour a day — during daytime (Shustov, 2010). It is no wonder that cries for help percolated through the scientific community and also reached ESO. In June 1992, the Council set up a working group *"to provide advice about ESO's future relations with C&EE astronomy under the recent changes in Europe"* (ESO press statement 11/92)[1]. The rationale for ESO involvement clearly illustrates that the organisation had accepted a wider role beyond providing top-of-the-line observing facilities for its Member State astronomers. It stated: *"The Council agreed that ESO, as the major European astronomy organisation, and with its many links to individual researchers, scientific institutes and observatories as well as to policy makers, is in an optimal position to assess objectively the very diverse needs and to provide support to C&EE astronomy in a non-bureaucratic and cost-effective way."* According to the report by the working group, there were about 60 institutes in the C&EE area with some 700 members of the IAU. At the time the American Astronomical Society (AAS) had raised some funds for support, as had the European Astronomical Society (EAS), but none were anywhere near sufficient to alleviate even the worst problems. On the other hand the European Community was preparing more comprehensive help, but it seemed clear that this programme would take time to become established and to achieve any substantial results. Therefore, the report concluded: *"Considering the present dramatic situation of astronomy in the C&EE countries and in line with ESO's historical obligations to European astronomy, it is highly desirable that ESO ... considers and executes a limited, but appropriate support programme as soon as possible, before a more substantial effort can be realised through more resourceful means."*

The three-year support programme began in early 1993 and had a yearly budget of half a million deutschmarks, to be dispensed partly as individual research grants, and partly to buy modern equipment (mainly computers, but also observational equipment, such as CCD detectors). Funds could also be allocated to support visits of astronomers from ESO Member States to C&EE institutes as well as to enable C&EE astronomers to participate in ESO conferences. It also enabled institutes to take out a free subscription to the journal *Astronomy & Astrophysics*. The programme was managed by ESO under a dedicated committee comprised of Nicolai Chugai (Moscow), James Lequeux (Paris), Georges Meylan (ESO), Giancarlo Setti (Bologna), Jean-Pierre Swings (Liège) and headed by Richard West (ESO). West knew the scientific communities of the countries in question well[2], spoke fluent Russian and had occasionally solved delicate questions of science diplomacy at the time of the Cold

[1] Members of the working group were Jan Bezemer, Jean-Pierre Swings, Gerhard Bachmann and Richard West.

[2] From 1982–85 West had served as General Secretary of the International Astronomical Union. As a young student, he had observed at the Abastumani Astrophysical Observatory — and brought back a Georgian wife.

War. If the response to the first call could be taken as a measure of the need, the numbers spoke for themselves: with 284 applications covering 936 applicants, the requested amount was eleven times bigger than the budget for the first year.

It is worthwhile dwelling a little on this activity, because it illustrates how what may often be seen as lofty political decisions were transformed into effective help "on the ground". This was driven by a genuine desire to help and enabled by what one can only describe as considerable personal courage. The articles in *The ESO Messenger* do not reveal anything of the dramatic circumstances under which the C&EE programme was actually carried out, probably because at the time, West was also editor of *The ESO Messenger.*

Delivering the programme also meant ensuring that the aid reached the right people. Conditions in the East were such that this often required delivery right to the front door. Since this could not be done safely through intermediaries, West himself travelled with what for people in that part of the world must have appeared to be large sums of money — in cash[3]. Personal grants amounted to 400 deutschmarks each, which at the time was the equivalent of a full year's salary. Since exchanging foreign currency was not simple, the money was packed in envelopes each with eight 50 deutschmark notes.

For security reasons, the local authorities occasionally provided armed protection during these missions, but within the ESO administration the unease was palpable and they welcomed the day when more orderly conditions again enabled regular bank transfers. After 1995, when the ESO programme folded, support was mainly taken over by INTAS[4], an organisation set up by the European Community and a number of countries and endowed with significant funds. The INTAS programme covered all areas of science and ran until 2006. It could be argued that the ESO support programme could only be but a drop in the ocean, but that probably does not do justice to it. The ESO programme was a "first aid" activity, delivered without delay in an emergency situation. The experience gathered was passed on to INTAS and other relief efforts[5]. Aside from the immediate relief, it sent an important signal to the scientists of the former Eastern Bloc and their societies, at a time when fundamental

[3] On at least one mission to Ukraine, he was accompanied by Georges Meylan.

[4] The International Association for the promotion of co-operation with scientists from the New Independent States of the former Soviet Union (NIS).

[5] West later advised the Soros Foundation regarding support for science in Eastern Europe.

science — and its practitioners — stood in low regard. It constituted a stamp of international recognition and approval, of political importance[6] and, of course, it was also a show of solidarity within the world's scientific community.

A further step towards engagement in broader policy matters came with ESO's participation in the Europa Ricerca event in Rome in June 1990, on the occasion of the 8th Ministerial Conference of EUREKA, and by invitation from its then President Antonio Ruberti. In 1993 Ruberti became European Commissioner for Research and began to think more seriously about how research in Europe could be integrated within a European framework. ESO also organised an exhibition at the Council of Europe between 26 September–12 October 1991. We shall return to the wider European policy perspectives shortly.

Learning the tools of the observing trade: young astronomers at the first ESO/OHP summer school in 1988 — Maria Grazia Franchini and Henri Boffin.

The training activities that had started with the summer school in Armenia changed character. In 1988, ESO and the Observatoire de Haute-Provence, under the directorship of Phillipe Veron, began a series of joint biannual summer schools for young astronomers. The rationale was that, with telescopes now being operated in faraway places, there were fewer options for young astronomers to become acquainted with modern observational techniques. The summer schools continued until 1996, when they were relaunched under the new name of NEON schools[7], thanks to an initiative by Michel Dennefeld and with financial support by the European Commission within the framework of the Marie Curie Actions.

[6] This was already seen in the report by the Council working group that stated: *"The recent experience has shown that C&EE institutes and individual scientists who are actively involved in ongoing international collaboration programmes have better chances of obtaining support by their national authorities and therefore of surviving scientifically."*

[7] NEON stands for the Network of European Observatories in the North. This originally meant the Osservatorio Astrofisico di Asiago in Italy, the Spanish/German Observatorio de Calar Alto and OHP, but soon brought the Canary Islands observatories as well as ESO into the partnership.

≈

In 1989, ESO introduced its own Research Student Programme, an idea of Harry van der Laan. With wonderful new research facilities becoming available, it was of course important to secure adequate recruitment to astronomy and to provide students with a perspective for their own development within this highly competitive, global science. The programme, initially with eight positions per year, was designed for postgraduate students preparing for a PhD. Since ESO is not a degree-granting institution, the idea was that the students would obtain their degree from a national university, but that they would spend 1–2 years at ESO, either in Garching or in Chile, as part of their doctoral work. At ESO, they would be aided by a local mentor, but also maintain contact with their formal supervisors back home. In *The ESO Messenger*, van der Laan presented the rationale: *"Having research students in our teams in addition to fellows, to visitors and to staff serves this purpose: Research students, bright, ambitious, naïvely demanding, contribute to the research mindedness of an institution.... Secondly, the ... programme serves to extend the linkage of ESO to Europe's universities.... A third aspect concerns the long-term quality and ambitions of European astronomy embodied in the next generation. The fellows and students who spend a year or two within ESO are better equipped to use its facilities for their personal research in future. In addition, and just as important, they will enable the institutes that employ them to use ESO telescopes and services to best advantage. They are the vanguard of VLT observers, training now and set to work for decades in the next century. Finally, and related to the preceding point, these youthful scientists establish patterns of professional and personal relations among themselves, relations that will guide their collaborations and projects of the future. The result will be ... growth of European excellence in astronomy. And that too is ESO's* raison d'etre." (van der Laan, 1989). The programme has been highly successful and up to the present day more than 200 doctoral students from almost 30 countries have benefitted from it. But the programme is not just about numbers, as the author experienced during a night stay at the VLT control room, chatting to two female astronomers: *"Here we're sitting, two young grad students, with the world's most advanced astronomical telescope at our control. It's fantastic!"* Indeed.

≈

By the end of the 20th century, ESO had established itself as a flagship organisation enabling European research in astronomy. Several countries were on their way to join while ESO itself was about to enter into an international collaborative project

(ALMA) with partners in Asia and the Americas. Further large projects[8] — some with potential ESO involvement — in combination with the overall changes in the European science policy landscape, that we shall discuss below, prompted the introduction of a Science Strategy Working Group by the ESO Council in June 2003. The committee structure of ESO already included the Scientific Technical Committee that addressed questions of the scientific orientation of ESO, for example, new instrumentation and major scientific activities, such as surveys. The SSWG, however, would instead address the overall policy aspects and develop paths for the future evolution of ESO within the broader European science landscape. Chaired by Ralf Bender, the initial membership of the SSWG comprised Claes Fransson, Gerry Gilmore, Franco Pacini, Jean-Loup Puget, Simon Lilly, Thomas Henning and Tim de Zeeuw as well as ESO staff members Bruno Leibundgut, Guy Monnet and Peter Quinn. ESO was not the only organisation to establish a dedicated body for such a purpose. In the face of the concentration in particle physics of mega-projects, both present and future, CERN had done the same, and in 2006, the CERN Council, building on the work of its working group, adopted a sweeping resolution regarding the future of particle physics in Europe. *"There is a fundamental need for an ongoing process to define and update the European strategy for particle physics,"* the text declared, *"Council, under Article II-2(b) of the CERN Convention, shall assume this responsibility, acting as a council for European particle physics*[9]*, ... Council will define and update the strategy based on proposals and observations from a dedicated scientific body that it shall establish for this purpose."* The question for the ESO Council was whether it should follow suit. It did not. Under its next chair, Tim de Zeeuw, the working group elaborated on three scenarios ranging from focussing on the projects currently in ESO's portfolio (mainly VLT, ALMA participation and E-ELT) to an all-encompassing role as the coordination body for all of European astronomy. Seeing that the strength of European astronomy derived from *"the synergy of a shared vision in large projects and the individual interests and innovation of member states in smaller projects"*[10], the ESO Council chose the middle way, reaffirming the notion that ESO should indeed focus on large projects beyond the capacity of any of the individual Member States. It essentially meant a continuation of the ideas that had underpinned the establishment of the organisation in the first place, the milestone of the 1998 Giacconi paper and into the era of the ELTs and, potentially, other large

[8] In December 2003 ESO organised a meeting of the OECD Global Science Forum in Munich, with a follow up meeting in April 2004 in Washington to discuss megaprojects in astronomy.

[9] Readers will remember that the acronym CERN stands for the Conseil Européen pour la Recherche Nucléaire.

[10] Report to the ESO Council, 14 March 2007.

infrastructure projects. The corollary of this was, inevitably, that ESO should not be responsible for the "small telescopes", even if they happened to be located at an ESO site. The discussion of small *vs.* large telescopes had been in progress for decades with many astronomers correctly emphasising the usefulness of the 1–2-metre-class telescopes, both in terms of the science that can be carried out with such facilities and their potential for training future generations of astronomers. But it would not be for ESO to secure the future for these telescopes.

≈

Of course, ESO was not the only research organisation that had achieved success in the global arena. CERN had long since grown into the leading organisation in the world for particle physics and in the area of space activities ESA had become the "NASA of Europe", providing Europe with an independent launch capacity and building and operating a host of satellites, many of which served scientific purposes. EMBL had been instrumental in building up research capacity in the new scientific discipline of molecular biology and demonstrated continued leadership in Europe. Other organisations at the European level had played important roles as well, each of them demonstrating the power of cooperation and collaboration between countries and the advantages of pooling resources, human as well as capital. While this happened two important political developments gained momentum: the general drive towards stronger political integration in Europe and the decision, in the spring of 2000, by the heads of states and governments to turn Europe into a strong[11], knowledge-based economy. This would clearly be significant for those who generate knowledge, namely scientists. The political framework that was created to foster this process became known as the European Research Area, an idea that harks back to the 1960s (and to which we have alluded on several occasions already) but was given a major boost by Europe's leaders, including increased funds for science and technology partly channelled through the Framework Programme of the European Union. In a very general way, everyone was now embracing some of the basic ideas that had underpinned the success of ESO CERN, and some of the other research organisations. But how did this affect ESO, which of course is not directly associated with the European Union? Firstly, the work of ESO's SSWG obviously took place within this context. Secondly, in the new spirit of the ERA, astronomical networks financed through the EU emerged. Three networks became important: OPTICON,

[11] The political ambition was to become the most competitive economy in the world, a goal that was quietly abandoned ten years later.

for optical astronomy, RadioNet, for radio astronomy, and finally ILIAS (and later ASPERA) for astroparticle research. OPTICON was perhaps especially attractive for astronomers in countries that were not yet members of ESO, since it included many institutes in these countries. However, ESO decided to become a full member of the network and has worked with it ever since. In this sense, OPTICON also provided a forum for discussions and for developing collaborative projects between institutes in ESO Member States and non-member states, but also involving ESO itself. Thirdly, a further initiative, in 2005, was ASTRONET, a cooperation between some of the main funding organisations in astronomy that also included ESO (and with ESA as an observer). In 2007, ASTRONET produced a common science vision document, followed in 2008 by the first European roadmap for astronomy covering both the ground and space segments — an equivalent to the Decadal Survey of US astronomy produced by the National Research Council Committee on Astronomy and Astrophysics[12].

≈

In the months before ERA was formally launched, some key players began to think of the shift in European policies, including José Mariano Gago, the Portuguese minister who had played an instrumental role in elevating support for science to the highest political levels, and Philippe Busquin, the new European Commissioner in charge of research and an ardent supporter of the ERA idea. Busquin began by talking to a number of key figures in the scientific community. At the suggestion of Curien, he interviewed Catherine Cesarsky, the newly appointed Director General of ESO, and the Director General of EMBL, Fotis Kafatos. Cesarsky pointed out the role that intergovernmental research organisations played in the European science landscape. Then, at an EU Council of Ministers meeting in Lisbon on 17–18 April, a select group of scientists was invited to present their views to the ministers, including Jean-Jacques Dordain from ESA (who later became Director General of that organisation) and Cesarsky. Having thought about the implications of the ERA proposal, Cesarsky realised that it was essential for the intergovernmental organisations, with their vast experience, not only to assist in the ERA process, but also to safeguard their own legitimate interests, and that this could best be done by joining forces. This led

[12] As already mentioned, several ESO Directors General had in the past expressed the desirability of having such a plan. As a substitute, ESO developed its own long-range plan, even if this could in no way cover all the needs of European astronomers. During her term of office as ESO DG, Catherine Cesarsky had therefore suggested to the European Astronomical Society (EAS) that it might consider taking up this task, but in the end the EAS declined.

to the formation of the EIROforum[13] partnership, which was formally sealed with the signing of the EIROforum Charter on 12 November 2002 in Brussels.

On 20 April 2005, the EIROforum partners published a joint vision paper for the future evolution of European science. Seen at the formal presentation at the Press Centre of the Berlaymont Building in Brussels (from left to right): Colin Carlile (ILL), Bill Stirling (ESRF), Robert Aymar (CERN) François Biltgen (President of the EU Competitiveness Council), David Southwood (ESA), Janez Potočnik (European Commissioner for Research), Catherine Cesarsky (ESO), Iain Mattaj (EMBL), Jerôme Pamela (EFDA-JET) and Katja Adler (BBC journalist and moderator). (Photo: European Commission)

The original EIROforum members were CERN, EFDA-JET (European Fusion Development Agreement-Joint European Torus), EMBL, ESA, ESO, ESRF (European Synchrotron Radiation Facility) and ILL (Institut Laue-Langevin)[14]. Between them they covered a wide spectrum of scientific disciplines including particle physics, fusion research, molecular biology, space science, astronomy and astrophysics, material science and neutron research. By virtue of their own scientific standing and involvement with multiple disciplines, the group could undertake actions and make pronouncements on important science policy issues, credibly and legitimately, in a

[13] EIROforum stands for the European Intergovernmental Research Organisations' Forum, a name that was suggested by Richard West during a meeting at Garching.

[14] In 2010, the partnership was augmented by the European X-ray Free Electron Laser (XFEL).

way that would have been beyond the bounds of any individual organisation. The partners also differed in size. ESO found itself in the middle, able to understand the specific problems of the small organisations but also to think big. Of course, ESO had traditionally maintained close scientific links to CERN and ESA, which were the largest members of the new partnership. All of this meant that ESO was ideally placed to play a key role as a motor for the collaboration for many years — strongly backed by Cesarsky, who had an in-depth understanding of the broader European science policy issues. The EIROforum was organised to have a Council, comprised of the Directors General[15] of the partner organisations, meeting twice a year, a Coordination Group and a number of Thematic Working Groups. The partnership established a rotating chairmanship with a term of one year. However, the first term, during which ESO held the chairmanship, was only for six months. This was thought to be appropriate since it was during the period of formation. Catherine Cesarsky held the EIROforum chair, with Peter Quinn serving as chair of the Coordination Group, and Richard West as the second ESO representative[16]. In October 2003, EIROforum signed a Statement of Intent of cooperation with the European Commission[17], opening up opportunities for continued mutual consultations and joint activities in support of the European Research Area.

≈

ESO's involvement in the ERA process, both on its own and through the EIROforum, was not based purely on altruistic motives. As mentioned, the ERA umbrella new groups and networks were formed, including in astronomy. Astronomers in the non-ESO member states in particular grasped these new opportunities. However, unlike the process that led to the creation of ESO, the ERA process was essentially a politically-driven, top-down process and Cesarsky realised the dangers this entailed. Thus on the European political stage, European astronomy became largely understood as those activities that enjoyed financial support from the European Union. ESO, on the other hand, whilst riding on the huge success of the VLT, was barely noticed at this level. The risk was that the new initiatives, however well intended and thought to foster collaboration and unity in European astronomical research, would actually cause the opposite — a fragmentation of efforts. Raising the awareness and the

[15] Or equivalent.

[16] The second term, from July 2008–June 2009, saw Tim de Zeeuw in the chair and the author chairing the Coordination Group.

[17] Renewed on 24 June 2010 with a formal signing ceremony at the ESO Headquarters.

profile of ESO at this political level was therefore necessary to bring European astronomy together on a much bigger scale, among other things manifested through the growth in the number of ESO Member States. At the same time, taking part in policy discussions at this level required resources that few in the astronomical community, aside from ESO, could muster.

Catherine Cesarsky and Philippe Busquin, then European Commissioner for Research, at Paranal in August 2003. In 2007, Busquin's successor, Janez Potočnik, also paid a visit to the observatory. (Photo by the author)

ESO at least had the capability to act as a conduit for the astronomical community in these matters. One example was a proposal by Piero Benvenuti that was meant to address the fragmentation of funding for certain research projects. In 2001, to overcome this problem (see also Chapter III-1), ESO pushed for a new fellowship scheme *vis-à-vis* the EU, under the name ELITE, meant to provide a one-stop funding process for important research projects in astronomy.

Although it was not implemented, the ELITE proposal, along with the study that underpinned it, became an important voice in the growing choir[18] that eventually succeeded in winning the argument for the creation of the European Research Council (ERC). The ERC is recognised today as perhaps the single most important new body within the European Research Area and is focussed exclusively on supporting excellence in research.

≈

We have seen how ESO became the natural partner in the quasi-global ALMA project. Given discussions in the past regarding expansion of its membership beyond Europe, it is therefore not surprising that this led to discussions with several overseas governments. At the time of writing, Brazil is at an advanced stage in the joining process. This book is about the past history and we shall leave speculations about the future to others. Yet, much has changed in the world both since the 1954 Leiden declaration, widely seen as ESO's birth certificate, and the 1962 Convention. Whilst its roots remain solidly European, in a globalised world, ESO has stepped onto a wider stage. Expressed differently: born in Europe, at home in the world.

[18] ESO was very much a voice in that choir, following the discussions that began with the first international meeting at the Nobel Foundation in the spring of 2002 and, together with more than 50 other organisations and associations, signing a declaration of support, published in *Science* in August 2004.

In July 2006, ESO organised a live videoconference with the Paranal Observatory at a major on-stage show at the central square in Munich, the Marienplatz, in connection with the German Wissenschaftssommer. *(Photo: Hans Hermann Heyer)*

Chapter IV-10

A Window to the Public

"The sharing of knowledge is quite simply a question of democracy and even of justice... ."

Prof. José Mariano Gago, Portuguese Minister for Science and Technology, RTD info, February 2005.

In the 100th issue of *The ESO Messenger*, Kurt Kjär, then technical editor of the magazine, recalled a somewhat bizarre incident at the ESO Office in Garching in 1977. He wrote: *"The ESO Headquarters was in its construction phase and the Garching staff (22 or 23 people) was provisionally housed in an apartment building in Garching. Only a small, discreet sign identified the place as ESO offices. One day, the manager of the publishing firm came to ESO to bring the blue-prints of* The ESO Messenger*... . To see someone from ESO, a visitor had first to ring the bell outside the building. Then a member of the ESO Secretariat, located in a flat on the ground floor of the building, opened the door and asked whom he wanted to see. Then the secretary informed the staff member concerned by telephone. The visitor went up to the apartment in which the staff member had his office. Here, he had to ring the bell again and wait a moment to be let in. This all must have appeared very strange and mysterious to the man. He said nothing while he was at ESO. But some weeks later, when the technical editor had matters to discuss with him at his place, he said: 'Now I know what ESO is all about — you can't fool me — astronomy is just a pretext. ESO is a secret-service agency!'* (Kjär, 2000). The oddness of the episode notwithstanding, Kjär's more sombre conclusion was clear: *"It shows that public relations activities should not be neglected in an organisation like ESO."*

Despite a well-developed tradition of the popularisation of science, mostly in the form of public lectures by distinguished scientists (Gregory & Miller, 1998), it is true that the European academic world had not given priority to public communication. In general, institutional arrangements to provide information and to support a broader public awareness were scarce and not held in particularly high regard in academia. As an organisation, ESO did not differ from the mainstream during the first couple of decades of its existence, despite the occasional *ad hoc* outreach activity. But change was on its way. Slowly, a new communication culture emerged, strongly inspired by the rather different approach taken by scientists in the US — in turn prompted by surveys conducted by the American Association for the Advancement of Science (AAAS) about public scientific literacy. At ESO, the wake-up call came in the mid-1980s, when the organisation moved towards the decision on the next generation telescope, the VLT. To prepare the ground, the representatives of the ESO Member States urged the organisation to undertake a dedicated effort in public outreach[1], and so in early 1986, what was originally called the Information and Photographic Service[2] was created, headed by Richard West. It was an excellent choice. West was a born communicator. In 1972 he had been awarded a prestigious communication prize in his native Denmark, the Rosenkjær Prize, given to scientists or persons from "cultural life" for their ability to communicate complex topics.

As already mentioned, the passage of Halley's Comet in 1986 provided an almost ideal opportunity to get into the "business". From the outset, the effort rested on two main pillars: press releases and exhibitions. Both contributed to building up public recognition of ESO, but the exhibitions also served to directly support attempts to broaden the Member State base. In fact, an ESO exhibition (or other public event) was often the first visible sign of a move towards a closer relationship between a non-member country and the organisation, e.g. in Portugal, Spain, Austria, etc. However, with the strong encouragement of the Director General of the time, Harry van der Laan, ESO soon added other activities to its outreach arsenal. The first was video production. The decision to build up competence and capacity in this field was originally driven by the desire to document the VLT project as it progressed. Video recordings were not just of historical importance but used actively to keep ESO staff informed in a timely fashion about ongoing developments, be it wind tunnel tests, production of mechanical pieces or construction activities — all activities that happened in different locations around Europe and the rest of the world. Video as a recording tool had been around for some years, but the important strategic (and bold at the time)

[1] In a statement by the Swiss government delegate at the ESO Council meeting in November 1984, he stressed that *"it was extremely important that an organisation like ESO makes headlines from time to time. This was most important for the tax payers, for parliamentarians, and government officials, particularly at a time where big decisions were lying ahead involving a substantial amount of money."* It is noteworthy, perhaps puzzling, that it did not come from the astronomers themselves.

[2] In various incarnations the department has existed ever since. Reflecting a slight switch in scope, in 1994 it became the Education and Public Relations Department, then the Public Affairs Department (also covering science policy and some international relations aspects) and, from 2008, the education and Public Outreach Department.

The inauguration of a major ESO exhibition at the Museon in the Hague in January 1989 also brought some of the key actors in ESO's history together including Jan Oort (left), one of the "founding fathers", Adriaan Blaauw (centre right), the second Director General, and Harry van der Laan, the fourth Director General of the organisation. (Photo by the author)

decision by ESO was — from the start — to choose a broadcast-quality solution on the principle that as ESO strove for excellence in its scientific and technical areas, it should do likewise in outreach and communication. This way, the recorded material could not only be used for internal purposes, but also be made available to television channels across the world (Madsen, 1993). From this nonetheless modest start, ESO's broadcast service has evolved into a highly professional unit able to work with top-of-the line television channels from CNN, Discovery Channel, NHK (of Japan), ABC (Australia) to all the major European channels whether public or private — and even Arab channels such as Al Jazeera[3]. As part of the broadcast service, ESO also began to develop advanced computer animations, both of astronomical phenomena and of new technology projects. The ability to deliver photo-realistic imagery became a powerful tool in promoting new projects and winning air time on television. Harry van der Laan's role in developing ESO's outreach activities was not limited to supporting the effort; he also decided to embed it in the Office of the Director General. This was an important and far-sighted move that made communication *Chefsache*

[3] While ESO produced and delivered significant amounts of footage to television, it also received camera teams from all over the world, including many from the most prestigious channels. In 2010, the first German-produced 3D documentary *Das Auge 3D*, about the VLT on Paranal, also became the first ever 3D documentary to be screened on German television — on the National Geographic Channel.

and enabled ESO to develop a high public profile, in fact to become one of the leading science communication efforts in Europe.

Later, as technology opened up for new media, based on the internet, ESO became increasingly present on these communication platforms as well. Over the years, to some limited extent, ESO became part of popular culture, featuring in commercials, films and even in stand-up comedy shows. For example, in 1990, the new NTT was used in a promotional campaign for a Swiss hotel chain (under the theme of "excellence") and later the VLT formed the backdrop for several car commercials (e.g., Audi). The acronym ESO could be found in crosswords, and occasionally ESO even appeared in science fiction movies on television or on the cinema screens, e.g., in 1985 in the German movie *Das Gespinst* (seemingly inspired by Fred Hoyle's 1957 novel *The Black Cloud*), which was shot partly at the ESO Headquarters in Garching, partly at La Silla. And in 2008, the Paranal Residencia was chosen for some scenes in the James Bond movie *A Quantum of Solace.*

Press releases formed the backbone of the outreach activities from the outset. Although the principle has always been followed that a press release would only be issued if there was real and solid content, ESO has continuously issued more press releases about astronomy than any other national research organisation in Europe and thus contributed significantly to giving European astronomy public visibility. The press releases have always been held in high regard by professional science journalists for their reliability, accuracy and the amount of information contained in them, as has been confirmed to ESO again and again. But press releases are — press releases; and media work can be hard and fast. For that reason, in 2005 ESO took a — possibly unique — initiative to publish an ethics statement, describing how ESO press releases are to be understood, how they are produced and what actions are taken should a scientific claim later turn out to be untenable[4].

As we have told the story of ESO in this book, we have occasionally also touched upon some of the many activities carried out by the group since 1986.

≈

While the department had grasped the opportunity of the Halley apparition and other astronomical events (such as supernova 1987A) and worked hard to promote

[4] http://www.eso.org/public/outreach/pressmedia/pr-proc.html

ESO's immediate goal of building the VLT, by 1993 it began to develop a more holistic view and strategic vision for its continued work. As a consequence it developed a separate strand of activities in the field of education, strongly encouraged by the Director General of the time, Riccardo Giacconi. Did ESO have an educational obligation? Giacconi clearly thought so. Unsurprisingly, he finishes his 2008 memoirs with a strong plea in favour of educational outreach: *"While our society relies increasingly on technology, based on science and reason, the effects on society of teaching rational thinking are not at all obvious.... I have pledged myself to do what I can still do ... to spread the light of reason."*

Beyond these high aims, ESO's educational activities dovetailed with wider efforts undertaken by the European Union and instigated by Antonio Ruberti, the energetic Italian scientist, politician and European Commissioner for Research, leading to the creation of a European Science Week. Under this umbrella, the European Commission agreed to co-fund scientific outreach activities with a "European dimension", including science education initiatives. For more than ten years, ESO became a prominent actor within this framework, often carrying out activities that were widely described as the flagships of the Science Week.

The first initiative was called The Future Astronomers of Europe and targeted pupils in upper secondary schools in 18 countries (17 in Europe[5] as well as Chile). They were invited to write essays with the title "A night with the VLT". The programme built on the premise that by the time the VLT would become operational many of these young people would be in the middle of or have finished their tertiary education. Drawing their attention to astronomy (and, of course, the VLT) before they made their career choice would be good. And, who knew, perhaps some would end up as astronomers and ultimately use the VLT themselves. As it was, this actually happened. So, in November 1993, 18 excited young people — national winners of the essay contest — arrived in Munich ready to undergo an astronomical crash course at the ESO Headquarters. A week later, the group departed for La Silla, where they had been given telescope time, first two nights on the Dutch 0.9-metre telescope and afterwards half a night on the NTT, to be used for observational programmes prepared by them in Garching. Shortly before their arrival at La Silla, however, three new supernovae (1993af, ag, ah) had been discovered, so the initial plans were overthrown and they instead had the opportunity of experiencing the excitement of observing such a unique event. And they were given their chance at the telescopes.

[5] Countries that were either Member States of ESO or the European Union.

Fortune favours the bold, as it is said. The young people carried out the observations, reduced the data at La Silla and saw their names published in IAU Circular 5897. Since the group had also been followed by a television crew, the excitement was not confined to the small group of young participants, but found its way back to a much larger audience in Europe.

For the 18 winners[6] of the Future Astronomers of Europe contest, the visit to La Silla was an experience they would never forget. Here, some of them are receiving instructions from Eva Grebel, as they prepare for a night's observations at the Dutch 0.9-metre telescope. (Photo by the author)

The Future Astronomers of Europe programme was a huge success, but it also revealed a serious problem. In spite of dedicated efforts to inform schools in all 18 countries about the programme, it turned out that none of the winners had actually heard about it through their schools. This at least suggested that an enhanced effort towards teachers might be indicated. During the next European Science Week, in 1994, ESO therefore organised a conference for interested science teachers with the title Astronomy — Science, Technology, Culture. At the conference, the participants adopted a

[6] Michael Bieglmayer (Austria), Nicolas Debry (Belgium), Edith Valenzuela (Chile), Thomas Greve (Denmark), Jere Veikko Kahanpää (Finland), Nicolas Leterrier (France), Sonja Schuh (Germany), Lambros Kourtis (Greece), Theo Honohan (Ireland), Carlo Farrigno (Italy), Alain Dondelinger (Luxembourg), Uri Naftali Shimron (The Netherlands), Niels Erland Haugen (Norway), João Pedro del Carvalho Saraiva (Portugal), Sergio Marco (Spain), Tyberiusz Moska (Sweden), Miroslava Zavadsky (Switzerland) and Christian Miller (United Kingdom). Several of them subsequently chose to study astrophysics.

Declaration on the Teaching of Astronomy in Europe's Schools, stating that *"Astronomy should contribute towards the consciousness that, in a complex society abounding in science and technology, a scientific education is essential for the choices that every citizen has to make in the democratic life. Students should feel that the Earth is a wonderful place in the Universe, and to be cared for and defended."* It set out some particular teaching goals, but also provided powerful arguments as to why and how astronomy can serve broader educational purposes. In this connection it stated that *"Astronomy teaching can contribute to an understanding of the physical laws, which start from the human level and reach the macro-cosmos to give a scientific organised outlook on our world and appreciate the uniqueness of the Earth for the human race. Astronomy locates our niche in space and time. Students should be aware of threats, from light pollution and radio interference, to our ability to observe the night sky."* It continued: *"Astronomy teaching conveys the fundamentals of the scientific method, including the associated doubt and lack of answers and the interplay between experiment and theory, thereby forcing students to adopt a critical attitude towards the many pseudo-sciences."*

The teachers' conference in Garching led to a long-term partnership between ESO and some of Europe's most dedicated science teachers. Seen here is Lucienne Gouguenheim giving a presentation at the meeting. (Photo: Hans Hermann Heyer)

The conference led, among other things, to the formation of a European Association for Astronomy Education, which became a very good partner for ESO's following educational activities. The meeting also laid the seeds of a much more substantial activity to support science teachers — the large Physics on Stage[7] and (later) Science on Stage Festivals, embedded in what became the EIROforum Science Teachers' Initiative (ESTI)[8]. Among the other activities to follow were Astronomy On-Line in 1996, a relatively early educational activity using the internet (and, in fact, leading many schools to become connected to the net, often with Government support), and Life in the Universe (2001), designed to create awareness about the level of serious

[7] Based on an initiative by ESO.

[8] It is estimated that these activities have exposed several tens of thousands of science teachers to new ideas about science teaching (Madsen, 2006).

science relating to exoplanets and the prospects for extraterrestrial life. Most of the activities had targeted *"the best and the brightest"* of young people. Catch a Star[9], on the other hand, was an internet-based programme with broader aims — trying to reach both younger children and those who might not have had enough confidence to enter into highly competitive programmes. Once again, the EAAE played a major role in assisting ESO in these activities.

Another activity deserves mention here: VT-2004, on the occasion of the transit of Venus in 2004 on 8 June of that year. In terms of reach, it was the largest programme ever undertaken by ESO, but while it exploited an astronomical event, just like the cases of Halley and the Shoemaker-Levy 9, this was a pure outreach event without new scientific results being expected. In a sense, the science had been done in the 18th and 19th centuries. Back then, the main scientific rationale lay in the opportunity — by means of triangulation — to use the crossing of Venus over the surface of the Sun, observed from different locations on the Earth, to determine the distance between the Earth and the Sun and thus also to determine the absolute scale of the Solar System. A Venus transit is a relatively rare event, the previous one having taken place in 1874. In 2004, the great opportunity lay in re-enacting the historical scientific experiment, but now involving a large number of people. Observing teams — this time amateurs and interested public — could observe the transit and report their data to a central hub at ESO. All in all more than 2700 teams took part, but many more simply chose to follow the event in the live webcast from ESO. On the day of the transit, ESO registered no less than 55 million web hits within an eight-hour period, and in total the website clocked 75 million hits during the programme. Perhaps not surprisingly, six million came from Germany. At the other end of the scale (but still somehow pleasing), six came from Rwanda. VT-2004 was led by

Children in Garching's central square, the Bürgerplatz, lining up to observe the Venus transit event on 8 June 2004, guided by volunteer ESO staff. (Photo: Heinz Kotzlowski)

[9] Proposed to ESO by Rosa Maria Ros, a Spanish astronomer based in Barcelona.

ESO, but also involved the EAAE, the Institut de Mecanique Celeste et de Calcul des Ephemerides (IMCCE)/Observatoire de Paris and the Astronomical Institute of the Academy of Sciences of the Czech Republic (Astronomický ústav Akademie věd České republiky) together with national nodes in 25 countries. This project was also co-funded by the European Commission.

≈

As time went by the educational activities grew in size and complexity. They also grew in budget and even with the funding from the European Commission, it was clear that more "shoulders" were needed. The activities initially had the character of small pilot programmes, but the need in Europe for action was such that critical mass became an issue, both in terms of conducting the activities and in order to achieve the improvement in quality and attractiveness of contemporary science education that all surveys demonstrated was so badly needed. ESO therefore initiated a close collaboration with its sister organisations, first CERN and ESA, and later, with all of the organisations that today form the EIROforum. This partnership in turn opened up the possibility of carrying out major activities such as ESTI — already mentioned — with science teaching festivals and a European science teaching journal entitled *Science in School*. The ESTI initiative was a unique and widely acclaimed activity, drawing in participants from far beyond the borders of Europe[10].

With the support of consecutive Directors General and the ESO Council, there is no doubt that among European science organisations ESO played a pioneering role with respect to outreach activities. Its activities were both based on the legitimate wish to promote its own programmes and projects and the fundamental recognition that ESO as a member of the worldwide knowledge community could make a unique contribution towards stimulating public awareness and interest in science in general and — of course — in astronomy in particular. *Noblesse oblige.*

This rationale also drove ESO's considerable involvement in the International Year of Astronomy (IYA2009), which ultimately involved almost 150 countries. ESO's engagement started in 2006, when the efforts of the IAU began in order to achieve the necessary resolution by the General Assembly of the United Nations declaring the year 2009 as the International Year of Astronomy, commemorating the

[10] Such as Canada, Chile and Russia.

quadrennial of the first use of an astronomical telescope by Galileo[11]. With its intergovernmental status, the fact that it was supported by a number of member states of the United Nations and the European origin of the IYA2009 proposal, ESO was able to act as a credible interlocutor on the international diplomatic stage[12], hand in hand with the IAU. Well aware of ESO's strength in this field, the IAU President asked the author to lead this political action in support of IYA2009, aided by Enikő Patkós. The approval of the IYA2009 at the UN General Assembly was obtained on 19 December 2007.

The IAU delegation that presented the IYA2009 initiative at the United Nations in New York on 1 November 2007. From left to right: The author, Beatriz Barbuy (IAU Vice President), Robert Williams (IAU President Elect), Catherine Cesarsky (IAU President), Luigi di Chiara (Italian UN Mission), Kevindran Govender (SALT), Nella Le Moli (INAF) and Leopoldo Benacchio (INAF).

Shortly afterwards, ESO's outreach activities were merged with the Hubble European Information Centre (also residing at ESO as part of the ESA/ESO European Coordination Facility for the Hubble Space Telescope). Together with the IYA2009

[11] The proposal to hold an International Year of Astronomy was originally made by the Italian astronomer (and former ESO Council President) Franco Pacini, first leading to a resolution by the International Astronomical Union at its 2003 General Assembly in Sydney and followed up with an endorsement by UNESCO at its General Conference in 2005.

[12] Including the United Nations General Assembly and the United Nations Committee for the Peaceful Uses of Outer Space, in which ESO has status as permanent observer.

secretariat, also hosted by ESO, an immensely powerful outreach group emerged, serving not just ESO, but the astronomical community as a whole. During IYA2009 ESO led four of the twelve IYA2009 Cornerstone projects, each a major activity with a global reach. Perhaps most spectacular was Around the World in 80 Telescopes, a live full 24-hour public webcast, featuring most of the research astronomical observatories in the world and attracting an estimated 200 000 viewers. The webcast, which included a lavish collection of stunning images, was hosted and produced at the ESO Headquarters, with astronomers at observatories around the globe participating. Another ESO-led Cornerstone project, launched in April 2009, was the Portal to the Universe, designed to serve as a "one-stop shop" for online astronomy information material. Finally, the Cosmic Diary was a site where professional scientists could blog about their work and their lives as scientists. No less than 14 ESO astronomers participated in the project. Another activity worthy of mention was the GigagalaxyZoom project, basically a triad of marvellous images of the Milky Way, which documented a virtual journey to the central dust band of our galaxy[13]. The GigagalaxyZoom project attracted about 500 000 visitors and was shown in 50 exhibitions in Europe. Yet all of this happened — so to speak — on the side, next to the normal tasks: keeping up with a steady stream of scientific press releases, videos and exhibitions, and many other activities. Making the most recent and exciting results from astronomical research available to the general public meant fulfilling a deep obligation to the millions of people, whose continued support for this great adventure decides whether it can go on or not.

[13] The first image, an 800-million-pixel panorama of the entire sky was produced by Serge Brunier, a French writer and astrophotographer, and Frédéric Tapissier in collaboration with ESO. The second image was a detailed view covering a 34 × 20 degree field of the constellations of Sagittarius and Scorpius, as observed with a 10-centimetre amateur telescope. This was produced by Stéphane Guisard, an engineer at the Paranal Observatory and also a keen amateur astronomer, and finally a 370-million-pixel view of NGC 6523, the Lagoon Nebula, obtained with the Wide Field Imager at the MPG/ESO 2.2-metre telescope at La Silla.

Epilogue

"Over the past half century, European astronomy has progressed from backwater to front-row player. . . ."

Johannes Andersen, 2011.

It is tempting to see the ESO story as an almost automatic, indeed irresistible, progression of science and technology, both interlinked and mutually enhancing. In this view, the politics and the human stories — what might be dismissed as "the many anecdotes" — could be understood merely as ripples on the pond. This would not necessarily be completely wrong since, at the end of the day, ESO is about scientific excellence. Science, understood as an ever-growing body of knowledge, is often seen this way. Yet this is too narrow a view, as the last chapters have aimed to demonstrate. For science is also a deeply human endeavour with all the implications that follow. While pursuing its goal of scientific excellence, ESO has both been influenced by and itself influenced wider developments in Europe. We have occasionally referred to the European Research Area. As an idea it can be traced back to the 1970s, yet as a general political concept, it first saw the light of day with the dawn of the new millennium. However, its basic tenets — free and unhindered cross-border collaboration, attainment of critical mass, development of common strategies and pooling of resources to reach shared goals, etc. — are exactly the features that have characterised ESO since its early days, fifty years ago. If, beyond those general ideas, one were to identify key lessons of the ESO story of interest to a wider audience, they would probably include:

The value of competition: We remember Heckmann's remarks about the dangers of a scientific monoculture. ESO was born at a time when Europe had fallen behind and the organisation was carried forward by the desire to regain lost ground. Yet science is not just about competition, it is as much about cooperation — a fact that has spurred the creation of the somewhat curious term of co-opetition. But a requirement for participation in joint projects at the appropriate level is the competence acquired by accepting the challenge of competition. Perhaps the best example is ESO's participation in the ALMA project — not as a junior partner, but on a par with the other participants.

Long-term thinking: When new science facilities, especially large and costly ones, are conceived, there is a natural tendency to focus on the initial investment. However, ensuring the necessary resources for operations and future upgrades are arguably as important, if not more so. Enabling long-term planning and execution, in turn, requires the proper institutional and legal structure, and ESO is deeply indebted to its founding fathers for providing exactly that, even if it took time.

Excellence, excellence, excellence, but...: Excellence is the fuel that makes the research engine run. Excellence in research, however, requires excellent facilities and telescopes. The NTT and the VLT with their advanced instruments provide ample examples of this. Excellence within a research infrastructure is of course no less important — among other things for its ability to recruit and retain expert staff. But the best is sometimes the enemy of the good. ESO occasionally struggled to complete technological projects in time, not because they were not finished (in a technical or contractual sense), but because enthusiastic engineers and technicians, convinced that they could do still better, were reluctant to let go of their piece of equipment. That there is a balance here is *because ...*

... *Time counts:* As every scientist knows, a discovery is only made once. Being second is no fun. But time, and *timing*, counts in the world of management and politics, as well — two worlds that play important roles in the realisation of large research facilities. In management, time equals money, in politics, timing can be a make or break question for the approval of a new project.

The tremendous impact of a major research infrastructure on the organisation of its research discipline: The creation of a major European facility in ground-based astronomy could only happen because the astronomers were willing and able to come together behind the idea. This may, at the outset, have required many a scientist to set aside his or her personal pet projects, but in retrospect, the result has been that the astronomical community is considered one of the best-organised research communities in Europe — perhaps in the world — and able to attract substantial funding for a range of activities. It is worth noting that out of this process a European astronomical community has emerged, where scientists work together across national borders as a matter of course. And it is a community that is self-confident, vibrant and ambitious in terms of its research objectives. The creation of this community is also due to the fact *that ...*

... *ESO acts as meeting point and breeding ground for talent:* Over the decades, ESO has provided a meeting place for astronomers, young and experienced, from all over the world. Friendships have been forged, ideas have been argued over and collaborative projects developed. Talented young scientists have worked for periods of time at ESO as students, as *co-opérants* or as fellows, before moving on to other institutes. A survey of the ESO fellowship scheme showed that 92% of those who worked as fellows in Garching stayed within their scientific field after leaving the organisation[1]. And a large number of scientists have simply met at the many conferences organised by ESO or been involved in various *ad hoc* working groups.

Growth: ESO has grown because of its projects, but ESO's projects have only been possible because ESO grew. Growth is not only a matter of increased resources. ESO's expansion into non-optical wavelengths, first with SEST and later on a huge scale through its involvement in the ALMA project is a case in point. The original European project, the LSA, was conceived outside ESO and even after the original project merged into ALMA it remained a partnership between a number of national European institutes for a while (together with the NSF/NRAO/NRC as their North American counterpart). But it became clear that if Europe was to act as an effective partner in this quasi-global project and to also secure wide backing within Europe, the European effort had to be unified and ESO was the right vehicle to realise this project.

≈

It might seem like a truism to state that Europe and ESO have come far since October 1962. But this has not happened for free. Substantial resources, both human and financial have been committed to allow European astronomers to catch up. As indeed they have. But time does not stand still, and so it is important to move on, to develop new projects and new methods of working. This is the challenge of the future. Without neglecting the scientific gains achieved with the current telescopes and thus the contribution made to increasing our knowledge, if we fail to meet that challenge, the past investment — undertaken to achieve a leading position for Europe's astronomers in the long term — will have been in vain. And so, inevitably, we arrive at the question of the E-ELT.

---

[1] ESO Annual Report, 2008.

What does the E-ELT mean to the Member States and their societies? First of all it means a major effort in the mobilisation of resources, financial and intellectual. In round numbers, the cost of the E-ELT is estimated to be about 1000 million euros. This equals the price of four Airbus A380 Superjumbos. More relevant to understand the figure is perhaps to look at the cost in comparison with that of the VLT, the project which marked the ESO's breakthrough into the international world of astronomical research. Adjusted for inflation, the 1999 cost was, as we will remember 990 million deutschmarks, or 506 million euros, to be shared amongst the eight Member States. If we now consider both the growth in the number of Member States and the growth in wealth over the last twenty-plus years, then the financial burden of the E-ELT on the 14 European Member States will amount to 80% of the VLT cost[2]. This book is written at the time when Brazil is in the process of joining ESO. If we add Brazil to the calculation, the financial burden on each Member State would fall to about 70% of the cost of the VLT. So much for cost. What about the gains? The E-ELT will need the involvement of high-tech industry on a scale hitherto unseen for an astronomical project. In turn, experience from the past suggests that there will be important spin-offs and market opportunities created that will reach far beyond the field of science[3]. But the E-ELT is first and foremost a "science machine".

What, then, will scientists get out of it? The honest answer, I believe, is that we don't know. Which is at the same time the best argument as to why we should do it. Yes, we know *what* to look for. We know *how* to look and we also know *why* we are looking. All of this is substantiated by our current knowledge and elaborated in thoroughly thought-out science cases. Yet, if on the eve of Galileo's first use of the telescope to study the heavens, the assembled scholars of the world had been asked where the intellectual voyage sparked by Galileo would eventually take us, would they have been right in their predictions? Some might have had an inkling of what was to come, for the human mind is awesome in its capability, but most would have been wrong. What we therefore can expect to find out there in the depths of space are surprises. Here lies the true fascination and power of science — its ability to harness the human intellect to unveil and help us understand those surprises and integrate this knowledge into the cultural edifice of humankind. If ESO can contribute to this endeavour, the efforts of its Member State societies will have been well invested.

[2] Expressed differently, together the 14 countries have a population of about 400 million. The E-ELT thus constitutes an investment of 2.5 euros per capita. Since this is spread over roughly 10 years, it means 25 cents per annum per capita.

[3] A study at CERN, the European laboratory for particle physics, showed that one euro of expenditure in high-tech procurement at CERN generates 3.7 times as much value in industry (Autio *et al.*, 2003).

References

Alloin, D. 2001, *ESO: Research Facilities in Santiago*, The ESO Messenger, 105
Andersen, J. 1993, *Ray Tracing Twenty Years at ESO*, The ESO Messenger, 72
Andersen, J. 1996, *Planning for La Silla in the VLT Era: What Came Out?*, The ESO Messenger, 83
Andersen, J. 2011, *Building a Strong, Unified European Astronomy*, in Lasota, J.-P. (ed.), Astronomy at the Frontier of Science, Springer Verlag
Ardeberg, A., Andersen, T., Owner-Petersen, M. & Jessen, N.-C. 1996, *A 25 m Live Optics Telescope*, Proceedings SPIE, 2871
Arnaboldi, M., Capaccioli, M., Mancini, D., Rafanelli, P., Caramella, R., Sedmak, G. & Vettolani, G. P. 1998, *VST: VLT Survey Telescope*, The ESO Messenger, 93
Arsenault, R., Hubin, N., Stroebele, S., Fedrigo, E., Oberti, S., Kissler-Patig, M., Bacon, R., McDermid, R., Bonaccini-Calia, D., Biasi, R., Gallieni, D., Riccardi, A., Donaldson, R., Lelouarn, M., Hackenberg, W., Conzelman, R., Delabre, B., Stuik, R., Paufique, J., Kasper, M., Vernet, E., Downing. M., Esposito, S., Duchateau, M., Franx, M., Myers, R. & Goodsell, S. 1986, *The VLT Adaptive Optics Facility Project: Telescope Systems*, The ESO Messenger, 123
ASTRONET 2008, *The ASTRONET Infrastructure Roadmap*
Autio, E., Bianchi-Streit, M. & Hameri, A.-P. 2003, *Technology Transfer and Technological Learning through CERN's procurement activity*, CERN
Baade, D., Giraud, E., Gitton, P., Glaves, P., Gojak, D., Mathys. G., Rojas, R., Storm, J. & Wallander, A. 1994, *Re-invigorating the NTT as a New Technology Telescope*, The ESO Messenger, 75
Babcock, H. W. 1953, *The Possibility of Compensating Astronomical Seeing*, Publications of the Astronomical Society of the Pacific, 65, 386
Balestra, A., Santin, P., Sedmak, G., Comin, M., Raffi, G. & Wallander, A. 1992, *NTT Remote Observing From Italy*, The ESO Messenger, 69
Banse, K. 2003, *The Munich Image Data Analysis System*, in Heck, A. (ed.), Information Handling in Astronomy — Historical Vistas, Kluwer Academic Publishers
Baranne, A., Queloz, D., Mayor, M., Adrianzyk, G., Knispel, G., Kohler, D., Lacroix, D., Meunier, J.-P., Rimbaud, G. & Vin, A. 1996, *ELODIE: A spectrograph for accurate radial velocity measurements*, Astronomy & Astrophysics Suppl. Ser., 119, 373
Bauersachs, W. 1982, *Dome for the 2.2 m Telescope Commissioned*, The ESO Messenger, 30
Beckers, J. M. 1988, *Increasing the size of the isoplanatic patch with multiconjugate adaptive optics*, in Ulrich, M.-H. (ed.), Very Large Telescopes and their Instrumentation, Proceedings of the ESO Conference held in Garching, March 21–24 1988
Beckers, J. M. 2003, *Interferometric Imaging in Astronomy: A Personal Retrospective*, Reviews in Modern Astronomy, Vol. 17
Behr, A. 1993, *Otto Heckmann 1901–1983*, The ESO Messenger, 33
Benvenuti, P. 1984, *The Space Telescope European Coordinating Facility Begins its Activity*, The ESO Messenger, 36
Beuzit, J. L. 2009, *The First Common-User AO Facility ADONIS*, ESO celebration "20 Years of Adaptive Optics at ESO", 27 November 2009
Beuzit, J. L. & Hubin, N. 1993, *ADONIS — a user-friendly adaptive optics system for the 3.6-m telescope*, The ESO Messenger, 73
Blaauw, A. 1974, *To all ESO Staff Members*, The ESO Messenger, 1
Blaauw, A. 1991, *ESO's Early History — The European Southern Observatory from Concept to Reality*, ESO, Garching, Germany
Blaauw, A. 2004, *My Cruise Through the World of Astronomy*, Annual Review, Astronomy & Astrophysics.
Block, D. L. 2011, *A Hubble Eclipse: Lemaître and Censorship*, Proceedings of the 80th Anniversary Conference held by the Faraday Institute, St. Edmund's College, Cambridge
Bonaccini Calia, D., Hackenberg, W., Cullum, M., Brunetto, E., Quattri, M., Allaert, E., Dimmler, M., Tarenghi, M., Van Kersteren, A., Di Chirico, C., Sarazin, M., Buzzoni, B., Gray, P., Tamai, R., Tapia, M. R., Davies, R., Rabien, S., Ott, T. & Hippler, S. 2001, *ESO VLT Laser Guide Star Facility*, The ESO Messenger, 105
Bonnet, H., Abuter, R., Baker, A., Bornemann, W., Brown, A., Castillo, R., Conzelmann, R., Damster, R., Davies, R., Delabre, B., Donaldson, R., Dumas, C., Eisenhauer, F., Elswijk, E., Fedrigo, E., Finger, G., Gemperlein, H., Genzel, R., Gilbert, A., Gillet, G., Goldbrunner, A., Horrobin, M., ter Horst, R., Huber, S., Hubin, N., Iserlohe, C., Kaufer, A., Kissler-Patig, M., Kragt, J., Kroes, G., Lehnert, M., Lieb, W., Liske, J., Lizon, J.-L., Lutz, D., Modigliani, A., Monnet, G., Nesvadba, N., Patig, J., Pragt, J., Reunanen, J., Röhrl, C., Silvio Rossi, S.,

Schmutzer, R., Schoenmaker, T., Schreiber, J., Ströbele, S., Szeifert, T., Tacconi, L., Tecza, M., Thatte, N., Tordo, S., van der Werf, P. & Weisz, H. 2004, *First Light of SINFONI at the VLT*, The ESO Messenger, 117

Bonnet, H., Bauvir, B., Wallander, A., Cantzler, M., Carstens, J., Caruso, F., Di Lieto, N., Guisard, S., Haguenauer, P., Housen, N., Mornhinweg, M., Nicoud, J.-L., Ramirez, A., Sahlmann, J., Vasisht, G., Wehner, S. & Zagal, J. 2006, *Enabling Fringe Tracking at the VLTI*, The ESO Messenger, 126

Booth, R. 1994, *A Southern Hemisphere Millimetre Array*, in Ishiguro, M. and Welch, W. J. (eds.), Astronomy with Millimeter and Submillimeter Wave Interferometry, ASP Conference series

Booth, R., Johansson, L. E. B. & Shaver, P. 1989, *SEST — the First Year of Operation*, The ESO Messenger, 57

Brand, J. 1982, *The Atmospheric Transmission at La Silla at 230 GHz*, The ESO Messenger, 2,

Breysacher, J. 1988, *Some Statistics about Observing Time Distribution on ESO Telescopes*, The ESO Messenger, 51

Breysacher, J. & Tarenghi, M. 1991, *Flexible Scheduling at the NTT, a New Approach to Astronomical Observations*, The ESO Messenger, 63

Brown, M. 1987, *European Telescope Will Be Biggest*, New York Times, December 22nd

Burrows, C. 2012, private communication with the author and STScI News Release 1996-02: *Disk around Star May Be Warped by Unseen Planet*

Calder, N. 1980, *The Comet Is Coming*, BBC

Cayrel, R., Spite, F., Spite, M., Hill, V., Primas, F., Andersen, J., Nordström, B., Beers, T. C., Bonifacio, P., Molaro, P., Plez, B. & Barbuy, B. 2001, *Measurement of stellar age from uranium decay*, Nature, 409, 691

Chaisson, E. 1994, *The Hubble Wars: Astrophysics Meets Astropolitics in the Two-Billion-Dollar Struggle Over the Hubble Space Telescope* Harvard University Press

Chauvin, G., Lagrange, A. M., Dumas, C., Zuckerman, B., Mouillet, D., Song, I., Beuzit, J.-L., Lowrance, P. 2004, *A Giant Planet Candidate near a Young Brown Dwarf*, Astronomy & Astrophysics, 425, L29

Chauvin, G., Lagrange, A.M., Dumas, C., Zuckerman, B., Mouillet, D., Song, I., Beuzit, J.-L. & Lowrance, P. 2004, *Giant Planet Companion to 2MASSW J1207334-393254*, Astronomy & Astrophysics, 425, L29

Danziger, J. & Bouchet, P. 2007, *SN 1987A at La Silla: The Early Days*, The ESO Messenger, 127

Dekker, H. 2009, *Evolution of Optical Spectrograph Design at ESO*, The ESO Messenger, 136

Dekker, H. & D'Odorico, S. 1992, *UVES, the UV-Visual Echelle Spectrograph for the VLT*, The ESO Messenger, 70

Dickmann, S. 1989, *Adaptive Optics shown on French telescope*, Nature, 341

Dierickx, P. & Ansorge, W. 1992, *Mirror Container and VLT 8.2-m Dummy Mirror Arrive at REOSC Plant*, The ESO Messenger, 68

Dierickx, P. 1993, *Manufacturing of the 8.2-m Zerodur Blanks for the VLT Primary Mirrors — a Progress Report*, The ESO Messenger, 71

Dierickx, P. & Zigmann, F. 1991, *Aluminium Mirror Technology at ESO: Positive Results Obtained with 1.8-m Test Mirrors*, The ESO Messenger, 65

D'Odorico, S. 1997 *A Deep Field with the Upgraded NTT*, The ESO Messenger, 90

D'Odorico, S., Enard, D., Lizon, J.-L., Ljung, B., Nees, W. & Ponz, D. 1983, *The ESO Echelle Spectrograph for the Cassegrain Focus of the 3.6-m Telescope*, The ESO Messenger, 33

D'Odorico, S., Beckers, J. & Moorwood, A. 1991, *A Progress Report on the VLT Instrumentation Plan*, The ESO Messenger, 65

D'Onofrio, M. & Carlo Burigana, C. (eds.) 2009: Massimo Capaccioli in Questions of Modern Cosmology — Galileo's legacy, Springer Verlag.

Duffner, R. W. 2009, *The Adaptive Optics Revolution*, University of New Mexico Press

Dürbeck, H. W., Osterbrock, D. E., Barrera, L. H. & Leiva, R. 1999, *Halfway from La Silla to Paranal — in 1909*, The ESO Messenger, 95

Ebersberger, J. & Weigelt, G. 1978, *Speckle Interferometry and Speckle Holography with the 1.5 m and 3.6 m ESO Telescopes*, The ESO Messenger, 18

Edvardsson, B., Andersen, J., Gustafsson, B., Lambert, D. L., Nissen, P. E. & Tomkin, J. 1993, *The Chemical Evolution of the Galactic Disk*, Astronomy & Astrophysics, 275, 101

Enard, D. 1987, *The VLT — Genesis of a Project*, The ESO Messenger, 50

Enard, D. 2002, *The early days of instrumentation at ESO*, The ESO Messenger, 109

ESO/Government of Chile Joint Committee for the Development of Astronomy 2006, *10 Years Exploring the Universe*

Ettlinger, E., Giordano, P. & Schneermann, M. 1999, *Performance of the VLT Coating Unit*, The ESO Messenger, 97
Fehrenbach, C. 1981, *The First Steps of the European Organization*, The ESO Messenger, 24
Finnish Ministry of Education (Opetusministeriön) 1999, *Tähtitieteen Suurhankkeet — Työrihmien Musitio*, 8, 1999
Fischer, R. & Walsh, J. 2009, *An Extension for ESO Headquarters*, The ESO Messenger, 135
Flensted Jensen, M. 2002, Deputy Chairman, *The Expert Group on the establishment of a European Research Council*, private conversation with the author
Fosbury, R. A. E. & Albrecht, R. 2002, *The Space Telescope — European Coordinating Facility*, The STScI Newsletter, 19, 3
Foy, R. & Labeyrie, A. 1985, *Feasibility of adaptive telescope with laser probe*, Astronomy & Astrophysics, 152, 2
Franchini, M., Molaro, P., Nonino, M., Pasian, F., Ramella, M., Vladilo, G., Centurion, M. & Bonifacio, P. 1992, *'Remote' Science with the NTT from Italy — Preliminary Scientific Results*, The ESO Messenger, 69
Fransson, C., Gilmozzi, R., Gröningsson, P., Hanuschik, R., Kjär, K., Leibundgut, B. & Spyromilio, J. 2007, *Twenty Years of Supernova 1987A*, The ESO Messenger, 127
Fusco, T. & Rousset, G. 2009, *NAOS-CONICA (a.k.a NACO) for the VLT*, Celebration meeting on the occasion of 20 years of AO at ESO, November 2009
Gehrels, N. 2005, *A short gamma-ray burst apparently associated with an elliptical galaxy at redshift z = 0.225*, Nature 437, 851
Gehring, G. & Rigaut, F. 1991, *New Year's Eve with Adaptive Optics*, The ESO Messenger, 63
Giacconi, R. 2008, *Secrets From The Hoary Deep — A Personal History of Modern Astronomy*, Johns Hopkins University Press, Baltimore, MD, USA
Gillessen, S., Eisenhauer, F., Trippe, S., Alexander, T., Genzel, R., Martins, F. & Ott, T. 2009, *Monitoring stellar orbits around the Massive Black Hole in the Galactic Center*, Astrophysical Journal, 692, 1075
Gillessen, S., Eisenhauer, F., Perrin, G., Brandner, W., Straubmeier, C., Perraut, K., Amorim, A., Schöller, M., Araujo-Hauck, C., Bartko, H., Baumeister, H., Berger, J.P., Carvas, P., Cassaing, F., Chapron, F., Choquet, E., Clenet, Y., Collin, C., Eckart, A., Fedou, P., Fischer, S., Gendron, E., Genzel, R., Gitton, P., Gonte, F., Gräter, A., Haguenauer, P., Haug, M., Haubois, X., Henning, T., Hippler, S., Hofmann, R., Jocou, L., Kellner, S., Kervella, P., Klein, R., Kudryavtseva, N., Lacour, S., Lapeyrere, V., Laun, W., Lena, P., Lenzen, R., Lima, J., Moch, D., Moratschke, D., Moulin, T., Naranjo, V., Neumann, U., Nolot, A., Paumard, T., Pfuhl, O., Rabien, S., Ramos, J., Rees, J. M., Rohloff, R. R., Rouan, D., Rousset, G., Sevin, A., Thiel, M., Wagner, K., Wiest, M., Yazici, S. & Ziegler, D. 2010, *GRAVITY: a four-telescope beam combiner instrument for the VLTI*, SPIE
Gilliotte, A. 1992, *Fine Telescope Image Analysis at La Silla*, The ESO Messenger, 68
Gilliotte, A., Ihle, G. & Lo Curto, G. 2005, *Improvements at the 3.6-m Telescope*, The ESO Messenger, 119
Gilmozzi, R. 2006, *Giant Telescopes of the Future*, Scientific American, May 2006
Gilmozzi, R. & Spyromilio, J. 2007, *The European Extremely Large Telescope (E-ELT)*, The ESO Messenger, 127
Glindemann, A., Bauvir, B., Delplancke, F., Derie, F; di Folco, E., Gennai, A., Gitton, P., Housen, N., Huxley, A., Kervella, P., Koehler, B., Lévêque, S., Longinotti, A., Ménardi, S., Morel, S., Paresce, F., Phan Duc, T., Richichi, A., Schöller, M., Tarenghi, M. & Wallander, A. 2001, *Light at the end of the tunnel — First Fringes with the VLTI*, The ESO Messenger, 104
Gray, P. M. 2000, *Assembly and Integration to First Light of the Four VLT Telescopes, in Telescope Structures, Enclosures, Controls, Assembly/Integration/Validation, and Commissioning*, Proceedings of SPIE, 4004
Gregory, J. & Miller S. 1998, *Science in Public: Communication, Culture and Credibility*, Basic Books
Grenon, M. 1990, *The Northern Chile Climate and its Evolution — A Pluridisciplinary Approach to the VLT Site Selection*, The ESO Messenger, 61
Grewing, M. & Guilloteau, S. 1997, *The LSA Study*, IRAM Newsletter 32
Grosbøl, P. & Biereichel, P. 2003, *IHAP — Image Processing and Handling*, in Heck, A (ed.), Information Handling in Astronomy — Historical Vistas, Kluwer Academic Publishers
Grothkopf, U. 2010, *ESO publication statistics*, as per March 2010
Gouiffes, C., Wampler, E. J., Baade, D. & Wang, L. F. 1989, *NTT Images of SN 1987A*, The ESO Messenger, 58

Guicharrousse, I. *Científico se alimentó sólo con agua y barro*, El Mercurio, 1 August 1987.

Hardy, J. W. 1994, *Adaptive Optics*, Scientific American, 270

Harwit, M, 1981: Cosmic Discovery: The Search, Scope, and Heritage of Astronomy, MW Books

Heckman, O. 1976, *Sterne, Kosmos, Weltmodelle*, R. Pieper & Co Verlag, Munich Zurich

Hjort, J., Watson, D., Fynbo, J. P. U., Price, P. A., Jensen, B. L., Jørgensen, U. G., Kubas, D., Gorosabel, J., Jakobsson, P., Sollerman, J., Pedersen, K. & Kouveliotou, C. 2005, *The optical afterglow of the short gamma-ray burst GRB 050709*, Nature, 437, 859

Hosfeld, R. 1954, *Comparisons of Stellar Scintillation with Image Motion*, Journal of the Optical Society of America, 44, 4

Ihle, G. 1995, *The Last Trip of the ESO GPO*, The ESO Messenger, 81

Ihle, G., Montano, N. & Tamai, R. 2007, *The 3.6-m Dome: 30 Years After*, The ESO Messenger, 129

Käufl, H.-U., Jouan, R., Lagage, P. O., Masse, P., Mestreau, P. & Tarrius, A. 1992, *TIMMI at the 3.6-m Telescope*, The ESO Messenger, 70

Käufl, H.-U. 1993, *Ground-Based Astronomy in the 10 and 20 μm Atmospheric Windows at ESO — Scientific Potential at Present and in the Future*, The ESO Messenger, 73

Käufl, H.-U. 1994, *N-Band Long-Slit Grism Spectroscopy with TIMMI at the 3.6-m Telescope*, The ESO Messenger, 78

Käufl, H.-U., Ageorges, N., Dietzsch, E., Hron, J., Relke, H., Scholz, D., Silber, A., Sperl, M., Sterzik, M., Wagner, R. & Weilenmann, U. 2000, *First Astronomical Light with TIMMI2, ESO's 2nd-Generation Thermal Infrared Multimode Instrument at the La Silla 3.6-m Telescope*, The ESO Messenger, 102

Kjär, K. 2000, *About the ESO Messenger*, The ESO Messenger, 100

Koehler, B. 1998, *ESO and AMOS Signed Contract for the VLTI Auxiliary Telescopes*, The ESO Messenger, 93

Krige, J. & Guzetti, L. (eds.) 1995, *History of European Scientific and Technological Cooperation*, Conference in Florence, Proceedings, European Commission, Brussels, Belgium

Krige, J. 2005, *Isidor I Rabi and CERN*, School of History, Technology & Society, Georgia Institute of Technology, Atlanta, GA, USA, Phys. Perspect., 7

Krügel, E. & Schulz, A. 1985, *Submillimetre Spectroscopy at La Silla*, The ESO Messenger, 40

Kuijken, K., Bender, R., Cappellaro, E., Muschielok, B., Baruffolo, A., Cascone, E., Iwert, O., Mitsch, W., Nicklas, H., Valentijn, E. A., Baade, D., Begeman, K. G., Bortolussi, A., Boxhoorn, D., Christen, F., Deul, E. R., Geimer, C., Greggio, L., Harke, R., Häfner, R., Hess, G., Hess, H. J., Hopp, U., Ilijevski, I., Klink, G., Kravcar, H., Lizon, J. L., Magagna, C. E., Müller, P. H., Niemeczek, R., De Pizzol, L., Poschmann, H., Reif, K., Rengelink, R., Reyes, J., Silber, A. & Wellem, W. 2002, *OmegaCAM: the 16k × 16k CCD Camera for the VLT Survey Telescope*, The ESO Messenger, 110

Kurz, R. & Shaver, P. S. 1999, *The ALMA Project*, The ESO Messenger, 96

Lagage, P. O., Pel, J. W., Authier, M., Belorgey, J., Claret, A., Doucet, C., Dubreuil, D., Durand, G., Elswijk, E. E., Girardot, P., Käufl, H.-U., Kroes, G., Lortholary, M. M., Lussignol, Y., Marchesi, M., Pantin, E., Peletier, R. R., Pirard, J. F., Pragt, J. J., Rio, Y., Schoenmaker, T., Siebenmorgen, R., Silber, A., Smette, A., Sterzik, M. & Veyssiere, C. 2004, *Successful Commissioning of Visir: The Mid-Infrared VLT Instrument*, The ESO Messenger, 117

Lagrange, A.-M., Chauvin, G., Ehrenreich, D., Rousset, G., Charton, J., Rabou, P., Montri, J., Lacombe, F., Mouillet, D., Gratadour, G., Rouan, D., Gendron, E., Fusco, T., Mugnier, L. & Allard, F. 2008, *A probable giant planet imaged in the β Pictoris disk. VLT/NACO Deep L-band imaging*, Astronomy & Astrophysics, 493, L21

Lauberts, A. 1982, *The ESO/Uppsala Survey of the ESO (B) Atlas*, ESO

Laustsen, S. 1974, Introduction, in Richter, W., Plathner, D., Roozefeld, J. F. and van der Ven, J: The Mechanical Design of the 3.6-m Telescope, ESO Technical Report 1

Laustsen, S. 1977, *Progress Report 3.6-m Telescope (1977)*, The ESO Messenger, 9

Laustsen, S. 2002, *How ESO got its Optics Group*, The ESO Messenger, 109

Laustsen, S., Madsen, C. & West, R. 1987, *Exploring the Southern Sky*, Springer Verlag

Leibundgut, B., Spyromillo, J., Walsh, J., Schmidt, B. P., Phillips, M. M., Suntzeff, N. B., Hamuy, M., Schommer, R. A., Aviles, R., Kirshner, R. P., Riess, A., Challis, P., Garnavich, P., Stubbs, C., Hogan, C., Dressler, A. & Ciaroullo, R. 1995,

Discovery of a Supernova (SN 1995K) at a Redshift of 0.478, The ESO Messenger, 81

Léna, P. 2008, *The Early Days of the VLTI*, in Richichi, A., Paresce, F. & Chelli, A. (eds.): The Power of Optical/IR Interferometry: Recent Scientific Results and 2nd Generation Instrumentation, Proceedings of the ESO Workshop held in Garching, April 4-8 2005, Springer

Léna, P. 2012, private communication with the author

Leroy, E. 1975, *Avanti for the Telescope*, The ESO Messenger, 3

Madsen, C. 1993, *Das Größte Teleskop der Welt — Konzept und Videodokumentation eines ehrgeizigen europäischen Technologieprojektes*, in Lehmann, R. G. (ed): Corporate Media — Handbuch der Audiovisuellen und multimedialen Lösungen und Instrumente, Verlag Moderne Industrie

Madsen, C. 2006, *Science on Stage — Towards a Rejuvenated Science Teaching in Europe*, Proceedings, the 9th International Conference on Public Communication of Science and Technology, Seoul, Korea

Madsen, C. 2010, *Scientific Europe: Policies and Politics of the European Research Area*, Multiscience, UK

Maitzen, H. M. & Hron, J. 2008, *Die Universitätssternwarte Wien — Pflanzstätte des Österrechischen ESO Beitritts*, Communications in Astroseismology, 149

Marcaide, J. 2011, private communication with the author

Mattila, K., Hämen-Antilla, A., Kajantie, K. & Valtonen, M. 1982, *Tähtitieteen asema Soumessa ja Soumen mahdollisuudet Osallistua kansainvällisiin tähtitieteen Hankkeisiin*

Mattila, K., Tornikoski, M., Tuominen, I. & Valtaoja, E. 2004, *Astronomy in Finland*, The ESO Messenger, 117

Mayor, M. 2012, private communication with the author

Mayor, M. & Queloz, D. 1995, *A Jupiter-mass companion to a solar-type star*, Nature, 378

McCray, W. P. 2004, *Giant Telescopes: Astronomical Ambition and the Promise of Technology*, Harvard University Press

McMullan, D. & Powell, R. 1977, *The electronographic camera*, New Scientist, 73, 1044

Mehra, J. 1973, *The Physicist's Conception of Nature*, Reidel, Dordrecht, 1973.

Mellier, Y., Maoli, R., Van Waerbeke, L., Schneider, P., Jain, B., Erben, T., Bernardeau, F., Fort, B., Bertin, E. & Dantel-Fort, M. 2000, *Cosmic Shear with Antu/FORS1: An Optimal Use of Service Mode Observation*, The ESO Messenger, 101

Merkle, F., Kern, P., Lena, P., Rigaut, F., Fontanella, J. C., Rousset, G., Boyer, C., Gaffard, J. P. & Jagourel, P. 1989, *Successful Tests of Adaptive Optics*, The ESO Messenger, 58

Menten, K. 2009, *Leo Brandt: Pionier der Funkmesstechnik und Initiator der Radioastronomie in Deutschland*, in Mittermaier, B. & Rusinek, B. A. (eds.): Leo Brandt (1908–1971) Ingenieur — Wissenschaftsförderer — Visionär, Proceedings, Scientific Conference on the occasion of the 100th anniversary of the Northrine-Westphalian Science politician and founder of the Forschungszentrum Jülich, Forschungszentrum Jülich

Monnier, J. D. 2003, *Optical Interferometry in Astronomy*, Reports on Progress in Physics 66, Institute of Physics

Morian, H. 2003, *Tiefer Blick ins All*, Schott Info, 1

Moorwood, A., Cuby, J.-G. & Lidman, C. 1998, *SOFI Sees First Light at the NTT*, The ESO Messenger, 91

Moorwood, A. F. M. 2003, *CRIRES Takes Shape*, The ESO Messenger, 114

Moorwood, A. F. M. 2009, *30 Years of Infrared Instrumentation at Eso: Some Personal Recollections*, The ESO Messenger, 136

Murtagh, F. & Fendt, M. 1994, *The SL-9/ESO Web Encounter*, The ESO Messenger, 77

Muñoz, H. 2008, *The Dictator's Shadow*, Basic Books

Müller, R., Höness, H., Espiard, J., Paseri, J. & Dierickx, P. 1993, *The 8.2-m Primary Mirrors of the VLT*, The ESO Messenger, 73

NASA 1998, *Hubble Takes First Image of a Possible Planet around Another Star and Finds a Runaway World*, NASA Release, 98-91

National Science Foundation 2006, *From the Ground Up: Balancing the NSF Astronomy Program — Report of the National Science Foundation*, Division of Astronomical Sciences, Senior Review Committee

Neufeld, D. A., Schilke, P., Menten, K. M., Wolfire, M. G., Black, J. H., Schuller, F., Müller, H. S. P., Thorwirth, S., Güsten, R. & Philipp, S. 2006, *Discovery of interstellar CF^+*, Astronomy & Astrophysics, 454, L37

Nilson, P. 1971, *The Uppsala General Catalogue of Galaxies*, Presented to the Royal Society of Science of Uppsala

Nordström, B., Mayor, M., Andersen, J., Holmberg, J., Pont, F., Jørgensen, B. F., Olsen, E. H., Udry, S. & Mowlavi, N. 2004, *The Geneva-Copenhagen survey of the Solar neighbourhood — Ages, metallicities, and kinematic properties of ~14 000 F and G dwarfs*, Astronomy & Astrophysics, 418, 3

Nordström, B., Mayor, M., Andersen, J., Holmberg, J., Pont, F., Jørgensen, B. F., Olsen, E. H., Udry, S. & Mowlavi, N. 2004b, *A Livelier Picture of the Solar Neighbourhood*, The ESO Messenger, 118

Otárola, A., Delgado, G., Booth, R., Belitsky, V., Urbain, D., Radford, S., Hofstadt, D., Nyman, L. Å., Shaver, P. & Hills, R. 1998, *European Site Testing at Chajnantor: a StepTowards the Large Southern Array*, The ESO Messenger, 94

Palouš, J. 2009, Česká republika v Evropské jižní observatory, Cs. cas. fyz., Vol. 59, No. 5, FZÚ AV ČR v.v.i

Pasachoff, J. 1979, *Astronomy: From the Earth to the Universe*, Saunders College Publishing

Pasquini, L., Avila, G., Blecha, A., Cacciari, C., Cayatte, V., Colless, M., Damiani, F., de Propris, R., Dekker, H., di Marcantonio, P., Farrell, T., Gillingham, P., Guinouard, I., Hammer, F., Kaufer, A., Hill, V., Marteaud, M., Modigliani, A., Mulas, G., North, P., Popovic, D., Rossetti, E., Royer, F., Santin, P., Schmutzer, R., Simond, G., Vola, P., Waller, L. & Zoccali, M. 2002, *Installation and Commissioning of FLAMES, the VLT Multifibre Facility*, The ESO Messenger, 110

Pasquini, L., Bonifacio, P., Randich, S., Galli, D. & Gratton, R. G. 2004, *Be in turn-off stars of NGC 6397: early Galaxy spallation, cosmochronology and cluster formation*, Astronomy & Astrophysics, 426, 2

Pedersen, H. & Cullum, M. 1982, *The CCD on La Silla*, The ESO Messenger, 30

Pedersen, H., Rigaut, F. & Sarazin, M. 1988, *Seeing Measurements with a Differential Image Motion Monitor*, The ESO Messenger, 53

Pedersen, H., Lindgren, H. & de Bon, C. 1991, *The Vaca Muerta Mesosiderite*, The ESO Messenger, 65

Peterson, B. A., D'Odorico, S., Tarenghi, M. & Wampler, E. J. 1991, *The NTT Provides the Deepest Look Into Space*, The ESO Messenger, 64

Physics World: *Boost for UK physics*, 13 May 1999

Plathner, D. 1976, *The 3.6-m Telescope on La Silla*, The ESO Messenger, 5

Quirrenbach, A. 2000, *Adaptive Optics with Laser Guide Stars: Basic Concepts and Limitations;* in Ageorges, N. and Dainty, J. C. (eds.): Laser guide star adaptive optics for astronomy, North Atlantic Treaty Organization, Scientific Affairs Division and Springer

Reimers, D. & Wisotzki, L. 1997, *Highlights From The Key Programme: 'A Wide-Angle Objective Prism Survey For Bright Quasars' — The Hamburg/ESO Survey*, The ESO Messenger, 88

Reipurth, B. 1995, *Astronomers in Chile Meet at ESO in Vitacura*, The ESO Messenger, 80

Reipurth, B. 2010, private communication

Reipurth, B. & Madsen, C. 1989, *Signposts of Low Mass Star Formation in Molecular Clouds*, The ESO Messenger, 55

Richichi, A. & Moorwood, A. 2006, *Second-generation VLTI Instruments: a First Step is Made*, The ESO Messenger, 125

Richter, W. 1976, *On the Vertical Support of Astronomical Research in Cassegrain Cages*, The ESO Messenger, 6

Richter, W. 1978, *Mechanical Constraints on Large Telescopes*, in Pacini, F., Richter, W. & Wilson, R. N. (eds.) 1978, ESO Conference on Optical Teescopes of the Future, Geneva, 12–15 December, 1977

Riess, A. G., Filippenko, A. V., Challis, P., Clocchiatti, A., Diercks, A., Garnavich, P. M., Gilliland, R. L., Hogan, C. J., Jha, S., Kirshner, R. P., Leibundgut, B., Phillips, M. M., Reiss, D., Schmidt, B. P., Schommer, R. A., Smith, R. C., Spyromilio, J. & Stubbs, C. 1998, *Observational Evidence from Supernovae for an Accelerating Universe and a Cosmological Constant,* The Astronomical Journal, 116, 1009

Roddier, F. & Sarazin, M. 1990, *The ESO differential image motion monitor*, Astronomy & Astrophysics, 227

Rule, B. H. & Sisson, G. M. 1965, *Types of Mounting*, in Crawford, D. L. (ed.), The Construction of Large Telescopes, Proceedings from Symposium no. 27 held in Tucson, Arizona, Pasadena and Mount Hamilton, California, USA, 5–12 April 1965, International Astronomical Union — Symposium no. 27, Academic Press, London

Rupprecht, G., Böhnhardt, H., Moehler, S., Møller, P., Saviane, I. & Ziegler, B. 2010, *Twenty Years of FORS Science Operations on the VLT*, The ESO Messenger, 140

Sack, M. 1980, *Det Ding muß sich drehn — Die astralen Zwillinge von München-Garching*, Die Zeit

Sarazin, M. 1986, *Seeing at La Silla: LASSCA 86*, The ESO Messenger, 44

Sarazin, M. 1988, *Site Evaluation for the VLT: Seeing Monitor No. 2 Tested in Garching*, The ESO Messenger, 52

Sarazin, M. 1992, *PARSCA 92: The Paranal Seeing Campaign*, The ESO Messenger, 68

Saviane, I., Ihle, G., Sterzik, M. & Kaufer, A. 2009, *La Silla 2010+*, The ESO Messenger, 136

Schilling, G. & Christensen, L. L. 2009, *Eyes on the Skies — 400 years of telescopic discovery*, Wiley VCH

Schmidt-Kaler, T. & Dachs, J. 1968, *The 61 cm Photometric Telescope of the Bochum University at La Silla,* The ESO Bulletin, 5

Schramm, J. 1996, *2010, Sterne über Hamburg — Die Geschichte der Astronomie in Hamburg*, Kultur und Geschichtskontor Bergedorf

Setti, G., Ulrich, M.-H., Dias de Deus, J. & Lago, T. 1988, *On the Situation of Astronomy in Portugal*, Report by JNICT to the Portuguese Government and to ESO

Shaver, P. A. 1989, *VLT Operations — a First Discussion*, The ESO Messenger, 57

Shaver, P. A. & Booth, R. S. 1991, *A Report on the SEST Users Meeting and Workshop on Millimetre-Wave Interferometry, 22-23 May 1991*, The ESO Messenger, 65

Shaver, P. A. & Booth, R. S. 1998, *The Large Southern Array*, The ESO Messenger, 91

Shustov, B. 2010, Private communication with the author.

Srianand, R., Chand, H., Petitjean, P. & Aracil, B. 2004, *Limits on the time variation of the electromagnetic fine-structure constant in the low energy limit from absorption lines in the spectra of distant quasars*, Physical Review Letters, 92:121302

Srianand, R., Noterdaeme, P., Ledoux, C. & Petitjean, P. 2008, *First detection of CO in a high-redshift damped Lyman-alpha system*, Astronomy & Astrophysics, 482, L39

Stock, J. 1968, *Astronomical observing conditions in Northern Chile*, The ESO Bulletin, 5

Soucail, G., Mellier, Y., Fort, B., Mathez, G. & Cailloux, M. 1988, *The giant arc in A 370: spectroscopic evidence for gravitational lensing from a source at $z = 0.724$*, Astronomy & Astrophysics, 191, LI9

Surdej, J., Swings, J.-P., Magain, P., Courvoisier, T. J.-L., Kuhr, H., Refsdal, S., Borgeest, U., Kayser, R. & Kellermann, K. 1987, *A new case of gravitational lensing*, Nature, 329, 695

Surdej, J., Arnaud, J., Borgeest, U., Djorgovski, S., Fleischmann, F., Hammer, F., Hursemekers, D., Kayser, R., Lefevre, O., Nottale, L., Magain, P., Meylan, G., Refsdal, S., Remy, M., Shaver, P., Smette, A., Swings, J.-P., Vanderriest, C., Van Drome, E., Veron-Cetty, M., Veron, P. & Weigelt, G. 1989, *Profile of a Key Programme: Gravitational Lensing*, The ESO Messenger, 55

Tammann, G. A. 1995, *The Role of ESO in European Astronomy*, in Krige, J. & Guzetti, L. (eds.): *History of European Scientific and Technological Cooperation*, Conference in Florence, November 1995, Proceedings, European Commission

Tammann, G. A. & Véron, P. 1985, *Halley's Komet*, Birkhäuser Verlag

Tanvir, N. R., Fox, D. B., Levan, A. J., Berger, E., Wiersema, K., Fynbo, J. P. U., Cucchiara, A., Krühlers, T., Hjorth, J., Jakobsson, P., Gehrels, N., Bloom, J. S., Greiner, J., Evans, P., Rol, E., Olivares, F., Farihi, J., Willingale, R., Starling, R. L. C., Cenko, S. B., Perley, D., Maund, J. R., Duke, J., Wijers, R. A. M. J., Adamson, A. J., Allan, A., Bremer, M. N., Burrows, D. N., Castro-Tirado, A. J., Cavanagh, B., de Ugarte Postigo, A., Dopita, M. A., Fatkhullin, T. A., Fruchter, A. S., Foley, R. J., Gorosabel, J., Holland, S. T., Kennea, J., Kerr, T., Klose, S., Krimm, H. A., Komarova, V. N., Kulkarni, S. R., Moskvitin, A. S., Mundell, C., Naylor, T., Page, K., Penprase, B. E., Perri, M., Podsiadlowski, P., Roth, K., Rutledge, R. E., Sakamoto, T., Schady, P., Schmidt, B. P., Soderberg, A. M., Sollerman, J., Stephens, A. W., Stratta, G., Ukwatta, T., Watson, D., Westra, E., Wold, T. & Wolf, C. 2009, *A γ-ray burst at a redshift of $z \approx 8.2$*, Nature, 461, 1254

Tarenghi, M., Gray, P., Spyromilio, J. & Gilmozzi, R. 1998, *The First Steps of UT1*, The ESO Messenger, 93

Tarenghi, M. 2010, Private communication with the author

UK Government 1998, *Modern Public Services for Britain: Investing in Reform — Comprehensive Spending Review*: New Public Spending Plans 1999–2002

van Belle, G., Sahlmann, J., Abuter, R., Accardo, M., Andolfato, L., Brillant, S., de Jong, J., Derie, F., Delplancke, F., Phan Duc, T., Dupuy, C., Gilli, B., Gitton, P., Haguenauer, P., Jocou, L., Jost, A., Di Lieto, N., Frahm, R., Mébardi, S., Morel, S., Moresmau, J.-M., Palsa, R., Popovic, D., Pozna, E., Puech, F., Lévêque, S., Ramirez, A., Schuhler,

N., Somboli, F. & Wehner, S. 2008, *The VLTI PRIMA facility*, The ESO Messenger, 134

van Boekel, R., Min, M., Leinert, C., Waters, L. B. F. M., Richichi, A., Chesneau, O., Dominik, C., Jaffe, W., Dutrey, A., Graser, U., Henning, T., de Jong, J., Köhler, R., de Koter, A., Lopez, B., Malbet, F., Morel, S., Paresce, F., Perrin, G., Preibisch, T., Przygodda, F., Schöller, M. & Wittkowski, M. 2004, *The building blocks of planets within the 'terrestrial' region of protoplanetary disks*, Nature, 432, 479

van der Kruit, P. 2012, private communication with the author

van der Laan, H. 1988, *Key Programmes on La Silla: a Preliminary Enquiry*, The ESO Messenger, 51

van der Laan, H. 1989, *The New Research Student Programme of the European Southern Observatory*, The ESO Messenger, 55

Véron, P. 1981, *The Inauguration of the ESO Headquarters Building at Garching*, The ESO Messenger, 24

von der Lühe, O., Bonaccini, D., Derie, F., Koehler, B., Lévêque, S., Manil, E., Michel, A. & Verola, M. 1997, *A New Plan for the VLTI*, The ESO Messenger, 87

von der Lühe, O., Berkefeld, T. & Soltau, D. 2005, *Multi-conjugate solar adaptive optics at the Vacuum Tower Telescope on Tenerife*, Comptes Rendus — Physique, 6, 10

Vreeswijk, P. M., Kaufer, A., Spyromilio, J., Schmutzer, R., Ledoux, C., Smette, A. & De Cia, A. 2010, *The VLT Rapid-Response Mode: implementation and scientific results*, SPIE Conference

West, R. M. 1974, *The Southern Sky Surveys — A review of the ESO Sky Survey Project*, The ESO Bulletin, 10

West, R. M. 1983, *Comets — Distant and Nearby*, The ESO Messenger, 32

West, R. M. 1985, *Comet Halley Observed at ESO*, The ESO Messenger, 41

West, R. M. 1994, *Comet Shoemaker-Levy 9 Collides with Jupiter — The Continuation of a Unique Experience*, The ESO Messenger, 77

West, R. M. 2002, *Memories of early times at ESO*, The ESO Messenger, 109

Wheeler, J. C. 2007, *Cosmic Catastrophes: Exploding Stars, Black Holes, and Mapping the Universe*, Cambridge University Press

Wilson, R. N. 1990, *Spherical Aberration in the HST: Optical Analysis of the Options Available*

Wilson, R. N. 1996, *Reflecting telescope optics I: basic design theory and its historical development*, Springer

Wilson, R. N. 1999, *Reflecting telescope optics II: basic design theory and its historical development*, Springer

Wilson, R. N. 2002, *First Astronomical Light at the NTT*, The ESO Messenger, 109

Wilson, R. N. 2003, *The History and Development of the ESO Active Optics System*, The ESO Messenger, 113

Wilson, R. N. 2010, Private communication with the author

Wilson, R. N. 2012, Private communication with the author

Wilson, R. N., Franza, F., Giordano, P., Noethe, L. & Tarenghi, M. 1988, *Active Optics: the NTT and the Future*, The ESO Messenger, 53

Woltjer, L. 1977, *The Case for Large Optical Telescopes*, in Pacini, F., Richter, W. & Wilson, R. N. (eds.) 1977, Optical Telescopes of the Future, Proceedings ESO Conference Geneva, December 12th-15th 1977

Woltjer, L. 1991, *The 'Discovery' of Paranal*, The ESO Messenger, 64

Woltjer, L. 2006, *Europe's Quest for the Universe*, EDP Sciences, Paris

Zago, L. 1992, *The Choice of the Telescope Enclosures for the VLT*, The ESO Messenger, 70

Appendices

Appendix 1

Important Milestones

Important milestones during the period covered in Part I

June 1953 — First discussions about a shared European observatory between Walter Baade and Jan Oort. Immediately thereafter, the subject is further discussed at the IAU Symposium No. 1 in Groningen.

26 January 1954 — The Leiden Declaration by leading astronomers from six European countries formally proposes that a joint European observatory be established in the southern hemisphere.

December 1955 — Site testing begins in South Africa to identify the best location for the ESO observatory.

5 October 1962 — Founding Members Belgium, France, Germany, the Netherlands and Sweden sign the ESO Convention at a ceremony in Paris.

1 November 1962 — Otto Heckmann becomes the first Director (General) of ESO, and provisional ESO offices are set up at the Hamburger Sternwarte.

November 1962 — Site testing in Chile begins.

7 November 1963 — Chile is selected to host the ESO observatory and the Agreement between Chile and ESO (the "*Convenio*") is signed.

15 November 1963 — Decision to establish the ESO Santiago office.

17 January 1964 — The ESO Convention enters into force following parliamentary ratification by France, Germany, the Netherlands and Sweden.

5 February 1964 — The ESO Council approves the Chile/ESO Agreement.

17 April 1964 — The Chilean Congress ratifies the Chile/ESO Agreement.

26 May 1964 — The ESO Council selects La Silla as the site for its observatory.

30 October 1964 — Acquisition of La Silla and land for the ESO Santiago Office in the Vitacura district.

March 1965 — ESO acquires a guesthouse in the Las Condes district of Santiago.

March 1965 — Construction of the La Silla Observatory begins.

24 March 1966 — Dedication ceremony for the road to the summit of La Silla.

30 November 1966 — First light for the ESO 1-metre telescope at La Silla, the first telescope at the observatory.

January 1967 — Construction of the ESO Santiago Office begins.

24 August 1967 — Denmark formally joins ESO as its 6th Member State.

2 October 1967 — Belgium, a founding member, ratifies the ESO Convention and thus formally joins the organisation.

July 1968 — First light for the GPO at La Silla.

July 1968 — First light for the ESO 1.52-metre telescope.

7 September 1968 — First light for the Bochum 0.6-metre telescope.

December 1968 — ESO helps found the journal *Astronomy and Astrophysics*.

February 1969 — First light for the Danish 0.5-metre telescope.

25 March 1969 — Inauguration of the La Silla Observatory by the President of the Republic of Chile, Eduardo Frei Montalva, and of the ESO Office in Santiago.

1 January 1970 — Adriaan Blaauw takes over as Director General of ESO.

16 September 1970 — ESO signs an agreement with CERN to collaborate on the realisation of the ESO 3.6-metre telescope, enabling ESO to establish offices at the CERN premises in Geneva and draw on CERN staff.

October 1970 — ESO's Telescope Division is set up at CERN.

December 1971 — First light for the ESO 0.5-metre telescope.

21 December 1971 — First light for the ESO 1-metre Schmidt telescope.

May 1974 — The first edition of *The ESO Messenger*, ESO's quarterly in-house magazine, is published.

1 January 1975 — Lodewijk Woltjer takes over as Director General of ESO.

10 November 1975 — First Light of the Swiss 0.4-metre telescope.

2 December 1975 — The ESO Council approves the proposal to locate ESO's permanent Headquarters building in Garching bei München, Germany.

7 November 1976 — First light for the ESO 3.6-metre telescope.

12–15 December 1977 — Major ESO Conference at CERN about future large telescopes leads to first ideas about the VLT.

1978 — Completion of the initial Quick Blue Survey of the southern sky, made with the 1-metre Schmidt telescope.

2 October 1978 — Construction of the new ESO Headquarters in Garching begins.

20 November 1978 — First light for the Danish 1.54-metre telescope.

31 January 1979 — An agreement is signed between ESO and the German Federal government as host state for the new ESO Headquarters.

March 1979 — First light for the Dutch 0.9-metre telescope.

8 August 1980 — First Light for the Swiss T70 telescope.

24–27 March 1981 — ESO Conference on the Scientific Importance of High Angular Resolution at Infrared and Optical Wavelengths. First milestone conference relating to large-scale optical interferometry at ESO.

5 May 1981 — First light of the 1.4-metre Coudé Auxiliary Telescope (CAT), and its Coudé Echelle Spectrometer (CES).

5 May 1981 — Inauguration of the new ESO Headquarters in Garching, Germany.

Important milestones during the period covered in Part II

1 March 1982 — Switzerland formally joins ESO as its 7th Member State.

24 May 1982 — Italy formally joins ESO as its 8th Member State. The joining fees of Switzerland and Italy are earmarked for the development of the NTT, a testbed for VLT-relevant technologies.

23 February 1983 — Agreement between ESA and ESO about the establishment of the Space Telescope European Coordinating Facility (ST-ECF) is signed.

16–19 May 1983 — Key meeting of European astronomers in Cargèse, Corsica, builds support for the VLT array concept.

June 1983 — First Light of the Cassegrain Echelle Spectrograph (CASPEC) at the ESO 3.6-metre telescope.

22 June 1983 — First light for the MPG/ESO 2.2-metre telescope.

1 March 1984 — ST-ECF begins operations at ESO Garching.

June 1984 — First tests at ESO for remote control of a telescope are carried out with the MPG/ESO 2.2-metre telescope.

November 1984 — First light of the *f*/35 chopping secondary system and the infrared photometers on the ESO 3.6-metre telescope.

November 1985 — First light of IRSPEC on the ESO 3.6-metre telescope.

29 September–2 October 1986 — The European scientific community supports ESO plans for the VLT at a major conference in Venice.

February 1987 — ESO plays a major role in the observations of Supernova 1987A in the Large Magellanic Cloud.

24 March 1987 — First light for the 15-metre Swedish–ESO Submillimetre Telescope (SEST).

8 December 1987 — Decision by the ESO Council to build the Very Large Telescope (VLT).

1 January 1988 — Harry van der Laan takes over as Director General of ESO.

July 1988 — First light of IRAC on the MPG/ESO 2.2-metre telescope.

October 1988 — The Chilean Government donates the land around Cerro Paranal to ESO.

23 March 1989 — First light of the 3.5-metre New Technology Telescope (NTT).

11 May 1989 — First light of the second ESO Faint Object Spectrograph and Camera (EFOSC2) instrument on the NTT.

16 April 1990 — First light of the Come-On instrument on the ESO 3.6-metre telescope.

4 December 1990 — Council approves Cerro Paranal as the site for the VLT.

Important milestones during the period covered in Part III

23 September 1991 — Construction of the Paranal Observatory begins.

30 May 1992 — First light of the IRAC2 instrument on the MPG/ESO 2.2-metre telescope.

21 July 1992 — First light of the Thermal Infrared MultiMode Instrument (TIMMI) on the ESO 3.6-metre telescope.

15 December 1992 — First light of the Come-On+ instrument on the ESO 3.6-metre telescope.

1 January 1993 — Riccardo Giacconi takes over as Director General of ESO.

16–22 July 1994 — The collision between Comet Shoemaker-Levy 9 and Jupiter draws worldwide attention to astronomy. La Silla telescopes play a major role in the observing campaign and the ESO Headquarters in Garching acts as focal point for the European media.

April 1995 — Site testing for the future Atacama Large Millimeter/submillimeter Array (ALMA) takes place in Chile together with National Radio Astronomy Observatory and National Astronomical Observatory of Japan.

18 April 1995 — Amendment to the *Convenio* between the Chilean Government and ESO is initialled.

5 September 1996 — The Chilean Senate ratifies the Amendment to the *Convenio* with ESO.

2 December 1996 — Amendment to the *Convenio* between the Chilean Government and ESO is signed.

4 December 1996 — Paranal Foundation Ceremony.

25–26 June 1997 — Memorandum of Understanding between ESO and NRAO about working towards a joint project in millimetre/submillimetre astronomy.

6 December 1997 — First light for the Son OF ISAAC instrument (SOFI) on the NTT.

11 February 1998 — First light for the second SUperb-Seeing Imager (SUSI2) on the NTT.

12 April 1998 — First light for the Swiss 1.2-metre Leonhard Euler telescope at La Silla.

25 May 1998 — First light for the first 8.2-metre VLT Unit Telescope (UT1).

15 September 1998 — First light for the first visual and near-UV FOcal Reducer and the low dispersion Spectrograph (FORS1) on VLT UT1.

6 October 1998 — First light of the Fibre-fed Extended Range Optical Spectrograph (FEROS) on the ESO 1.52-metre telescope.

16 November 1998 — First light of the Infrared Spectrometer And Array Camera (ISAAC) instrument on VLT UT1.

18 December 1998 — The work of two independent research teams, which includes observations at La Silla, are described as the "Breakthrough of the Year" by *Science*. Their results indicate that the expansion of the Universe is accelerating. In 2011 the Nobel Prize in Physics was awarded to the principal investigators for this discovery.

15 January 1999 — First light with the 67-million-pixel Wide Field Imager (WFI) camera on the MPG/ESO 2.2-metre telescope.

1 March 1999 — First light for VLT UT2.

5 March 1999 — Official inauguration of the Paranal Observatory. The four UTs are given the names Antu (UT1), Keuyen (UT2), Melipal (UT3) and Yepun (UT4).

Important milestones during the period covered in Part IV

June 1999 — The choice of Llano de Chajnantor is endorsed by ESO Council as the site for ALMA.

1 September 1999 — Catherine Cesarsky takes over as Director General of ESO.

27 September 1999 — First light for the Ultraviolet Visual Echelle Spectrograph (UVES) on Kueyen.

29 October 1999 — First light for FORS2 on Kueyen.

26 January 2000 — First light for VLT UT3, Melipal.

4 September 2000 — First light for VLT UT4, Yepun.

11 October 2000 — First light of the second Thermal Infrared MultiMode Instrument (TIMMI2) on the ESO 3.6-metre telescope.

17 March 2001 — First light for the Very Large Telescope Interferometer (VLTI).

5 April 2001 — ESO signs an agreement with representatives from North America on Phase one of the ALMA project.

7 May 2001 — Portugal formally joins ESO and becomes the 9th Member State.

25 November 2001 — First light for NAOS/CONICA instrument on VLT UT4, Yepun.

26 February 2002 — First light of the VIsible Multi-Object Spectrograph (VIMOS) on VLT UT3, Melipal.

1 April 2002 — First light of the Fibre Large Array Multi Element Spectrograph (FLAMES) on UT2, Kueyen.

24 June 2002 — The United Kingdom joins ESO as its 10th Member State.

December 2004 — The ESO Council accords the highest priority to ESO leading the European efforts to build en Extremely Large Telescope (ELT).

15 December 2002 — First light of the MID-infrared Interferometric instrument (MIDI) on the VLTI.

11 February 2003 — First light of the High Accuracy Radial Velocity Planet Searcher (HARPS) at the ESO 3.6-metre telescope.

25 February 2003 — ESO and the US National Science Foundation (NSF) sign a bilateral agreement on the Phase two (construction) of ALMA.

18 April 2003 — First light of the MACAO–VLTI facility.

24 June 2003 — The 0.6-metre Rapid Eye Mount (REM) telescope starts operations at La Silla.

25 July 2003 — The Republic of Chile grants concession of the land on Chajnantor for the ALMA project.

6 November 2003 — Ground-breaking ceremony at the site of the ALMA Operational Support Facility.

24 January 2004 — First light for the first 1.8-metre Auxiliary Telescope (AT1).

21 March 2004 — First fringes of the Astronomical Multi-BEam combineR (AMBER) on the VLTI.

30 April 2004 — First light for the VLT Imager and Spectrometer in the InfraRed (VISIR) on UT3, Melipal.

7 July 2004 — Finland formally joins ESO as the 11th Member State.

9 July 2004 — First light for the Spectrograph for INtegral Field Observation in the Near-Infrared (SINFONI) on the VLT's UT4, Yepun.

10 September 2004 — The VLT obtains the first-ever image of a planet outside the Solar System.

14 September 2004 — Agreement is reached between ESO, the US National Science Foundation and the National Institutes of Natural Sciences, Japan, concerning the construction of ALMA.

2 February 2005 — First light for the second Auxiliary Telescope (AT2).

4 July 2005 — As with the case of the collision between Comet Shoemaker-Levy 9 and Jupiter, ESO plays an important role, both scientifically and with respect to the media, in connection with the Deep Impact mission, which collided a dedicated space probe with Comet 9P/Tempel 1.

14 July 2005 — First light for the submillimetre Atacama Pathfinder Experiment (APEX).

1 November 2005 — First light for the third Auxiliary Telescope (AT3).

7 December 2005 — ESO signs the European contract for the production of up to 32 ALMA antennas. This is the largest ever ESO contract for industrial work on a ground-based astronomy project.

28 January 2006 — First light of the VLT laser guide star, on UT4, Yepun.

28 February 2006 — Decision to host the ALMA Santiago Central Office at ESO Vitacura.

4 June 2006 — First light for the CRyogenic high-resolution InfraRed Echelle Spectrograph (CRIRES) at UT1.

15 September 2006 — The robotic TAROT–South telescope starts work at La Silla.

11 December 2006 — The ESO Council agrees to proceed with studies for the European Extremely Large Telescope (E-ELT).

27 November–1 December 2006 — The European scientific community supports ESO's ELT plans at a major conference in Marseille.

15 December 2006 — First light for the fourth Auxiliary Telescope (AT4).

14 February 2007 — Spain formally joins ESO as its 12th Member State.

25 March 2007 — First light for the Multi-conjugate Adaptive optics Demonstrator (MAD) at the visitor focus of Melipal.

30 April 2007 — The Czech Republic formally joins ESO as the 13th Member State.

5 August 2007 — First light for the LArge BOlometer CAmera (LABOCA) instrument on APEX.

22 August 2007 — First light for the High Acuity, Wide field *K*-band Imaging (HAWK-I) instrument on UT4, Yepun.

1 September 2007 — Tim de Zeeuw takes over as Director General of ESO.

20 December 2007 — The International Year of Astronomy is approved by the United Nations designating UNESCO and the IAU together with ESO and other "associations and groups in astronomy" as implementing bodies.

8 September 2008 — First light for the Phase Referenced Imaging and Microarcsecond Astrometry (PRIMA) instrument on the VLTI.

9 November 2008 — First light for X-shooter on the VLT.

10 December 2008 — Several ESO telescopes were used in a 16-year-long study to obtain the most detailed view ever of the surroundings of the supermassive black hole at the Galactic Centre.

18 December 2008 — The ALMA Observatory is equipped with its first antenna.

1 July 2009 — Austria formally joins ESO as its 14th Member State.

6 July 2009 — Start of construction of the new ALMA headquarters at the ESO premises in Santiago's Vitacura district.

17 September 2009 — First ALMA antenna arrives at 5000-metre-altitude Chajnantor site.

25 November 2009 — First three ALMA antennas are successfully linked (phase closure) at 5000-metre-altitude Chajnantor site.

11 December 2009 — VISTA, the new infrared survey telescope, starts work.

Appendix 2

List of ESO Council Presidents and Directors General

Year	President of Council	Director General
1962	Jan Oort	Otto Heckmann
1963	Jan Oort	Otto Heckmann
1964	Jan Oort	Otto Heckmann
1965	Jan Oort/Bertil Lindblad	Otto Heckmann
1966	Gösta Funke	Otto Heckmann
1967	Gösta Funke	Otto Heckmann
1968	Gösta Funke	Otto Heckmann
1969	Henk Bannier	Otto Heckmann
1970	Henk Bannier	Adriaan Blaauw
1971	Henk Bannier	Adriaan Blaauw
1972	Augustin Alline	Adriaan Blaauw
1973	Augustin Alline	Adriaan Blaauw
1974	Augustin Alline/Henk Bannier	Adriaan Blaauw
1975	Bengt Strömgren	Lodewijk Woltjer
1976	Bengt Strömgren	Lodewijk Woltjer
1977	Bengt Strömgren	Lodewijk Woltjer
1978	Jean-François Denisse	Lodewijk Woltjer
1979	Jean-François Denisse	Lodewijk Woltjer
1980	Jean-François Denisse	Lodewijk Woltjer
1981	Jean-François Denisse	Lodewijk Woltjer
1982	Paul Ledoux	Lodewijk Woltjer
1983	Paul Ledoux	Lodewijk Woltjer
1984	Paul Ledoux	Lodewijk Woltjer
1985	Kurt Hunger	Lodewijk Woltjer
1986	Kurt Hunger	Lodewijk Woltjer
1987	Kurt Hunger	Lodewijk Woltjer
1988	Per-Olof Lindblad	Harry van der Laan
1989	Per-Olof Lindblad	Harry van der Laan
1990	Per-Olof Lindblad	Harry van der Laan
1991	Franco Pacini	Harry van der Laan
1992	Franco Pacini	Harry van der Laan
1993	Franco Pacini	Riccardo Giacconi
1994	Peter Creola	Riccardo Giacconi
1995	Peter Creola	Riccardo Giacconi
1996	Peter Creola	Riccardo Giacconi
1997	Henrik Grage	Riccardo Giacconi
1998	Henrik Grage	Riccardo Giacconi
1999	Henrik Grage	Riccardo Giacconi/Catherine Cesarsky

2000	Arno Freytag	Catherine Cesarsky
2001	Arno Freytag	Catherine Cesarsky
2002	Arno Freytag	Catherine Cesarsky
2003	Piet van der Kruit	Catherine Cesarsky
2004	Piet van der Kruit	Catherine Cesarsky
2005	Piet van der Kruit	Catherine Cesarsky
2006	Richard Wade	Catherine Cesarsky
2007	Richard Wade	Catherine Cesarsky/Tim de Zeeuw
2008	Richard Wade	Tim de Zeeuw
2009	Laurent Vigroux	Tim de Zeeuw
2010	Laurent Vigroux	Tim de Zeeuw
2011	Laurent Vigroux	Tim de Zeeuw
2012	Xavier Barcons	Tim de Zeeuw

Appendix 3

List of Interviewees

Johannes Andersen, Phillip Auer, Dietrich Baade, Klaus Banse, Xavier Barcons, Jacques Beckers, Fernando Bello, Adriaan Blaauw, Henri Boffin, Domenico Bonaccini Calia, Albert Bosker, Roy Booth, Bob Brown, Massimo Capaccioli, Mark Casali, Catherine Cesarsky, Ian Corbett, Phil Crane, John Danziger, Michel Dennefeld, Sandro D'Odorico, Philippe Dierickx, Hedwig Dröll, Jan Doornenbal, Robert Fischer, Bob Fosbury, Roland Geyl, Riccardo Giacconi, Alain Gilliotte, Andreas Glindemann, Roberto Gilmozzi, Flavio Gutierrez, Thijs de Graauw, Peter Gray, Henrik Grage, Michael Grewing, Preben Grosbøl, Olivier Hainaut, Daniel Hofstadt, Norbert Hubin, Henning Jørgensen, Bertrand Koehler, Luboš Kohoutek, Christine Sachs-Kohoutek, Joachim Krautter, Richard Kurz, Harry van der Laan, Teresa Lago, Svend Laustsen, Pierre Léna, Hans Michael Maitzen, Jon Marcaide, Kalevi Mattila, Alistair McPherson, Jorge Melnick, Suzanne Messerlian, Hans Morian, Jeremy Mould, Paul Murdin, Birgitta Nordström, Lars-Åke Nyman, Michael Olberg, Angel Otárola, Peter Quinn, Jan Palouš, Christian Patermann, Jutta Quentin, Bo Reipurth, Gero Rupprecht, Hans Rykaczewski, Marc Sarazin, Dominik Schenkirz, Hans-Emil Schuster, Michael Schneermann, Boris Shustov, Chris Sterken, Jean-Pierre Swings, Massimo Tarenghi, Ewine van Dishoeck, Paul Vanden Bout, Oskar von der Lühe, Richard West, Ray Wilson and Lodewijk Woltjer.

Additional information was provided by:
Mary Bauerle, Leopoldo Benacchio, Elly Berkhuijsen, Daniel Bonneau, Bernard Delabre, Frédéric Derie, Dieter Engels, Mart de Groot, Erik Høg, Gerardo Ihle, Masato Ishiguro, Max Kraus, Jet Merkelijn-Katgert, Bruno Leibundgut, Lothar Noethe, Maximilian Metzger, Franco Pacini, Jørgen Otzen Petersen, Valentina Rodriguez, Ulrich Schwarz, Josef Strasser, Jason Spyromilio, Will Sutherland, Piet van der Kruit, Huug van Woerden, Laura Ventura, Tom Wilson and Tim de Zeeuw.

Appendix 4

List of Acronyms

2MASS	Two Micron All Sky Survey
A&A	Astronomy and Astrophysics
AAAS	American Association for the Advancement of Science
AAS	American Astronomical Society
AAT	Anglo Australian Telescope
ABB	Asea-Brown Boveri
ACS	Advanced Camera for Surveys (HST)
ADONIS	Adaptive optics system (3.6-metre)
AEM	ALMA construction consortium
AIV	Assembly Integration and Verification
ALFA	Adaptive optics with Laser guide star For Astronomy
ALMA	Atacama Large Millimeter/submillimeter Array
alt-az	altitude-azimuth
AMBER	Astronomical Multiple BEam Recombiner (VLTI)
AMOS	Advanced Mechanical and Optical Systems (Belgium)
AO	Adaptive Optics
AOF	Adaptive Optics Facility
AOS	Array Operations Site (ALMA)
APEX	Atacama Pathfinder Experiment
ARC	ALMA Regional Centre
ASPERA	European Strategy for Astroparticle Physics
ASTE	Atacama Submillimeter Telescope Experiment
ASTEC	Australian Science and Technology Council
AstroGrid	Research and development project to produce a virtual observatory based on grid technologies
AT	Auxiliary Telescope
ATC	UK Astronomy Technology Centre
ATF	ALMA Test Facility
AU	Astronomical Unit
AUI	Associated Universities Inc.
AURA	Association of Universities for Research in Astronomy (US)
AUSTRALIS	Consortium of Australian institutes involved in OzPoz
AVO	Astrophysical Virtual Observatory
B&C	Boller & Chivens
BTA	Bolshoi Teleskop Alt-azimutalnyi (Large Altazimuth Telescope)
BUS	BackUp Structure (ALMA)
C&EE	Central and Eastern European
CAMCAO	Camera for Multiconjugated Adaptive Optics (VLT)
CARSO	Carnegie Southern Observatory
CASPEC	Echelle spectrograph for the 3.6-metre telescope
CAT	Coudé Auxiliary Telescope
CAUP	Centro de Astrofisica (Porto)
CCD	Charge Coupled Device
CDFS	Chandra Deep Field South
CDS	Centre de Données Astronomiques de Strasbourg
CEA	Commissariat à l'énergie atomique et aux énergies alternatives (Atomic Energy Commission, France)
CEPAL	United Nations Economic Commission for Latin America
CERGA	Centre d'Etudes et de Recherches Géodynamiques et Astronomiques
CERN	Conseil Européen pour la Recherche Nucléaire (the European Organization for Nuclear Research)
CES	Coudé Echelle Spectrometer
CfA	Harvard Smithsonian Center for Astrophysics
CFHT	Canada France Hawaii Telescope
CGE	Compagnie Générale d'Électricité
CMB	Cosmic Microwave Background
CNAA	Consorzio Nazionale per l'Astronomia e l'Astrofisica
CNR	Consiglio Nazionale delle Ricerche (Italian National Research Council)
CNRS	Centre National de la Recherche Scientifique
CO	Carbon Monoxide
COBE	Cosmic Background Explorer
Come-On	Adaptive optics prototype (3.6-metre)
Come-On+	Adaptive optics prototype (3.6-metre)

CONICA	Coudé Near-Infrared Camera (VLT)
CONICYT	Comisión Nacional de Investigación Científica y Tecnológica
CORALIE	Echelle Spectrograph (Swiss 1.2-metre Leonard Euler telescope)
CORAVEL	Spectrovelocimeter (La Silla & OHP)
COSTAR	Corrective Optics Space Telescope Axial Replacement (HST)
CRIRES	CRyogenic high-resolution InfraRed Echelle Spectrograph (VLT)
CSIC	Consejo Superior de Investigaciones Científicas
CSIRO	Commonwealth Scientific and Industrial Research Organisation
CTIO	Cerro Tololo Inter-American Observatory
DARPA	Defense Advanced Research Projects Agency
DG	Director General
DIMM	Differential Image Motion Monitor
DLR	Deutsche Luft- und Raumfahrtforschung
DSM	Direction des Sciences de la Matière (CEA)
EAAE	European Association for Astronomy Education
EAS	European Astronomical Society
EBU	European Broadcasting Union
EC	European Commission
EEAB	ELT External Advisory Board
E-ELT	European Extremely Large Telescope
EFDA	European Fusion Development Agreement
EFOSC	ESO Faint Object Spectroscopic Camera (3.6 metre)
EIE	European Industrial Engineering
EIROforum	European Intergovernmental Research Organisations' Forum
ELDO	European Launcher Development Organisation
ELODIE	Cross-dispersed echelle spectrograph (OHP)
ELT	Extremely Large Telescope
EMBL	European Molecular Biology Laboratory
EMMI	ESO Multi-Mode Instrument (NTT)
ENO	European Northern Observatory
ERA	European Research Area
ERC	European Research Council
ESA	European Space Agency
ESAC	European Science Advisory Committee (for ALMA)
ESE	ELT Science and Engineering working group
ESFRI	European Strategy Forum for Research Infrastructures
ESO	European Southern Observatory (European Organisation for Astronomical Research in the Southern Hemisphere)
ESPRESSO	Echelle SPectrograph for Rocky Exoplanet- and Stable Spectroscopic Observations (VLT)
ESRC	ELT Standing Review Committee
ESRF	European Synchrotron Radiation Facility
ESRO	European Space Research Organisation
ESTI	EIROforum Science Teachers' Initiative
EU	European Union
EURAB	European Research Advisory Board
EUREKA	Pan-European network encouraging market-oriented, collaborative research and development (R&D) projects
EVA	Extra-Vehicular Activity
FC	Finance Committee
FEIC	Front-End Integration Centre (ALMA)
FEROS	Fibrefed Extended Range Optical Spectrograph
FIERA	CCD controller electronics
FINITO	Fringe-tracking Instrument of NIce and Torino (VLTI)
FLAMES	Multi-object, intermediate and high resolution spectrograph (VLT)
FOC	Faint Object Camera (HST)
FORS1	FOcal Reducer and low dispersion Spectrograph (VLT)
FORS2	FOcal Reducer and low dispersion Spectrograph 2 (VLT)
FOSC	Faint Object Spectroscopic Camera
FP6	Sixth EC Framework Programme
FSU	Friedrich-Schiller-Universität (Jena)
FUEGOS	Multi-Fibre Spectrograph (VLT)
FWHM	Full Width at Half Maximum
GB	Gigabyte
GCSII	Guide Star Catalog II
GDP	Gross Domestic Product
GHz	Gigahertz
GIRAFFE	Medium-high resolution spectrograph (VLT)

GMS	Gamma-ray burst Monitoring System (La Silla)
GMT	Giant Magellan Telescope
GOODS	Great Observatories Origins Deep Survey
GPO	Grand Prism Objectif
GRAAL	GRound-layer Adaptive optics Assisted by Lasers (AOF)
GRAVITY	Adaptive optics assisted, near-infrared VLTI instrument (VLTI)
GRB	Gamma-ray burst
GTC	Gran Telescopio Canarias
HARPS	High Accuracy Radial velocity Planet Searcher (3.6-metre)
HAWK-I	High Acuity Wide field K-band Imager (VLT)
HDF	Hubble Deep Field
HES	Hamburg/ESO Survey
HH	Herbig–Haro (objects)
HIRES	Keck 10-metre telescope spectrograph
HRS	High Resolution Spectrograph (HST)
HST	Hubble Space Telescope
IAA	Institute of Astrophysics of Andalucía
IAC	Instituto de Astrofísica Canarias
IAP	Institut Astrophysique de Paris
IAU	International Astronomical Union
IC	Index Catalogue
ICSU	International Council of Scientific Unions
IGN	Instituto Geográfico Nacional
IHAP	Image Handling and Processing System
ILIAS	Integrated Large Infrastructures for. Astroparticle Science
ILL	Institut Laue-Langevin
IMCCE	Institut de Mecanique Celeste et de Calcul des Ephemerides
INAF	Istituto Nazionale di Astrofisica (Italy)
INNSE	Innocenti-Santeustacchio (Italy)
INTAS	The International Association for the promotion of co-operation with scientists from the New Independent States of the former Soviet Union
IR	Infrared
IRAC	InfraRed Array Camera (MPG/ESO 2.2-metre)
IRACE	Infrared Detector High Speed Array Control and Processing Electronics
IRAF	Image Reduction and Analysis Facility
IRAM	Institut de Radioastronomie Millimétrique
IRSPEC	Infrared spectrometer for the NTT
ISAAC	Infrared Spectrometer And Array Camera (VLT)
ISO	Infrared Space Observatory (ESA)
ISOCAM	Infrared Camera (ISO)
ISR	Intersecting Storage Rings (CERN project)
IT	information Technology
IUE	International Ultraviolet Explorer
IYA2009	International Year of Astronomy 2009
JAO	Joint ALMA Office
JCMT	James Clerk Maxwell Telescope
JET	Joint European Torus
JIF	Joint Infrastructure Fund
JNICT	Junta Nacional de Investigação Cientifica e Tecnológica
JPL	Jet Propulsion Laboratory
KMOS	*K*-band Multi-Object Spectrograph (VLT)
KPNO	Kitt Peak National Observatory
LABOCA	Large APEX BOlometer CAmera
LAM	L'Observatoire Astronomique Marseille-Provence
LAMA	Large Active Mirrors in Aluminum
LAMOST	Large Sky Area Multi-Object Fibre Spectroscopic Telescope
LAOG	Laboratoire d'Astrophysique de Grenoble
LASSCA	La Silla Seeing Campaign
LBT	Large Binocular Telescope
LESS	LABOCA Survey of the Extended Chandra Deep Field South
LETI	Laboratoire d'électronique et de technologie de l'information (Grenoble)
LGS	Laser Guide Star
LGSF	Laser Guide Star Facility
LMC	Large Magellanic Cloud
LSA	Large Southern Array
LSST	Large Synoptic Survey Telescope
LZOS	Lytkarino Optical Glass Factory
M#n	Mirror #n
MACHO	Massive Compact Halo Objects
MAD	Multi-conjugate Adaptive optics Demonstrator
mas	Milliarcseconds
MATISSE	Multi AperTure mid-Infrared SpectroScopic Experiment (VLTI)

MATRA	French automotive and defence company
MFAS	Multi-Fibre Area Spectrograph (VLT)
MIDAS	Munich Image Data Analysis System
MIDI	MID-infrared Interferometric instrument (VLTI)
MIT	Massachusetts Institute of Technology
MLE	Maximum Likely Earthquake
MMA	Millimeter Array
MMB	Mirror Maintenance Building
MMT	Multiple Mirror Telescope
MPA	Max-Planck-Institut für Astrophysik
MPE	Max-Planck-Institut für extraterrestrische Physik
MPG	Max-Planck-Gesellschaft
MPIfR	Max-Planck-Institut für Radioastronomie
MPP	Max-Planck-Institut für Plasmaphysik
MPQ	Max-Planck-Institut für Quantenoptik
msl	metres above sea level
MUSE	Multi Unit Spectroscopic Explorer (VLT)
NACO	NAOS-CONICA (VLT)
NAOJ	National Astronomical Observatory of Japan
NAOS	Nasmyth Adaptive Optics System (VLT)
NASA	National Aeronautics and Space Administration (US)
NATO	North Atlantic Treaty Organization
NEON	Network of European Observatories in the North
NFR	Swedish Natural Science Research Council
NFRA	Netherlands Foundation for Research in Astronomy
NGC	New General Catalogue
NGST	Next Generation Space Telescope
NICMOS	Near Infrared Camera and Multi-Object Spectrometer (HST)
NINS	National Institutes of Natural Sciences
NIRMOS	Near-InfraRed Multi-Object Spectrograph
NIS	New Independent States (Soviet Union)
NNI	Net National Income
NNTT	National New Technology Telescope
NOAO	National Optical Astronomy Observatory
NOT	Nordic Optical Telescope
NOVA	The Netherlands Research School for Astronomy (Nederlandse Onderzoekschool voor Astronomie)
NRAO	National Radio Astronomy Observatory (US)
NRC	National Research Council
NSF	National Science Foundation (US)
NTT	New Technology Telescope
OB	Observing Block
OBE	Operating Basis Earthquake
OECD	Organisation for Economic Co-operation and Development
ÖGA[2]	Austrian Society for Astronomy and Astrophysics
OHP	Observatoire de Haute-Provence
OmegaCAM	VST survey camera
ONERA	Office National D'Etudes et de Recherche Aerospatiale
OPC	Observing Programmes Committee
OPD	Optical Path Difference
OPTICON	Optical Astronomy Network (FP5)
OSF	Operations Support Facility (ALMA)
OSO	Onsala Space Observatory
OWL	OverWhelmingly Large Telescope
OzPoz	Multi-fibre positioner feeding GIRAFFE and FLAMES
PARSCA 92	Paranal Seeing Campaign
PARSEC	Paranal Artificial Reference Source for Extended Coverage
PI	Principal Investigator
PPARC	Particle Physics and Astronomy Research Council
PRIMA	Phase Referenced Imaging and Microarcsecond Astrometry facility (VLTI)
QBS	Quick Blue Survey
QSO	Quasi-Stellar Object
QUEST	QUasar Equatorial Survey Team
R&D	Research and Development
RadioNet	Radio Astronomy network (FP5)
REM	Rapid Eye Mount (La Silla)
REOSC	French optical company
RMS	root mean square
SAG	Servicio Agrícola Ganadero (Agricultural and Livestock Service)
SALT	Southern African Large Telescope
SEA	Sociedad Española de Astronomía
SEST	Swedish-ESO Submillimetre Telescope

SRON	Netherlands Foundation for Space Research
SHARP	Camera on NTT
SINFONI	Spectrograph for INtegral Field Observations in the Near Infrared (VLT)
SKA	Square Kilometre Array
SL-9	Shoemaker-Levy 9
SMA	Submillimeter Array
SODAR	SOund Detection And Ranging
SOFI	Son OF ISAAC (NTT)
SOFIA	Stratospheric Observatory for Infrared Astronomy
SOIMI	Società Impianti Industriali
SPHERE	Spectro-Polarimetric High-contrast Exoplanet REsearch (VLT)
SPIE	International society for optics and photonics
SPIFFI	Spectrometer for Infrared Faint Field Imaging (VLT)
SPS	Super Proton Synchrotron (CERN project)
SRC	Science Research Council (UK)
SSWG	Science Strategy Working Group
STC	Scientific Technical Committee
ST-ECF	Space Telescope European Coordinating Facility
STFC	Science and Technology Facilities Council (UK)
STScI	Space Telescope Science Institute (US)
SUSI	SUperb Seeing Imager (NTT)
SUSI2	Superb Seeing Imager 2 (NTT)
TAROT	Télescope à Action Rapide pour les Objets Transitoires (La Silla)
TB	Terabyte
TEKES	Finnish Agency for Technology and Innovation
TIMMI	Thermal Infrared Multimode Instrument (3.6-metre)
TMT	Ten Meter Telescope
TMT	Thirty Meter Telescope
TNO	Netherlands Organisation for Applied Scientific Research
TP	Telescope Project (Division)
TPD	Netherlands Institute of Applied Physics
TRS	Technical Research Support
UC	Users Committee
UCLA	University of California, Los Angeles
UK	United Kingdom
UKIDSS	UKIRT Infrared Deep Sky Survey
UKIRT	United Kingdom Infrared Telescope
UKST	UK Schmidt Telescope
ULTRACAM	High-speed camera (NTT)
UN	United Nations
UNESCO	United Nations Educational, Scientific and Cultural Organization
UT	Unit Telescope (VLT)
UT1–4	VLT Unit Telescopes 1–4: Antu, Kueyen, Melipal and Yepun
UTC	Coordinated Universal Time
UV	ultraviolet
UVES	UV-Visual Echelle Spectrograph (VLT)
VIMOS	VIsible MultiObject Spectrograph (VLT)
VINCI	VLT INterferometer Commissioning Instrument (VLTI)
VIRCAM	VISTA IR Camera
VIRGO	Gravitational Wave Detector at the European Gravitational Observatory
VIRMOS	Visible Infra-Red Multi-Object Spectrograph (VLT)
VISA	VLTI Sub-Array
VISIR	VLT Imager and Spectrometer for mid-InfraRed
VISTA	Visible and Infrared Survey Telescope for Astronomy
VLA	Very Large Array
VLT	Very Large Telescope
VLTI	Very Large Telescope Interferometer
VO	Virtual Observatory
VST	VLT Survey Telescope
WDR	Westdeutscher Rundfunk
WFC3	Wide Field Camera 3 (HST)
WFCAM	Infrared wide field camera for the UK Infrared Telescope on Mauna Kea
WFI	Wide-Field Imager (MPG/ESO 2.2-metre)
WMAP	Wilkinson Microwave Anisotropy Probe
WTT	Wide Terrestrial Telescope
XFEL	European X-ray Free Electron Laser
XMM-Newton	ESA's X-ray space observatory
X-shooter	Multi-wavelength (ultraviolet-infrared) medium resolution spectrograph (VLT)

Appendix 5

Index of Names

C

D

E

F

G

H

R

S

T

U

V

W

Y

Z

Appendix 6

Subject Index

F

G

H

I

M

N

O

P

Q

R

S

T

U

V

W

X

Y

Z